Handbook of Industrial Chemistry

Handbook of Industrial Chemistry

Editor: Cory Simmons

NY RESEARCH
P R E S S

New York

Published by NY Research Press
118-35 Queens Blvd., Suite 400,
Forest Hills, NY 11375, USA
www.nyresearchpress.com

Handbook of Industrial Chemistry
Edited by Cory Simmons

International Standard Book Number: 978-1-63238-653-3 (Hardback)

Cataloging-in-Publication Data

Handbook of industrial chemistry / edited by Cory Simmons.
    p. cm.
Includes bibliographical references and index.
ISBN 978-1-63238-653-3
1. Chemistry, Technical. 2. Chemical engineering. I. Simmons, Cory.
TP145 .H36 2019
660--dc23

Contents

Preface

Industrial chemistry is the study of chemical processes in industry. It also involves the development of chemical products through the selection of raw materials, production optimization and process improvement. It integrates the principles of chemistry, chemical engineering and environmental science to develop products for the pharmaceutical industry, polymer manufacturing, hydrocarbon processing, food processing, etc. The objective of this book is to give a general view of the different areas of industrial chemistry and their applications in modern industries. It is a valuable compilation of topics, ranging from the basics to the most complex advancements in this field. With state-of-the-art inputs by acclaimed experts in this field, this book will be an essential guide for both academicians and researchers.

The information contained in this book is the result of intensive hard work done by researchers in this field. All due efforts have been made to make this book serve as a complete guiding source for students and researchers. The topics in this book have been comprehensively explained to help readers understand the growing trends in the field.

I would like to thank the entire group of writers who made sincere efforts in this book and my family who supported me in my efforts of working on this book. I take this opportunity to thank all those who have been a guiding force throughout my life.

Editor

The influence of La_2O_3-doping on structural, surface and catalytic properties of nano-sized cobalt–manganese mixed oxides

Abdelrahman A. Badawy[1] · Shaimaa M. Ibrahim[2,3]

Abstract Structural, textural and catalytic activity of CoO–Mn_2O_3 system as being influenced by La_2O_3-doping (0.75–3 mol%) and calcination temperatures (300–500 °C) were investigated. The techniques employed were XRD, N_2-adsorption–desoprtion at −196 °C, EDX and catalysis of H_2O_2-decomposition in aqueous solution at 30–50 °C. The results revealed that the investigated system consisted of nano-sized Co_2MnO_4 as a major phase together with un-reacted portion of Co_3O_4 and γ-Mn_3O_4. Doping with the smallest amount of La_2O_3 greatly increased the surface molar ratio of Mn/Co (88 %) for the solid calcined at 500 °C with subsequent increase of the catalytic activity more than 12-fold for solids calcined at 500 °C. The increase of La_2O_3-dopant above 0.75 mol% decreased progressively the surface molar ratio of Mn/Co with subsequent decrease in the catalytic activity which still measured higher values than that measured for the un-doped catalyst.

Keywords Cobalt manganite · La_2O_3-dopant · Coprecipitation · Catalytic decomposition of H_2O_2

✉ Abdelrahman A. Badawy
aabadawy107@yahoo.com

[1] Physical Chemistry Department, National Research Centre, Dokki, Cairo, Egypt

[2] Chemistry Department, Faculty of Education, Ain Shams University, Cairo, Egypt

[3] Present Address: Department of Chemistry, Faculty of Science, Qassim University, Buraidah, Saudi Arabia

Introduction

Manganese oxides are reported to be considered as environment-friendly materials. MnO_2 and Mn_3O_4 were found to be active and stable catalysts for the combustion of organic compounds [1, 2].

Nanocrystalline manganese oxide powders were synthesized in previous study by an inert gas condensation technique [3]. The manganese oxide, which is prepared, is a mixture of MnO and Mn_3O_4. The particle size of manganese oxides is greatly dependent on their preparation conditions. Dimesso et al. [4] claimed that manganese oxides can be prepared by an inert gas condensation technique followed by annealing in air and oxygen at various temperatures. The predominant phase of MnO and Mn_3O_4 are obtained after annealing in air at 400 °C.

Mixed oxides containing transition metal oxides are used to design the catalytic materials to replace noble metal catalysts. Lahousse et al. [1] have found that γ-MnO_2 and Pt/TiO_2 catalysts measured high catalytic activities as compared to noble metals catalysts.

Mixing manganese with transition metal oxides in many catalytic systems modify the catalytic activity of individual components [5, 6]. Mixed oxide materials are active for oxidation–reduction reactions and combustion processes. For example, cobalt–zinc manganites, manganese–CeO_2 mixed oxides and Co-containing mixed oxides prepared from hydrotalcite-like precursors were active catalysts in the reduction of nitrous oxide [7–9]. Also, Co–Mn mixed oxides were found to be active catalysts for oxidation of ethanol [10] and conversion of synthesis gas to light olefins [11]. However, Ag–Mn, Ag–Co and Ag–Ce composite oxides supported on Al_2O_3 have been reported as catalysts for oxidation of volatile organic compounds [12]. In a previous study, Mn–Cu mixed oxides have been reported to be catalytically more active towards ethanol oxidation as compared to individual Mn_2O_3 and CuO [6]. Both copper and manganese mixed oxides catalysts were found to be more active catalysts in many industrial oxidation processes, such as CO oxidation by O_2, combustion of toluene, methanol, ethylene, ammonia, NO_2 and other combustion reactions [13–16]. However, Mn–Zr mixed oxide samples are active towards dehydrogenation of isopropanol, giving rise to acetone with high selectivity at partial conversion [17].

Hydrogen peroxide and its solutions find use as antiseptic in medicine [18, 19], other applications such as bleach in the textile and paper/pulp industry, in treatment of waste water [20]. However, the literature survey reveals that mixed oxide catalyst is more active in H_2O_2 decomposition. These catalysts have attracted much attention of chemists due to their application as low-cost fuel cells, their stability and high activity [21, 22]. The decomposition of hydrogen peroxide in presence of some metal oxides as $LaMnO_3$ at room temperature and nanocrystalline $LaCrO_3$ was investigated by Khetre et al. [23]. They found that the catalytic activity was increased by increasing both the amount of the catalyst and pH. The probable reaction mechanism has been suggested in which an intermediate surface complex is thought to be responsible for the enhancement of the decomposition of hydrogen peroxide.

The present work aimed at studying the effect of La_2O_3-doping of CoO/Mn_2O_3 system prepared by coprecipitation method on its structural, surface and catalytic properties. The techniques employed were XRD, EDX, N_2-adsorption isotherms carried out at -196 °C and catalytic decomposition of H_2O_2 in aqueous solution at 30–50 °C.

Experimental

Materials

Equimolar proportions of CoO/Mn_2O_3 were prepared by coprecipitation method of their mixed hydroxides from their nitrates solution using 1 M NaOH solution at pH 8 and a temperature of 70 °C. The carefully washed precipitate was dried at 110 °C till constant weight, and then subjected to heating at 300, 400, and 500 °C for 4 h. Three La_2O_3-doped samples were prepared by impregnating a given dry weight of the mixed hydroxides with calculated amount of lanthanum nitrate dissolved in the least amount of distilled water sufficient to make pastes. The pastes were dried at 110 °C and then calcined at 300, 400 and 500 °C for 4 h. The dopant concentrations in the calcined solids were 0.75, 1.5, and 3 mol% La_2O_3.

Techniques

X-ray powder diffractograms of various investigated samples calcined at 300, 400 and 500 °C were determined using a Bruker diffractometer (Bruker D 8 advance target). The patterns were run with copper K_α with secondly monochromator ($\lambda = 1.5405$ Å) at 40 kV and 40 mA. The scanning rate was $0.8°$ in 2θ min^{-1} for phase identification and line broadening profile analysis, respectively. The crystallite size of the phases present in pure and variously La_2O_3-doped solids was determined using the Scherrer equation [24]:

$$d = K\lambda/\beta_{1/2}\cos\theta,$$

where d is the mean crystallite diameter, λ is the X-ray wave length of the X-ray beam, K is the Scherrer constant (0.89), $\beta_{1/2}$ is the full width at half maximum (FWHM) of the main diffraction peaks of the investigated phases, in radian and θ is the diffraction angle.

Energy dispersive X-ray analysis (EDX) was carried out on a Hitachi S-800 electron microscope with a Kevex Delta system attached. The parameters were as follows: -15 kV accelerating voltage, 100 s accumulation time, 8 μm window width. The surface molar composition was determined by the Asa method (Zaf-correction, Gaussian approximation).

Different surface characteristics, namely specific surface area (S_{BET}), total pore volume (V_p), mean pore radius (r^-) and pore volume distribution curves ($\Delta v/\Delta r$) of various solids were determined from nitrogen adsorption–desorption isotherms measured at -196 °C using NOVA Automated Gas sorbometer. Before undertaking such measurements, each sample was degassed under a reduced pressure of 10^{-5} Torr for 3 h at 200 °C. The values of V_p were computed from the relation:

$$V_p = 15.45 \times 10^{-4} \times V_{st}\, cm^3/g,$$

where V_{st} is the volume of nitrogen adsorbed at P/P^0 tends to unity. The values of r^- were determined from the equation:

$$r^- = \frac{2V_p}{S_{BET}} \times 10^4 Å.$$

The catalytic activities of pure and variously La_2O_3-doped solids were determined by studying the decomposition of H_2O_2 in their presence at temperatures within 30–50 °C using 25, 50 and 100 mg of a given catalyst sample with 0.5 ml volume of H_2O_2 of known concentration diluted to 20 ml with distilled water (initial concentration of $H_2O_2 = 0.01$ mol/L). The reaction kinetics was monitored by measuring the volume of oxygen liberated at different time intervals until no further O_2 was liberated. The volume of the liberated oxygen was recalculated under STP.

Results and discussion

X-ray investigation of various solids

X-ray diffractograms of un-doped and variously La$_2$O$_3$-doped solids calcined at 300–500 °C were determined from the recorded diffractograms of these solids are illustrated in Figs. 1, 2, 3 for the solids calcined at 300, 400 and 500 °C, respectively. The different structural characteristics of the solids investigated are given in Table 1. Table 1 includes the peak area of the main diffraction lines of different phases present and the crystallite size of CoMn$_2$O$_4$ phase formed calculated from the Scherrer equation.

Examination of Figs. 1, 2, 3 and Table 1 shows the following: (1) pure and variously La$_2$O$_3$-doped solids calcined at 300–500 °C consisted of nano-sized cobalt manganite (Co$_2$MnO$_4$) (02-1061-JCPDS-ICDD, Copyright, 2001) as a major phase together with un-reacted Co$_3$O$_4$ (42-1467-JCPDS-ICDD, Copyright, 2001) and γ-Mn$_3$O$_4$ (18-0803-JCPDS-ICDD, Copyright, 2001) phases. (2) The peak area of the main diffraction lines of Co$_2$MnO$_4$ phase decreases progressively by increasing the amount of La$_2$O$_3$-added in different solids. For example, the peak area of the main diffraction peak of cobalt manganite for the solids calcined at 500 °C measured 73.5, 55.1, 37.7 and 28.9 (a.u.) for pure sample and those treated with 0.75, 1.5 and 3 mol% La$_2$O$_3$, respectively. (3) The crystallite size of the produced Co$_2$MnO$_4$ varies between 21.3 and 75.7 nm depending on the dopant concentration and calcination temperature. (4) The increase in calcination temperature of various solids investigated within 300–500 °C increased progressively the peak area of the main diffraction line corresponding to the produced Co$_2$MnO$_4$. (5) No diffraction peak of lanthanum or lanthanum-manganite and

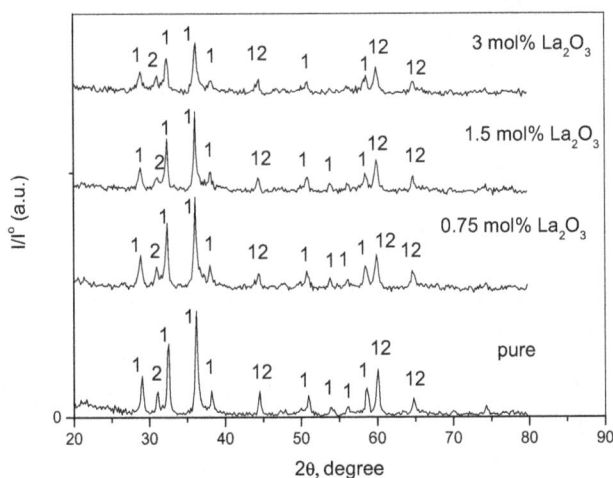

Fig. 2 X-ray diffractograms of pure and variously doped solids calcined at 400 °C. *Lines 1* refer to Co$_2$MnO$_4$, *lines 2* refers to Co$_3$O$_4$

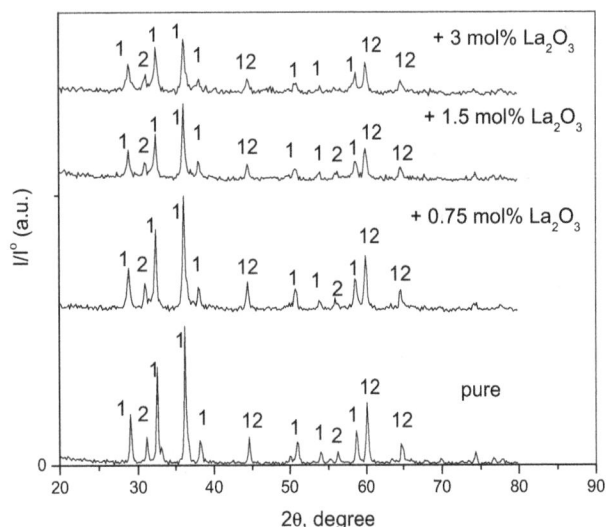

Fig. 3 X-ray diffractograms of pure and variously doped solids calcined at 500 °C. *Lines 1* refer to Co$_2$MnO$_4$, *lines 2* refers to Co$_3$O$_4$

lanthanum-cobaltite composites were detected in the diffractograms. This finding suggested clearly that La$_2$O$_3$ acted only as a doping agent.

These results show clearly the role of La$_2$O$_3$ in hindering the solid–solid interaction between cobalt and manganese oxide to yielding cobalt manganite. The increase in calcination temperature within 300–500 °C stimulated the formation of Co$_2$MnO$_4$. The formation of Co$_2$MnO$_4$ took place according to the reaction:

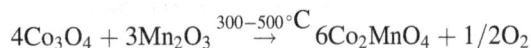

$$4Co_3O_4 + 3Mn_2O_3 \overset{300-500\,°C}{\rightarrow} 6Co_2MnO_4 + 1/2O_2$$

The addition of smallest amounts of La$_2$O$_3$ (0.75–3 mol%) followed by heating at 300–500 °C hindered Co$_2$MnO$_4$ formation to an extent proportional to its amount added.

Fig. 1 X-ray diffractograms of pure and variously doped solids calcined at 300 °C. *Lines 1* refer to Co$_2$MnO$_4$, *lines 2* refers to Co$_3$O$_4$

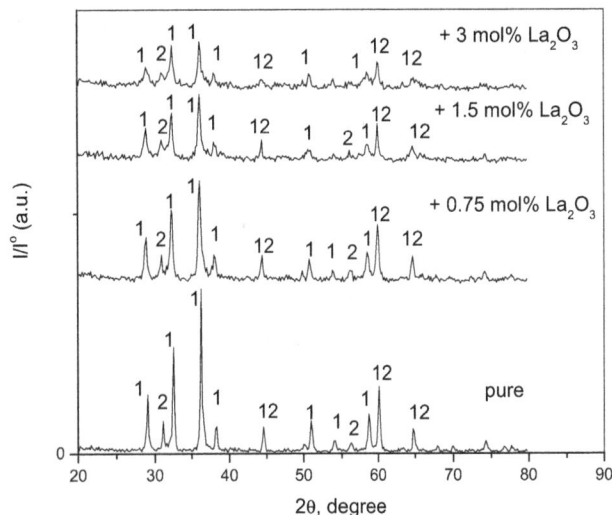

Table 1 Peak area of different phases present in pure and variously doped Co_3O_4–Mn_3O_4 solids calcined at 300–500 °C and the crystallite size of Co_2MnO_4 phase

Solids	Calcination temperature (°C)	Peak area (a.u.) of diffraction lines of			Crystallite size (nm)
		Co_2MnO_4 phase 2.48 Å (100 %)	Co_3O_4 phase 2.85 Å (34 %)	γ-Mn_3O_4 phase 3.08 Å (90 %)	Co_2MnO_4
Pure (0.5 mol CoO + 0.5 mol Mn_2O_3)	300	47.5	11.2	17.8	51.3
	400	70.3	13.9	27	59.7
	500	73.5	14.5	26.5	75.7
+0.75 mol% La_2O_3	300	39	7	10.8	26.3
	400	45.9	11.7	20.5	27.5
	500	55.1	14.3	21.3	31.5
+1.5 mol% La_2O_3	300	31.6	11.1	15.4	28.7
	400	31.7	10.8	15.4	30.4
	500	37.7	8.6	15.3	31.5
+3 mol% La_2O_3	300	25.2	9.3	11.6	21.3
	400	21.3	7.8	9.8	23.5
	500	28.9	9.8	15.2	30.8

The retardation effect of La_2O_3 might be attributed to dissolution of some of La_2O_3 (La^{3+}) cation in the lattices of Co_3O_4 and/or Mn_3O_4 assuming that Co_3O_4 consisted of (2CoO and Co_2O_3, i.e., cobalt cations existed in di- and/or tetra-valent states. The dissolution of La_2O_3 in Co_3O_4 lattices could be simplified adopting Kröger's notations [25] according to:

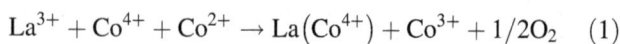

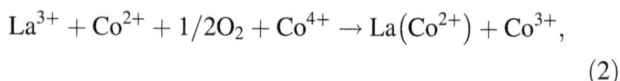

$$La^{3+} + Co^{4+} + Co^{2+} \rightarrow La(Co^{4+}) + Co^{3+} + 1/2O_2 \quad (1)$$

$$La^{3+} + Co^{2+} + 1/2O_2 + Co^{4+} \rightarrow La(Co^{2+}) + Co^{3+}, \quad (2)$$

where La (Co^{4+}) is lanthanum cation located in the position of host tetravalent cobalt cation lattice and La (Co^{2+}) is lanthanum cation located in the position of host divalent cobalt lattice cation in Co_3O_4. It is clear from Eqs. 1 and 2 that incorporation of La^{3+} in cobaltic oxide lattice decreased the concentrations of divalent and tetravalent cobalt cations converting them into Co^{3+} ions. The trivalent cobalt cations did not participate directly in the formation of $CoMn_2O_4$ which involved divalent cobalt and trivalent manganese cations. So, the incorporation of lanthanum cation in Co_3O_4 lattice according to mechanism 1 decreased the number of divalent cobalt involved in the formation of $CoMn_2O_4$ phase. Consequently, the retardation of La_2O_3 dopant in $CoMn_2O_4$ formation could be understood.

Surface characteristics of various prepared solids

Nitrogen adsorption–desorption isotherms measured at −196 °C were determined for pure and variously La_2O_3-doped Co_3O_4/Mn_3O_4 solids calcined at 300, 400 and 500 °C. All isotherms belong to type II of Branuer classification [26] having hysteresis loops of small areas closing at P/P^0 at about 0.5. Figure 4 depicts representative N_2-adsorption–desorption isotherms measured over pure and variously doped solids calcined at 500 °C. Figure 5 depicts $\Delta v/\Delta r$ curves of various solids calcined at 300 and 500 °C. Analysis of the recorded adsorption–desorption isotherms permitted us to calculate different surface characteristics, namely specific surface area (S_{BET}), total pore volume (V_p) and mean pore radius (r^-). The computed values of S_{BET}, V_p and r^- are given in Table 2.

Examination of Table 2 and Fig. 5 shows the following: (1) the S_{BET} and V_p values of pure and variously doped solids increased progressively by increasing the calcination temperature within 300–500 °C. The increase was, however, more pronounced for pure mixed solids which attained 113 and 295 % for S_{BET} and V_p, respectively [27]. (2) The r^- values cited in the last column of Table 2 show that the investigated solids are mesoporous adsorbents measuring (r^-) values varying between 30 and 88 Å. (3) La_2O_3-doping decreased effectively the S_{BET} to an extent directly proportional to its amount present. The doping process did not affect the r^- values which remained almost constant for heavily doped sample (3 mol% La_2O_3). (4) Most of the investigated solids exhibit bimodal pore volume distribution curves except the heavily doped sample (3 mol% La_2O_3) exhibited tri-modal distribution curves (c.f. Figure 5). The maximum hydraulic pore radius was located at 11.5 and 27.5 Å for pure solids and found at 10 and 23 Å for solids doped by 0.75 mol% La_2O_3 and 9.5 and 30 Å for solids doped by 1.5 mol% La_2O_3 and 10, 16 and 26 Å for solids doped by 3 mol% La_2O_3 being calcined at 500 °C.

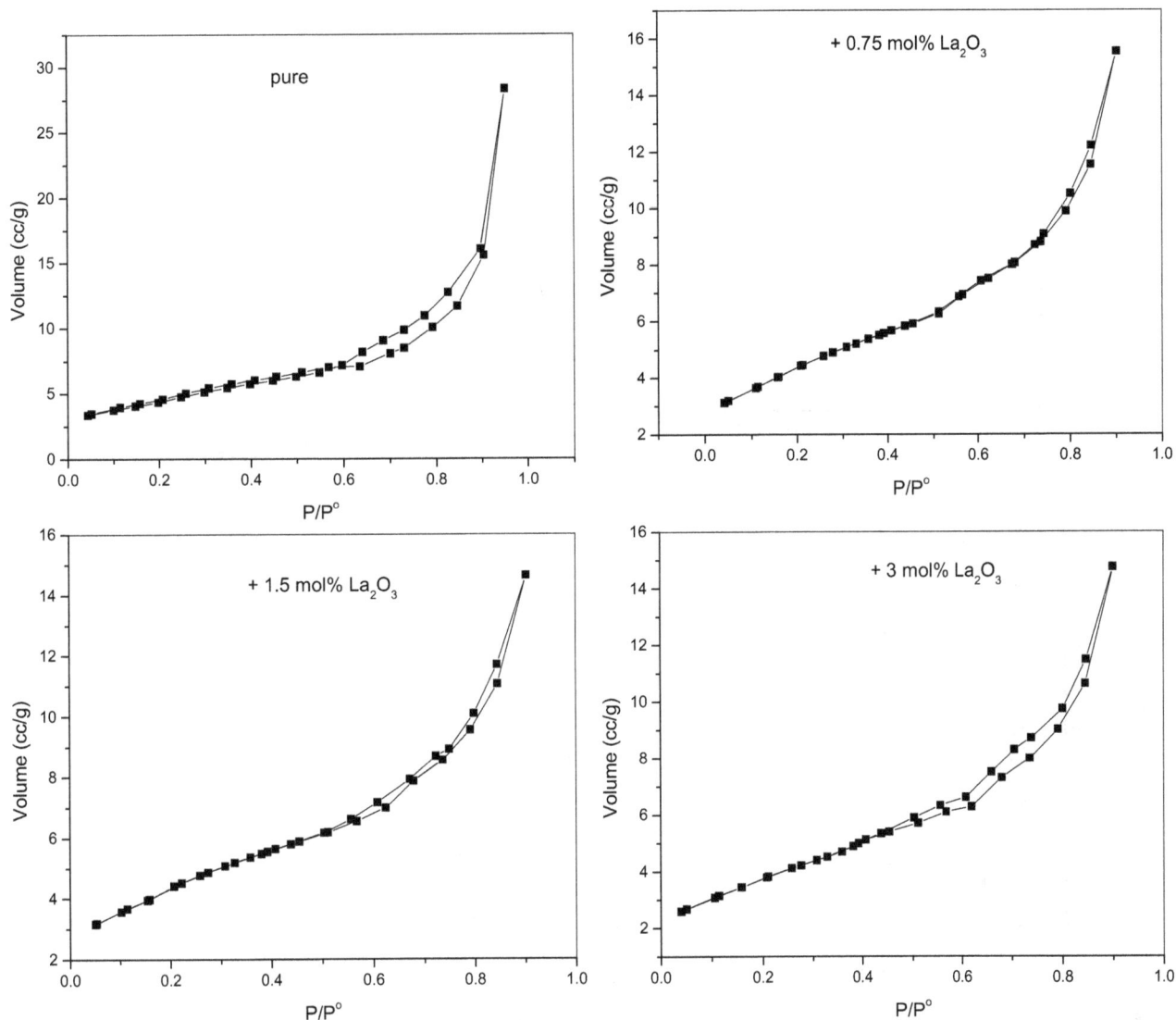

Fig. 4 N_2-adsorption–desorption isotherms measured over pure and variously doped solids calcined at 500 °C

These results show clearly the role of La_2O_3-doping in modifying the various surface characteristics of the system investigated. Comparison of (r^-) values of pure and variously doped solids calcined at 300–500 °C decreased effectively the calculated (r^-) values. This decrease might be followed by an increase in the S_{BET} values opposite to what was found. So, one might expect that the dopant process decreased the concentration of the narrowest pore located at 11.5 Å.

Energy dispersive X-ray analysis of various solids

EDX investigation of pure and doped solids calcined at 300–500 °C was determined. The relative atomic abundance of manganese, cobalt, oxygen and lanthanum species present in the uppermost surface layers of the calcined solids is given in Table 3. It is well known that EDX technique supplies an accurate determination of relative atomic concentration of different elements present on their outermost surface layers [28–34]. In fact, this technique (EDX) has been successfully employed in determining the surface composition of a big variety of catalytic systems such as CuO/Mn_2O_3 [28], CuO/ZnO [29, 31, 33], TiO_2/Al_2O_3 [30], CuO/NiO [32] and Co_3O_4/Fe_2O_3 [34]. The thickness of these layers is bigger than those measured by using XPS technique. XPS is a well-known surface-sensitive technique that supplies very accurate relative atomic abundance of cationic and anionic species on the surfaces of investigated solids. Surface and bulk compositions of various solids are given in Table 3. Inspection of the results given in Table 3 reveals the following: (1) the surface composition of pure and variously doped solids is different from those of their bulk. (2) The surface concentration of cobalt species is always smaller than that presents in the

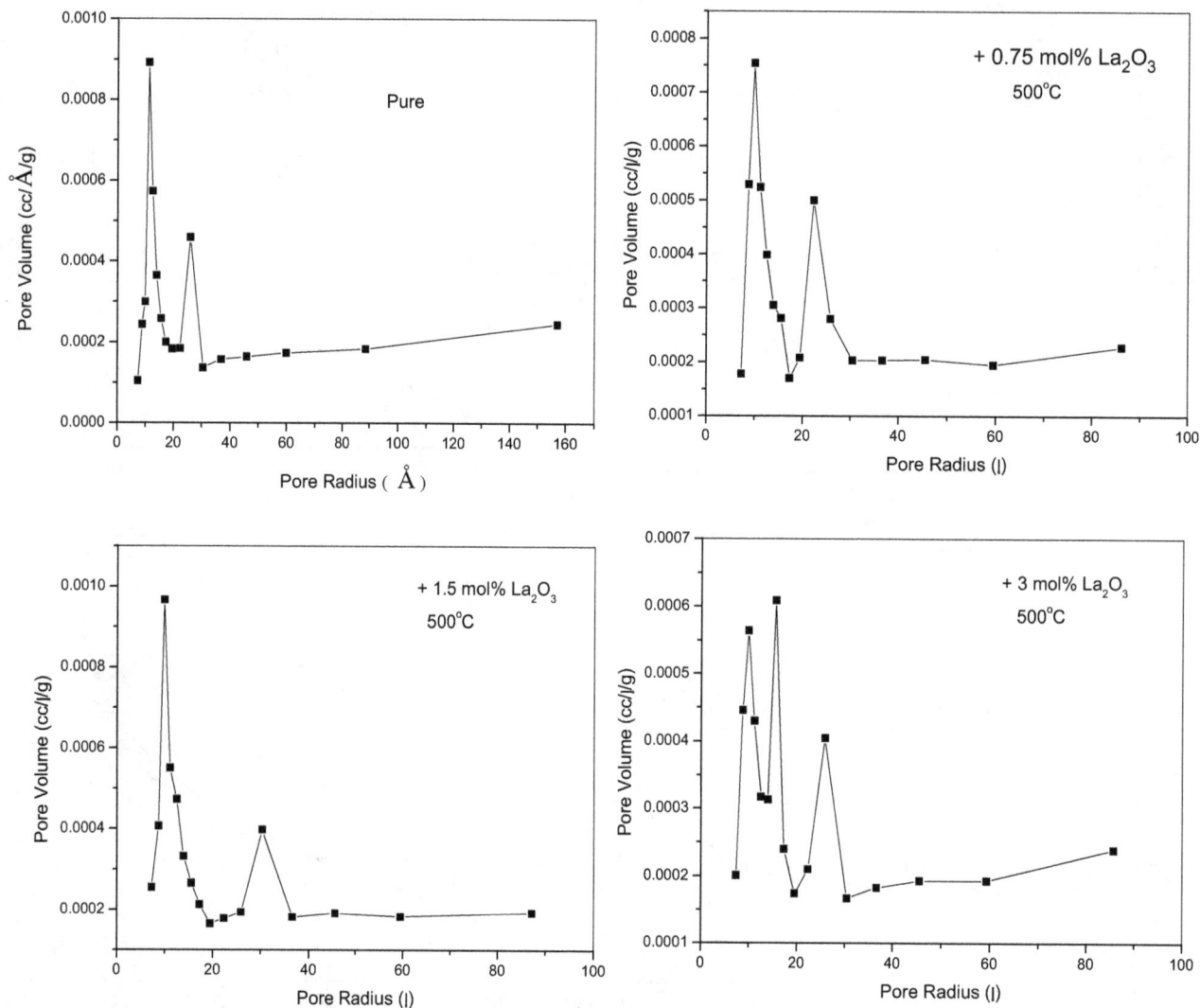

Fig. 5 Pore volume distribution curves $\Delta v/\Delta r$ for pure and variously doped solids calcined at 500 °C

Table 2 Surface characteristics of pure and variously treated solids being calcined at 300–500 °C	Solids	Calcination temperature (°C)	S_{BET} (m^2/g)	Total pore volume V_p (cc/g)	Mean pore radius r^- (Å)
	Pure (0.5 mol CoO +0.5 mol Mn$_2$O$_3$)	300	16	0.038	48
		400	17	0.045	52
		500	34	0.15	88
	+0.75 mol% La$_2$O$_3$	300	13	0.021	32
		400	14	0.022	31
		500	16	0.024	30
	+1.5 mol% La$_2$O$_3$	300	12	0.011	18
		400	14	0.015	21
		500	15	0.016	21
	+3 mol% La$_2$O$_3$	300	10	0.016	32
		400	12	0.020	33
		500	14	0.023	33

Table 3 Surface molar composition of pure and variously treated solids determined by EDX

Solid	Calcination temperature (°C)	Element	Atomic abundance (at.%)		Mn/Co ratio (surface)
			Bulk	Surface	
Pure (0.5 mol CoO + 0.5 mol Mn$_2$O$_3$)	300	Mn	28.57	37.0	33.6
		Co	14.29	1.1	
		O	57.14	62.0	
	400	Mn	28.57	38.6	35.1
		Co	14.29	1.1	
		O	57.14	60.3	
	500	Mn	28.57	56.0	37.3
		Co	14.29	1.5	
		O	57.14	42.5	
+0.75 mol% La$_2$O$_3$	300	Mn	28.39	38.3	47.9
		Co	14.19	0.8	
		O	57.21	60.64	
		La	0.21	0.26	
	400	Mn	28.39	48.2	53.5
		Co	14.19	0.9	
		O	57.21	50.54	
		La	0.21	0.36	
	500	Mn	28.39	63.3	70.3
		Co	14.19	0.9	
		O	57.21	35.44	
		La	0.21	0.4	
+1.5 mol% La$_2$O$_3$	300	Mn	28.21	38.0	46.5
		Co	14.11	0.8	
		O	57.26	60.74	
		La	0.42	0.46	
	400	Mn	28.21	38.9	47.4
		Co	14.11	0.82	
		O	57.26	59.74	
		La	0.42	0.54	
	500	Mn	28.21	42.2	48.0
		Co	14.11	0.88	
		O	57.26	56.34	
		La	0.42	0.58	
+3 mol% La$_2$O$_3$	300	Mn	27.85	37.6	41.8
		Co	13.93	0.9	
		O	57.38	60.4	
		La	0.84	1.1	
	400	Mn	27.85	38.6	42.0
		Co	13.93	0.92	
		O	57.38	59.28	
		La	0.84	1.2	
	500	Mn	27.85	40.9	43.1
		Co	13.93	0.95	
		O	57.38	56.95	
		La	0.84	1.2	

bulk of pure and variously doped solids. (3) The concentration of surface manganese was several folds that of cobalt. This finding might suggest that cobalt hydroxide was precipitated much earlier than manganese hydroxides. This conclusion seems logical because of the significant difference between the values of the solubility products of $Co(OH)_3$ and $Mn(OH)_2$ which measured 5.9×10^{-15} and 1.6×10^{-44}, respectively. (4) Lanthanum species present in the uppermost surface layers of all doped solids calcined at temperature within 300–500 °C is bigger than the amount present in the bulk of solids. This finding is expected because all doped solids were prepared by wet impregnation method [35, 36]. Furthermore, the surface concentration of lanthanum increases by increasing the calcination temperature of the doped solids. (5) The addition of the smallest amount of La_2O_3 (0.75 mol%) much increased the surface concentration of manganese present in all solids calcined at 300–500 °C. The increase in surface concentration of manganese species attained 4, 25 and 13 % for the solids calcined at 300, 400 and 500 °C, respectively. The increase of the dopant concentration above 0.75 mol% La_2O_3 decreased the surface concentration of manganese which still remained bigger than that measured for the un-doped samples calcined at the same temperatures.

It is well known that manganese species present in the uppermost surface layers in pure and doped solids are considered as the most catalytically active constituent involved in H_2O_2-decomposition. This assumption is based on the possible presence of manganese cations in different oxidation states varying between di- and hepta-valence states leading to a significant increase in the concentration of manganese ion pairs participating in the catalytic decomposition process. So, the bigger the surface concentration of manganese the bigger will be the concentration of the most catalytically active constituent and the bigger the catalytic activity. This speculation will be confirmed in the next section of the present work dealing with the catalytic decomposition of H_2O_2 on pure and variously La_2O_3-doped solids calcined at 300–500 °C.

Catalytic activity of pure and variously doped solids

The catalytic decomposition of H_2O_2 in aqueous solution was studied at 30, 40 and 50 °C over pure and variously doped solids precalcined at 300, 400 and 500 °C. First-order kinetics was observed in all cases. In fact, straight line was found upon plotting $\ln a/a - x$ against the time intervals t, where a is the initial concentration of H_2O_2 and $a - x$ is its concentration at time t. The slopes ($d \ln a/a - x/dt$) of plots determine the reaction rate constant (k) for the reaction conducted at a given temperature over a given catalyst sample. The reaction kinetics were monitored using

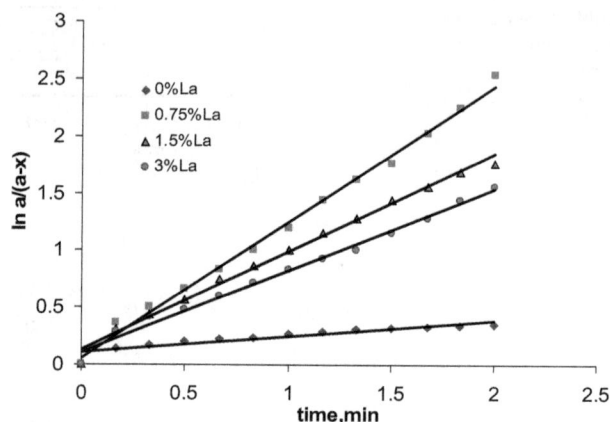

Fig. 6 First-order plots for catalytic decomposition of H_2O_2 at 50 °C, over pure and variously doped solids calcined at 500 °C

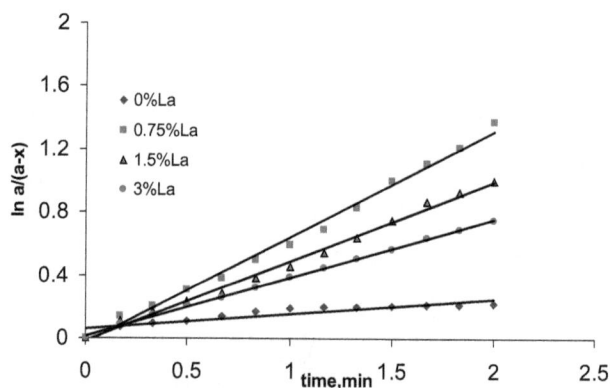

Fig. 7 First-order plots for catalytic decomposition of H_2O_2 at 50 °C, over pure and variously doped solids calcined at 300 °C

25, 50 and 100 mg of catalyst sample. The results, not given, showed that k value increases linearly by increasing the catalyst mass indicating the absence of any possible solid gas diffusion of liberated oxygen. Figures 6 and 7 depict representative first-order plots of the catalyzed reaction conducted at 30 and 50 °C over pure and variously La_2O_3-doped solids pre-calcined at 300 and 500 °C, respectively. The values of the reaction rate constant per unit mass for the reaction carried out at 30, 40 and 50 °C were calculated from the slope of the first-order plots. Table 4 includes only the computed values of $k_{50 °C}$. Inspection of Table 4 reveals the following: (1) the catalytic activity, expressed as reaction rate constant per unit mass, measured for pure and variously doped solids increased by increasing the calcination temperatures of all solids investigated calcined at 300–500 °C. (2) The presence of the smallest amount of La_2O_3 (0.75 mol%) increased considerably the (k) values. The increase reached about 10^3-fold for the catalytic reaction carried out at 30 and 50 °C. So, the comparison between the role of ZrO_2, as shown in our

Table 4 Reaction rate constant per gram catalyst, $k^*_{50°C}$ (min^{-1} - g^{-1}) $\times 10^3$ of catalytic decomposition of H_2O_2, activation energy (ΔE) for the catalytic reaction carried out at 50 °C over pure and variously La_2O_3-doped solids calcined at 300–500 °C

Solid	Calcination temperature (°C)	$k^*_{50°C}$ (min^{-1} g^{-1})	$\Delta E \times 103$ (kJ mol^{-1})
Pure (0.5 mol CoO + 0.5 mol Mn$_2$O$_3$)	300	1.24	5.2
	400	1.4	
	500	1.66	
+0.75 mol% La$_2$O$_3$	300	13.6	1.5
	400	20.0	
	500	22.4	
+1.5 mol% La$_2$O$_3$	300	10.2	2.0
	400	14.0	
	500	16.4	
+3 mol% La$_2$O$_3$	300	7.2	2.7
	400	12.0	
	500	13.6	

previous studied [27], and La_2O_3-doping of Co_3O_4–Mn_3O_4 system showed clearly that the increase in the catalytic activity due to the doping process was much more pronounced by doping with La_2O_3 as compared to ZrO_2-doping [27]. (3) Increasing the dopant concentration above 0.75 mol% decreased the catalytic activity to an extent proportional to the amount of La_2O_3 added. (4) The observed increase in the catalytic activity of pure and variously doped solids by increasing their calcination temperature within 300–500 °C might be attributed to the observed increase in surface concentration of manganese species (considered as the most catalytically active constituent) and the observed increase in the specific surface areas (c.f. Tables 2, 3). (5) The observed significant increase in the catalytic activity due to doping with 0.75 mol% La_2O_3 could be also attributed to the observed decrease in the crystallite size of Co_2MnO_4 phase (c.f. Table 1). (6) The decrease in the catalytic activity of the heavily doped solids might result from the observed increase in the crystallite size of Co_2MnO_4 (the major phase present in pure and doped solids) and also due to the observed decrease in surface concentration of manganese species besides the significant decrease in the S_{BET} (c.f. Tables 1, 2, 3).

In order to throw more light about the role of both calcination temperature and dopant concentration (La_2O_3) in the mechanism of the catalyzed reaction the activation energy of which (ΔE) was determined for pure and doped solids calcined at 300–500 °C. ΔE values were calculated from the values of k measured for the reaction carried out at 30, 40 and 50 °C by direct application of the Arrhenius equation. The computed ΔE values are given in Table 4. Examination of Table 4 shows that ΔE values for pure and

variously doped solids decreased progressively as a function of dopant concentration. This trend ran parallel to observed increase in the catalytic activity, expressed as $k_{50 °C}$ values (c.f. Table 4). This finding expresses the observed increase in the catalytic activity of sample of 0.75 mol% La2O3-doping. On the other hand, increasing the dopant concentration above this limit increased ΔE values which remained almost smaller than ΔE values measured for the pure catalyst samples.

Conclusions

The main conclusions derived from the results obtained can be summarized as follows:

1. The role of La_2O_3-doping on structural, textural, surface composition and catalytic activity was investigated.
2. The prepared mixed solids consisted of nano-sized Co_2MnO_4 as a major phase together with un-reacted portion of Co_3O_4 and γ-Mn_3O_4.
3. Doping with La_2O_3(0.75 mol%) much increased both surface molar ratio of Mn/Co with (0.75 mol%) about 88 % with subsequent increase in the catalytic activity more than 12-fold.
4. The apparent activation energy of the catalyzed reaction measured 5.2, 1.5, 2.0 and 2.7 kJ/mol for the un-doped and catalyst doped with 0.75, 1.5 and 3 mol% La_2O_3.

References

1. Lahousse C, Bernier A, Grange P, Delmon B, Papaefthimiou P, Ioannides T, Very Kios X (1998) Evaluation of γ-MnO$_2$ as a VOC removal catalyst: comparison with a noble metal catalyst. J Catal 178:214–225
2. Baldi M, Escribano VS, Amores JMG, Milella F, Busca G (1998) Characterization of manganese and iron oxides as combustion catalysts for propane and propene. Appl Catal B 17:L175
3. Chen C-Y, Lin C-K, Tsai N-H, Tsay C-Y, Lee P-Y, Chen G-S (2008) Characterization of nanocrystalline manganese oxide powder prepared by inert gas condensation. Ceram Int 34:1661
4. Dimesso L, Heider I, Hahn H (1999) Synthesis of nanocrystalline Mn-oxides by gas condensation. Solid State Ion 123:39–46
5. Li W, Lin Y, Zhang Y (2003) Promoting effect of water vapor on catalytic oxidation of methane over cobalt, manganese mixed oxides. Catal Today 83:239–245
6. Morales MR, Barbero BP, Cadús LE (2006) Total oxidation of ethanol and propane over Mn–Cu mixed oxide catalysts. Appl Catal B 67:229
7. Fierro G, Jacono ML, Inversi M, Dragone R, Ferraris G (2001) Preparation, characterization and catalytic activity of Co–Zn-based manganites obtained from carbonate precursors. Appl Catal B. 30:173
8. Chmielarz L, Kustrowski P, Rafalska-Lasocha A, Majda D, Dziembaj R (2002) Catalytic activity of Co–Mg–Al, Cu–Mg–Al and Cu–Co–Mg–Al mixed oxides derived from hydrotalcites in SCR of NO with ammonia. Appl Catal B 35:195

9. Qi G, Yang RT, Change R (2004) MnOx–CeO2 mixed oxides prepared by co-precipitation for selective catalytic reduction of NO with NH3 at low temperatures. Appl Catal B. 51:93

10. Kovanda F, Rojka T, Dobešová J, Machovič V, Bezdička P, Obalová L, Grygar T (2006) Mixed oxides obtained from Co and Mn containing layered double hydroxides: Preparation, characterization, and catalytic properties. J Solid State Chem 179:812

11. Mirzaei AA, Faizi M, Habibpour R (2006) Effect of preparation conditions on the catalytic performance of cobalt manganese oxide catalysts for conversion of synthesis gas to light olefins. Appl Catal A: Gen 306:98–107

12. Luo M-F, Yuan X-X, Zheng X-M (1998) Catalyst characterization and activity of Ag–Mn, Ag–Co and Ag–Ce composite oxides for oxidation of volatile organic compounds. Appl Catal A: Gen 175:121–129

13. Liu Y, Luo MF, Wei ZB, Xin Q, Ying PL, Li C (2001) Catalytic oxidation of chlorobenzene on supported manganese oxide catalysts. Appl Catal B. 29(1):61

14. Ferrandon M, Bjornbom E (2001) Hydrothermal stabilization by lanthanum of mixed metal oxides and noble metal catalysts for volatile organic compound removal. J Catal 200(1):148–159

15. Alvarez-Galvan MC, O'Shea VADP, Fierro JLG, Arias PL (2003) Alumina-supported manganese- and manganese–palladium oxide catalysts for VOCs combustion. Catal Commun 4(5):223–228

16. Zimowska M, Michalik-Zym A, Janik R, Machej T, Gurgul J, Socha RP, Podobiński J, Serwicka EM (2007) Catalytic combustion of toluene over mixed Cu–Mn oxides. Catal Today 119:321

17. López EF, Escribano VS, Resini C, Gallardo-Amores JM, Busca G (2001) A study of coprecipitated Mn–Zr oxides and their behaviour as oxidation catalysts. Appl Catal B 29:251

18. Khetre SM, Jadhav HV, Bangale SV, Jagdale PN, Bamane SR (2011) Use of mixed metal oxide as a catalyst in the decomposition of hydrogen peroxide. Ind J Chem 2(2):252–259

19. Shivankar VS, Thakkar NV (2006) Chiral mixed ligand Co (II) and Ni (II) complexes: synthesis and biological activity. Ind J Chem 46A:382–387

20. Wood A (2004) Life after ACC: little change for most firms' EH&S efforts. Chem Week 166:27

21. Afsin B, Roberts MW (1992) Surface structure and the instability of the formate overlayer at a Pb(110) surface. Catal Lett 13(3):277–282

22. Kga Y, Oho Y, Tsukanoto K, Nakajima T (1990) Relationships between the gas permeabilities and the microstructures of plasma sprayed oxide layers. Solid State Ion 40/41:1000

23. Khetre SM, Jadhav HV, Bangale SV, Jagdale PN, Bamane SR (2011) Synthesis, characterization and hydrophilic properties of nanocrystalline ZnCo2O4 oxide by combustion route. Der Chemica Sinica 2(4):303–311

24. Cullity BD (1978) Publishing Cos, 2nd edn. Addison-Wesley, Reading

25. Kröger FA (1964) Chemistry of imperfect crystals. North-Holland, Amsterdam

26. Rouquerol F, Rouquerol J, Sing K (1999) Adsorption by powders and porous solids: principles, methodology and applications. Academic Press, San Diego

27. Ibrahim SM, Badawy AA, El-Shobaky GA, Mohamed HA (2014) Structural, surface and catalytic properties of pure and ZrO2-doped nanosized cobalt–manganese mixed oxides. Can J Chem Eng 92:676–684

28. El-Shobaky GA, El-Shobaky HG, Badawy AA, Fahmy YM (2011) Physicochemical, surface and catalytic properties of nanosized copper and manganese oxides supported on cordierite. Appl Catal A: Gen 409–410:234

29. El-Shobaky GA, Hassan HMA, Yehia NS, Badawy AA (2010) Effect of CeO2-doping on surface and catalytic properties of CuO–ZnO system. J Non-Cryst Solids 356:32–38

30. El All SA, El-Shobaky GA (2009) Structural and electrical properties of γ-irradiated TiO2/Al2O3 composite prepared by sol–gel method. J Alloys Compd 479:91–96

31. El-Shobaky GA, Yehia NS, Hassan HMA, Badawy AA (2009) Catalytic oxidation of CO by O2 over nanosized CuO–ZnO system prepared under various conditions. Can J Chem Eng 87:792–800

32. El-Shobaky GA, Radwan NRE, El-Shall MS, Turky AM, Hassan HMA (2009) Physicochemical, surface and catalytic properties of nanocrystalline CuO–NiO system as being influenced by doping with La2O3. Colloids Surf A: Physicochem Eng Asp 345:147–154

33. El-Shobaky GA, Yehia NS, El-Hendawy AA, Abo-Elenin RMM, Badawy AA (2009) Effects of preparation conditions on surface and catalytic properties of copper and zinc mixed oxides system. Open Catal J 2:45–53

34. Fagal GA, Badawy AA, Hassan NA, El-Shobaky GA (2012) Effect of La2O3-treatment on textural and solid–solid interactions in ferric/cobaltic oxides system. J Solid State Chem 194:162–167

35. Radwan NRE, Fagal GA, El-Shobaky GA (2001) Effects of CeO2-doping on surface and catalytic properties of CuO/Al2O3 solids. Colloids Surf A 178:277

36. El-Shobaky GA, Fagal GA, Mokhtar M (1997) Effect of ZnO on surface and catalytic properties of CuOAl2O3 system. Appl Catal A: Gen 155:167

2

Dynamics of inhibition patterns during fermentation processes-Zea Mays and Sorghum Bicolor case study

Neba F. Abunde[1] · N. Asiedu[2] · Ahmad Addo[1]

Abstract Recently ethanol production involved the processing and fermentation of sorghum and maize extracts. Sorghum and maize are cheaper, locally available and a substitute to imported barley malt. Large scale ethanol fermentation systems are usually hampered by instability, in the form of oscillations resulting from ethanol inhibition and the lag response of yeast cells to this inhibition. There is limited information regarding the mathematical nature of such inhibitions in the fermentation of sorghum and maize extracts. In the present work, mathematical models are developed to determine the nature of ethanol inhibition during the fermentation of sorghum and maize extracts. The models were sets of coupled ordinary differential equations based on a Monod type cell growth kinetic model that accounts for product inhibition. The Inhibition patterns considered were; Linear, Sudden Growth Stop and Exponential. The results obtained showed that there is product inhibition during ethanol fermentation using sorghum extracts, with inhibition patterns being Linear and Exponential. However, the results obtained from ethanol fermentation of maize extract also showed that there is product inhibition during ethanol fermentation using maize extracts, with inhibition patterns being Linear and Sudden Growth Stop. The obtained models described with high accuracy, 99% Confidence Interval the dynamics of substrate utilization, product formation and cell growth. These inhibitions which affect the high ethanol yields can be minimized by setting up an optimal control problem using the developed models and solved to determine the control variables that minimize the effect of such inhibitions during the fermentation of sorghum and maize extracts.

Keywords Alcoholic fermentation · Mathematical modeling · Ethanol inhibition · Maize extracts sorghum extracts

List of symbols

μ_{max}	Maximum specific growth rate (h^{-1})
q_{pmax}	Maximum rate of product formation (h^{-1})
P_{xmax}	Product concentration when product formation ceases (g/100 g)
P_{pmax}	Product concentration when cell growth ceases (g/100 g)
K_{ix}	Product inhibition coefficient on cell growth
K_{ip}	Product inhibition coefficient on product formation
K_{isx}	Substrate inhibition coefficient on cell growth
K_{isp}	Substrate inhibition coefficient on product formation
K_{sx}	Substrate saturation (Monod) constant for cell growth (g/100 g)
K_{sp}	Substrate saturation (Monod) constant for product formation (g/100 g)
Y_x	Yield coefficient of cell based on substrate utilization (g/g)
Y_p	Yield coefficient of cell based on substrate utilization (g/g)
G_s	Yield coefficient of cell based on substrate utilization (g/g h)

✉ N. Asiedu
nasiedusoe@yahoo.co.uk

[1] Department of Agricultural Engineering, College of Engineering, Kwame Nkrumah University of Science and Technology, Kumasi, Ghana

[2] Department of Chemical Engineering, College of Engineering, Kwame Nkrumah University of Science and Technology, Kumasi, Ghana

M_s Cell growth coefficient on substrate (g/g h)

Introduction

In several studies regarding the alcoholic fermentation oscillations in batch fermenters resulting from ethanol inhibition and the lag response to yeast cells to this inhibition has been observed and reported. It is often conventional during the modeling of ethanol fermentation to predefine an inhibition pattern but the success of this practice is based on probabilities, since such patterns vary based on the type and strength of the fermentation wort. Sorghum, a cereal which belongs to the family Graminae was first used as a brewing adjunct during the Second World War and is now used in most breweries as locally available alternative to imported barley malt. In recent years, the search for cheaper locally available substitutes to imported barley malt rekindled the involvement of most firms in expensive experiments regarding beer production from various materials and today, most of the more successful firms use maize and sorghum in their beer production process. In a generalized view of processing sorghum and maize for beer production, though involves several unit operations, the fermentation step is regarded as the heart of the entire production where a near optimal environment is desired for microorganisms to grow, multiply and produce the desired product, Alford [1]. However, the fermentation of sorghum and maize extracts at large scale is usually hampered by sub optimal conditions including instability, in the form of oscillations resulting from ethanol inhibition and the lag response of yeast cells to this inhibition, Chen and McDonald [2, 3], Beuse et al. [4], Fengwu [5]. These inhibitions observed results in an increase in residual sugar at the end of the fermentation, which decreases raw material consumption and correspondingly, decreases the ethanol yield if no economically acceptable attenuation strategies are developed, Fengwu [5]. In a typical procedure for modeling ethanol fermentation, if inhibition is considered, it is often conventional to predefine the inhibition pattern and this practice increases uncertainties in the model since ethanol inhibition pattern varies depending on the type of microorganism, and on the type and strength of fermentation wort, Russell [6]. This increase the unreliability of process controllers and simulators since these automatic tools are usually based on a mathematical representation of the considered system, Alford [1]. Dynamic models were developed, incorporating three effects of product inhibition into the kinetic model and simulation resulted in interesting findings.

Model development

The fermentation process kinetics was described with a Monod type cell growth model that accounts for substrate and product inhibition.

Modeling kinetics of growth and product formation

Starting from the Monod Equation for cell growth and product formation, Eq. (1), three inhibition patterns were considered in modeling product inhibition; linear, Sudden growth stop and exponential as shown in Table 1 below.

$$\mu(S) = \frac{\mu_{max}S}{K_{sx} + S} \tag{1}$$

$$q_p(S) = \frac{q_{pmax}S}{K_{sp} + S} \tag{2}$$

Introducing the effect of product inhibition on the Monod equation, using the respective inhibition factors, the following kinetic models were obtained:

Kinetics with Linear Product Inhibition, -Hinshelwood–Dagley model [7]

$$\mu(S,P) = (1 - K_{ix}P)\frac{\mu_{max}S}{K_{sx} + S} \tag{3}$$

$$q(S,P) = (1 - K_{ip}P)\frac{q_{max}S}{K_{sx} + S} \tag{4}$$

Kinetics Sudden Growth Stop Product Inhibition, -Ghose and Tyagi [8]

$$\mu(S,P) = \left(1 - \frac{P}{P_{max}}\right)\frac{\mu_{max}S}{K_{sx} + S} \tag{5}$$

$$q(S,P) = \left(1 - \frac{P}{P_{pmax}}\right)\frac{q_{max}S}{K_{sx} + S} \tag{6}$$

Kinetics with Exponential Product Inhibition,- Aiba and Shoda model [9]

$$\mu(S,P) = exp(-K_{ix}P)\frac{\mu_{max}S}{K_{sx} + S} \tag{7}$$

$$q(S,P) = exp\left(-K_{ip}P\right)\frac{q_{max}S}{K_{sx} + S} \tag{8}$$

Table 1 Mathematical expressions for the product inhibition factors

Product inhibition factor	Mathematical expression
Linear	$(1 - K_2P)$
Sudden growth stop	$\left(1 - \frac{P}{P_{max}}\right)$
Exponential	$e^{-K_1 P}$

Material Balance and development of Differential Equations

The dynamic equations describing the cell growth, product formation and substrate utilization can be developed by applying the principle of conservation of mass. The following system of first order ordinary differential equations that present cell growth, product formation and substrate utilization are in Eqs. (9)–(11)

$$\frac{dX}{dt} = \mu X \tag{9}$$

$$\frac{dP}{dt} = qX \tag{10}$$

$$\frac{dS}{dt} = -\frac{1}{Y_x}\frac{dX}{dt} - \frac{1}{Y_p}\frac{dP}{dt} - G_sX - M_sX \tag{11}$$

Using the batch kinetic models developed above and substituting μ and q in Eq. (5) with each of their product inhibition expressions, the approximate representation of the fermentation process in each inhibition scenario could be described by the following equations;

Dynamics with linear product inhibition:

$$\frac{dX}{dt} = (1 - K_{ix}P)\frac{\mu_{max}S}{K_{sx} + S}X \tag{12}$$

$$\frac{dP}{dt} = (1 - K_{ip}P)\frac{q_{max}S}{K_{sp} + S}X \tag{13}$$

$$\frac{dS}{dt} = -\frac{1}{Y_x}\frac{dX}{dt} - \frac{1}{Y_p}\frac{dP}{dt} - G_sX - M_sX \tag{14}$$

Dynamics with sudden growth stop product inhibition:

$$\frac{dX}{dt} = \left(1 - \frac{P}{P_{max}}\right)\frac{\mu_{max}S}{K_{sx} + S}X \tag{15}$$

$$\frac{dP}{dt} = \left(1 - \frac{P}{P_{pmax}}\right)\frac{q_{max}S}{K_{sp} + S}X \tag{16}$$

$$\frac{dS}{dt} = -\frac{1}{Y_x}\frac{dX}{dt} - \frac{1}{Y_p}\frac{dP}{dt} - G_sX - M_sX \tag{17}$$

Dynamics with exponential product inhibition:

$$\frac{dX}{dt} = \exp(-K_{ix}P)\frac{\mu_{max}S}{K_{sx} + S}X \tag{18}$$

$$\frac{dP}{dt} = \exp\left(-K_{ip}P\right)\frac{q_{max}S}{K_{sp} + S}X \tag{19}$$

$$\frac{dS}{dt} = -\frac{1}{Y_x}\frac{dX}{dt} - \frac{1}{Y_p}\frac{dP}{dt} - G_sX - M_sX \tag{20}$$

Parameter Estimation and Model Statistical Validity

The identification of model parameters for each of the three systems of equations was made with Matlab and the ode45 solver used to simulate the differential equations. This was done by minimizing the overall sum of squared error, Eq. (11) between the model simulation and experimental data points of the process variables (Biomass, Substrate and Product).

$$\varepsilon = \min \sum (X(k_1,k_2,\ldots,k_n) - X^e)^2 + (S(k_1,k_2,\ldots,k_n) - S^e)^2 + (P(k_1,k_2,\ldots,k_n) - P^e)^2$$

For that purpose the Matlab routine "fmincon" was applied. Here $k_i, i = 1 \div n$ was vector of model parameters to be determined as output of minimization procedure. Once the model parameters were estimated, the capability of the mathematical model to describe the ethanol fermentation process was tested statistically using the F- tests and this was done using STATA at a confidence interval of 99%, to find out the confidence level for the developed mathematical model.

Results and discussion of sorghum extract

Table 2 presents the parameters for the four different models used to describe the dynamics of fermentation, while Figs. 1, 2, 3 and 4 presents the fitting of the models with respect to the experimental data. The results led to the following interesting conclusions; alcoholic fermentation of sorghum extracts using Saccharomyces cerevisiae shows the existence of product inhibition, and the patterns of inhibition could be described as linear or exponential models which showed lowest error. It is shown in this work that the errors for the models showing product inhibition

Table 2 Model Parameters for fermentation using Sorghum Extracts

	Inhibition			
	No Inhibition	Linear	SGS	Exponential
Model parameters				
μ_{max}	0.3887	0.9557	0.6582	2.2628
q_{pmax}	17.4649	2.2101	4.7227	9.0145
P_{xmax}			7.9885	
P_{pmax}			9.8916	
K_{ix}		0.1294		0.4210
K_{ip}		0.1004		0.1030
K_{sx}	249.9922	125.433	81.8551	179.2911
K_{sp}	199.9980	29.213	74.8407	199.8710
Y_x	0.1001	0.1086	0.1368	1.0000
Y_p	0.6085	1.8895	1.2710	0.5936
G_s	0.0010	0.0502	0.0345	0.0010
M_s	0.0100	0.0564	0.0423	0.0100
Model error	5.4262	0.3407	0.4054	0.3270

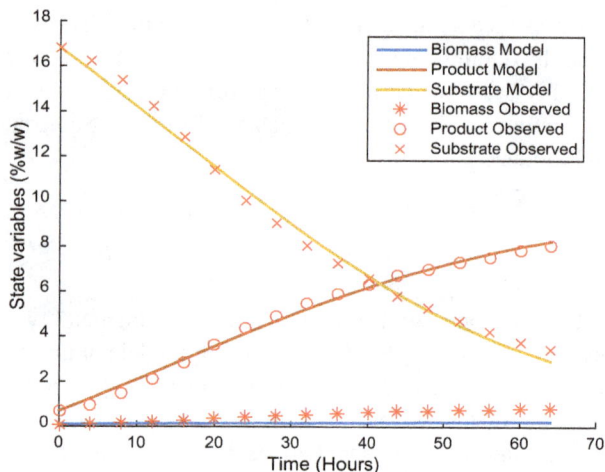

Fig. 1 Experimental results and model fitting, case of no inhibition (Monod-model)

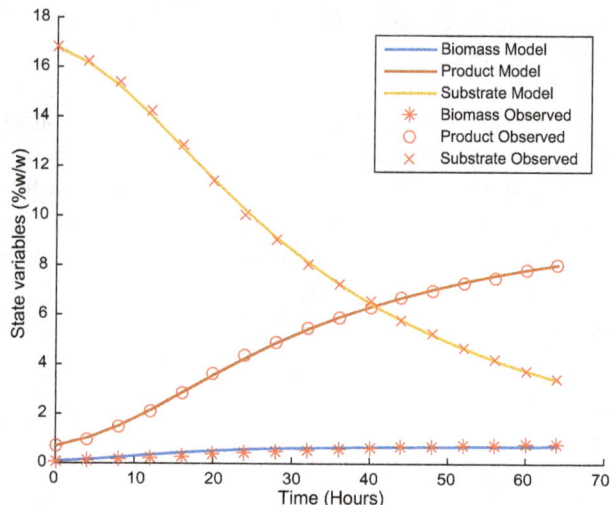

Fig. 2 Experimental results and model fitting, case of linear inhibition model

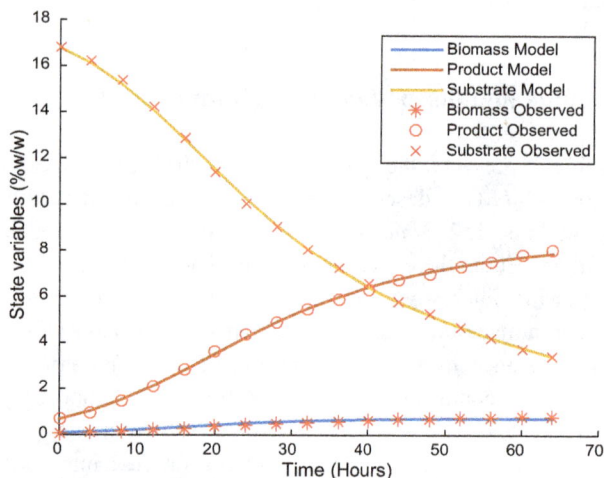

Fig. 3 Experimental results and model fitting, case of Exponential inhibition model

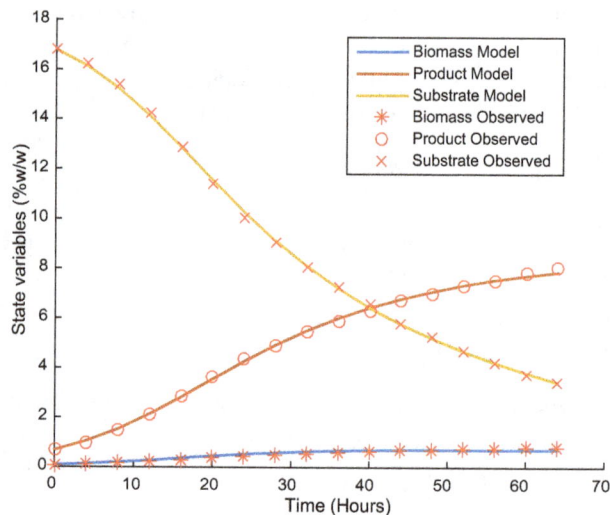

Fig. 4 Experimental results and model fitting, case of Sudden Growth Stop inhibition model

were all lower than that of the Monod which was used as a control. The results confirmed previous observations by Chen and McDonald, 1990a,b; Beuse et al., 1998, 1999; Ingledew, 1999 and Fengwu, 2007 that results ethanol, whether produced by yeast cells during fermentation or externally added into a fermentation system, can trigger inhibitions once its concentration approaches inhibitory levels. Also important in fermentation kinetics is the product yield, growth and maintenance coefficients.

The linear model showed a relatively high product yield coefficient with a relatively low growth and maintenance coefficient compared to the exponential model. This suggest that even though both models showed similar accuracy in describing the fermentation dynamics, designing a control policy with the Linear model will result in more substrate being converted into extracellular product and lesser amount for cell growth and maintenance hence

higher productivity and yield. Tables 3 and 4 presents the model statistical validity using two sample F-test for variance and the results show that at a 99% confidence interval, the states prediction of the linear and exponential models showed no significant difference with the experimental data.

Also important in a fermentation process is how the substrate and product vary in the fermenter as the cells grow. The linear and exponential inhibition models were both used to simulate this variation and both showed a decrease in substrate concentration and an increase in product concentration with cell growth. However, with the linear model the process arrives at steady state within the time used for the simulation as shown in Fig. 5 by the curve at the edges of the plots. This behavior was not

Table 3 Model Statistical Validity with kinetics of linear inhibition, two sample F-test for variance (Biomass)

	Biomass		Product		Substrate	
	Experimental (Xobs)	Model (Xpred)	Experimental (Pobs)	Model (Ppred)	Experimental (Sobs)	Model (Spred)
Mean	0.508	0.533	4.903	4.893	9.126	9.107
Standard Error	0.059	0.055	0.610	0.605	1.114	1.105
Standard Deviation	0.244	0.228	2.514	2.495	4.595	4.558
Observations	17	17	17	17	17	17
Confidence Interval	0.990		0.990		0.990	
F	0.8664		0.9842		0.9840	
Pr(F < f) Two-tailed	0.7778		0.9750		0.9747	

Table 4 Model Statistical Validity with kinetics of Exponential inhibition, two sample F-test for variance

	Biomass		Product		Substrate	
	Experimental (Xobs)	Model (Xpred)	Experimental (Pobs)	Model (Ppred)	Experimental (Sobs)	Model (Spred)
Mean	0.508	0.549	4.903	4.903	9.126	9.105
Standard error	0.059	0.050	0.610	0.610	1.114	1.1061
Standard deviation	0.244	0.207	2.514	2.514	4.595	4.561
Observations	17	17	17	17	17	17
Confidence interval	0.990		0.990		0.990	
F	0.7209		0.9994		0.9850	

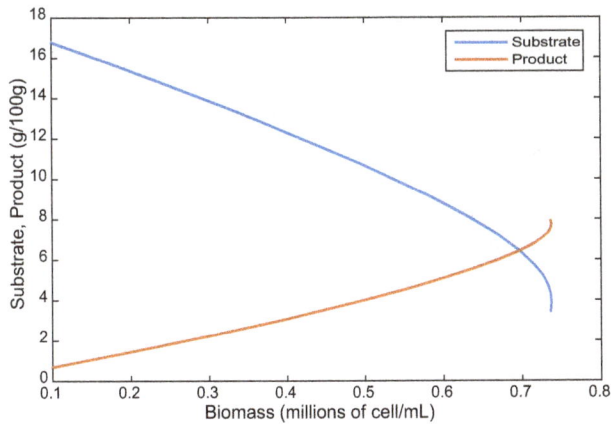

Fig. 5 Simulation of substrate and Product variation as a function of biomass during fermentation using linear inhibition kinetics

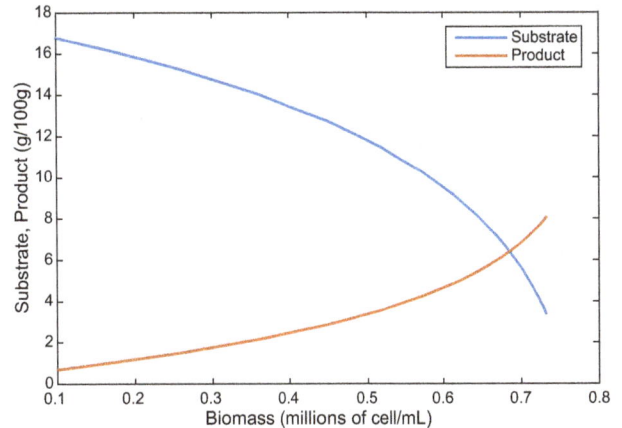

Fig. 6 Simulation of substrate and Product as a function of biomass during fermentation using exponential inhibition kinetics

observed in the exponential inhibition model as shown in Fig. 6, confirming our earlier assertions with the yield coefficients that using the linear model in a control policy will result in high productivity. 3D profiles using the proximal interpolant method implemented using the Matlab curve fitting tool also revealed interesting findings regarding the fermentation process. This was to observe the formation of product as cells grow and consume substrate.

Figure 7 presents product variation with cell growth and substrate consumption using the linear model and Fig. 8 using the exponential model. It can be found that with the linear model, as ethanol accumulated in the fermenter up to a certain concentration, nonlinearities, described as insta-bilities were observed in the product profile. This can be attributed to the higher product yield coefficient of the linear model compared to the exponential model which

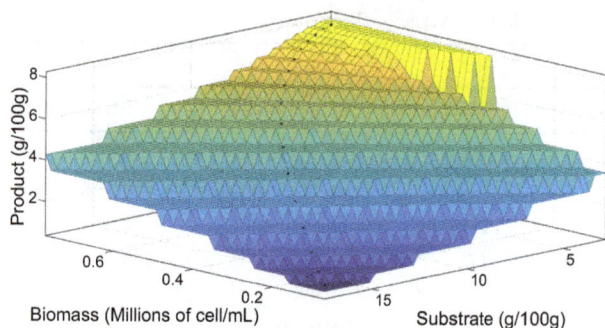

Fig. 7 3D Proximal Interpolant Simulation of Product variation as a function of substrate and biomass using Linear inhibition kinetic Model

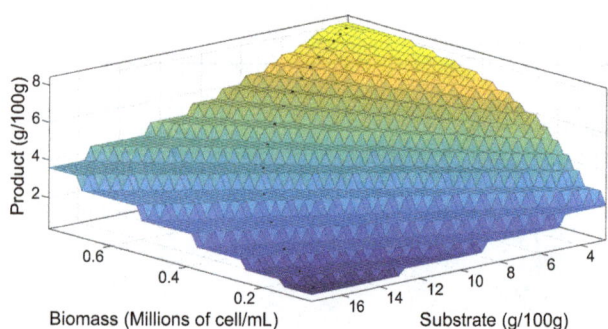

Fig. 8 3D Proximal Interpolant Simulation of Product variation as a function of substrate and biomass using Exponential inhibition kinetic Model

resulted in ethanol accumulating faster in the fermenter and rapidly reaching inhibitory levels, resulting in transient instabilities in the fermenter showed by the nonlinearity, Ingledew [10], Fengwu [5]. This high ethanol concentration is inhibitory to the yeast cell by disrupting the integrity of the cell membrane, Russell [6], Sutton [11]. Figures 9,

10, 11 and 12 show comparisons between models developed for linear, sudden growth stop and exponential inhibitions during the fermentation process.

Results and discussion maize extract

Table 5 presents the parameters for the four different models used to describe the dynamics of fermentation while Figs. 1, 2, 3 and 4 presents the fitting of the models with respect to the experimental data. The results led to the following interesting conclusions; there exist ethanol inhibition in the alcoholic fermentation of maize extracts, and the patterns of inhibition could be described as linear decrease on the inhibitory ethanol concentration or sudden growth seizure at inhibitory concentration (models which showed lowest error). It is shown in this work that the errors for the models showing product inhibition were all lower than that of the Monod which was used as the control. The results confirmed previous observations by Chen and McDonald [2, 3], Beuse et al. [4], Ingledew [10], Fengwu [5], that ethanol either produced by yeast cells during fermentation or externally added into a fermentation system, can trigger inhibitions once its concentration approaches inhibitory levels. Also important in fermentation kinetics is the rate of ethanol accumulation in the fermentation broth, which depicts the maximum achievable rate of product. Even though the linear and sudden growth stop models described the dynamics of ethanol inhibition, the sudden growth stop model showed a very high maximal rate of ethanol accumulation compared to the exponential model. This suggests that designing a control policy with this model will result in high process productivity.

Tables 6 and 7 presents the model statistical validity using two sample F-test for variance and the results show that at a 99% confidence interval, the states prediction of

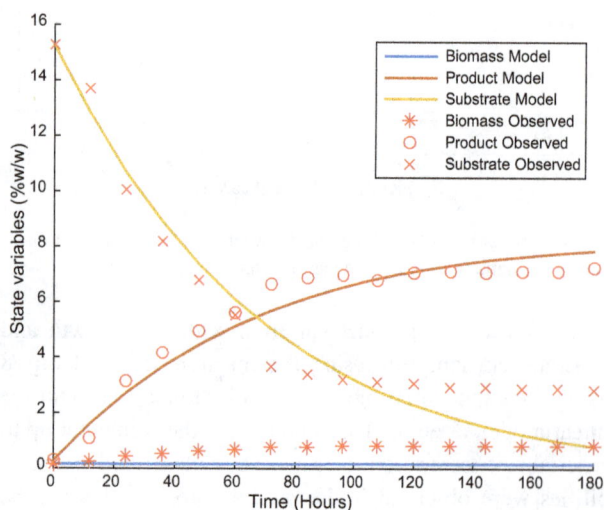

Fig. 9 Experimental results and model fitting, case of no inhibition (Monod-model)

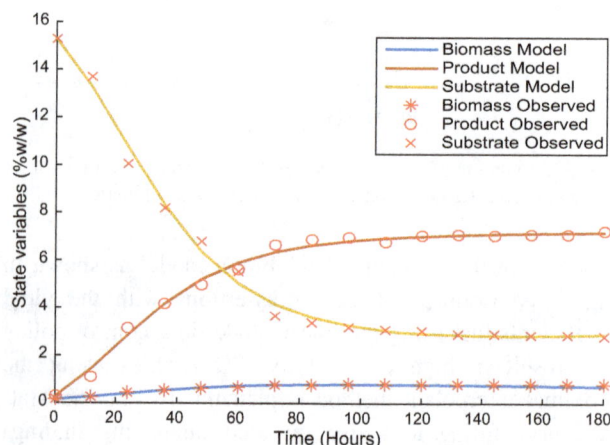

Fig. 10 Experimental results and model fitting, case of linear inhibition model

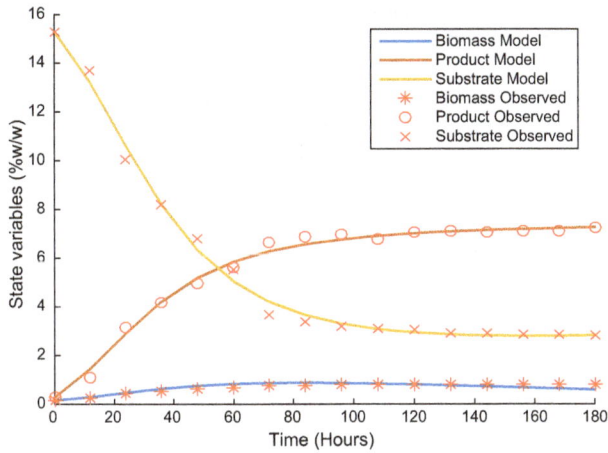

Fig. 11 Experimental results and model fitting, case of Sudden Growth Stop inhibition model

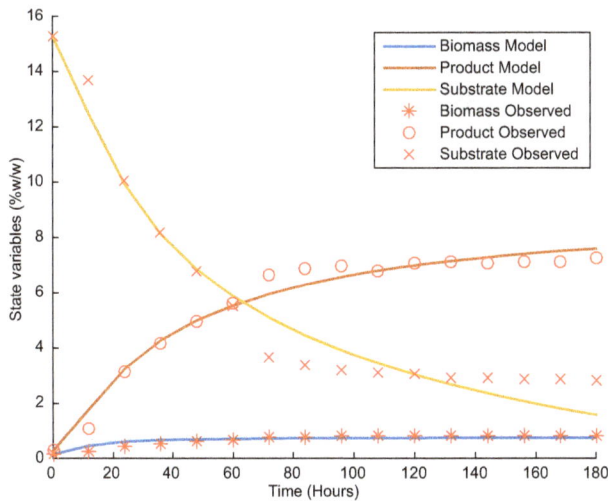

Fig. 12 Experimental results and model fitting, case of Exponential inhibition model

Table 5 Model Parameters for beer fermentation using maize extracts

| | Inhibition | | | |
	No inhibition	Linear	SGS	Exponential
Model parameters				
μ_{max}	0.0100	0.0567	0.0630	2.9922
q_{pmax}	10.4778	0.7784	0.9338	8.2438
P_{xmax}	–	–	6.7249	–
P_{pmax}	–	–	7.3560	–
K_{ix}	–	0.1459	–	0.7179
K_{ip}	–	0.1375	–	0.2631
K_{sx}	250.000	1.2621	1.9085	231.9805
K_{sp}	200.00	7.1081	11.7803	199.9890
Y_x	0.1000	0.1084	0.1108	1.0000
Y_p	0.5370	1.2548	1.3938	0.6210
G_s	0.0010	0.0018	0.0051	0.0010
M_s	0.010	0.0108	0.0141	0.0100
Model error	27.9231	2.0381	2.0941	11.340

the linear and exponential models showed no significant difference with the experimental data.

The linear and sudden growth stop models inhibition models were both used to simulate this variation and both showed a decrease in substrate concentration and an increase in product concentration with cell growth. What was intriguing is the curve-like behavior that was observed at the end of the profiles, as shown in Figs. 13 and 14. There are two theories to the explanation of this behavior: Ethanol inhibition and stationary/decline phases in batch growth kinetics. In the case of ethanol inhibition, accumulation of ethanol to inhibitory levels leads to disruption of the cell membrane and correspondingly nonlinearity in the substrate consumption and product formation profiles, Chen and McDonald [2, 3], Beuse et al. [4], Fengwu [5], Sutton [11]. The more intense curvature observed with the sudden growth model is due to its relatively high maximal

rate of product formation, hence rapidly accumulates ethanol to inhibitory concentration. Regarding the stationary and decline phase theory in batch growth kinetics, the curve-like behavior observed in the linear model can be attributed to the fact the cells grow and get to the stationary and decline phases resulting and hence no longer consume substrate and produce ethanol as in the exponential phase, resulting in the observed nonlinearity. These theories are further confirmed by the Figs. 15 and 16 which, respectively, simulates cell growth and product formation throughout the duration of fermentation. It can be observed that cells in the fermenter start declining after a certain time of fermentation buttressing the attainment of ethanol inhibition and disruption of cell membrane leading to cell death. The nonlinear patterns observed at the start of the fermentation can be attributed to high sugar concentrations encountered immediately after hydrolysis which exert osmotic stress on yeast leading to nonlinearity in their pattern of growth and product formation Russell [6], Sutton [11].

Conclusions and recommendations-sorghum

A mathematical approach to study the product inhibitor on the kinetics of alcoholic fermentation of sorghum and maize has been presented. The results of sorghum extract fermentation showed that there exist both ethanol inhibition in the alcoholic fermentation of sorghum extracts and can be described at a 99% confidence interval as being a

Table 6 Model Statistical Validity with kinetics of linear inhibition, two sample F-test for variance

	Biomass		Product		Substrate	
	Experimental (Xobs)	Model (Xpred)	Experimental (Pobs)	Model (Ppred)	Experimental (Sobs)	Model (Spred)
Mean	0.673	0.670	5.580	5.592	5.644	5.641
Standard error	0.054	0.052	0.566	0.556	1.023	1.017
Standard deviation	0.218	0.206	2.263	2.223	4.091	4.069
Observations	16	16	16	16	16	16
Confidence interval	0.990		0.990		0.990	
f	0.9002		0.9647		0.9890	
Pr(F < f) two-tailed	0.8414		0.9455		0.9832	

Table 7 Model Statistical Validity with kinetics of Sudden Growth Stop inhibition, two sample F-test for variance

	Biomass		Product		Substrate	
	Experimental (Xobs)	Model (Xpred)	Experimental (Pobs)	Model (Ppred)	Experimental (Sobs)	Model (Spred)
Mean	0.673	0.684	5.579	5.592	5.644	5.641
Standard error	0.054	0.052	0.566	0.555	1.023	1.016
Standard deviation	0.218	0.209	2.263	2.221	4.091	4.066
Observations	16	16	16	16	16	16
Confidence interval	0.990		0.990		0.990	
f	0.9194		0.9630		0.9878	
Pr(F < f) two-tailed	0.8728		0.9427		0.9814	

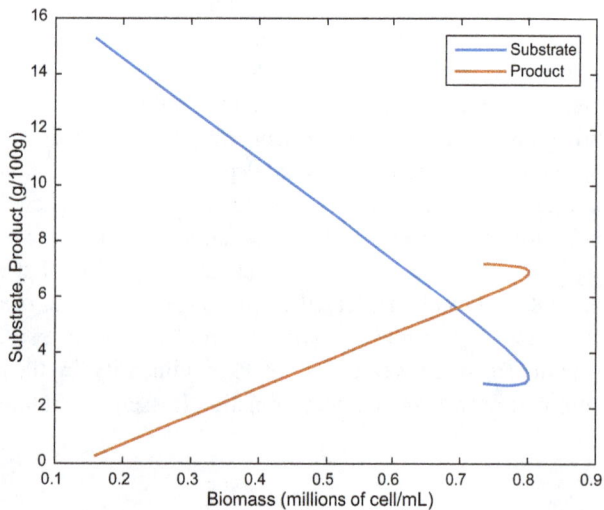

Fig. 13 Simulation of substrate and Product variation as a function of biomass during fermentation of maize extracts using Linear Inhibition kinetic model

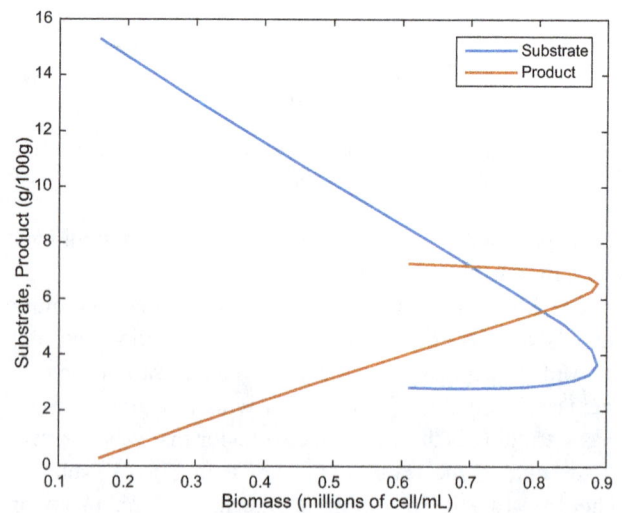

Fig. 14 Simulation of substrate and Product as a function biomass during fermentation, using Sudden Growth Stop Inhibition kinetic

linear and or an exponential. The results of maize extract fermentation showed that there exist ethanol inhibition which can be described with a 99% confidence interval as being either a linear or sudden growth at the inhibitory

ethanol concentration. The sudden growth stop model resulted in a relatively high maximal rate of ethanol accumulation hence rapidly approaching inhibitory levels. The paper, therefore, recommends for further research to study optimal control problems that can be formulated

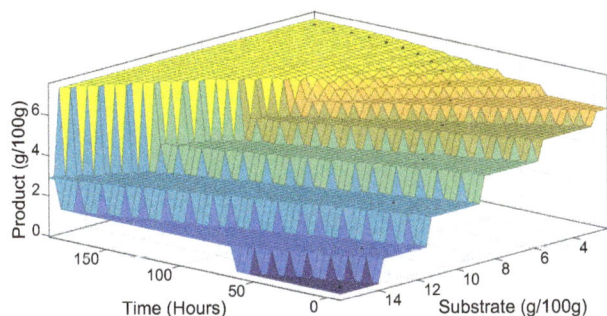

Fig. 15 3D Proximal Interpolant Simulation of product dynamics during fermentation using linear inhibition kinetic Model. Nearest neighbor interpolant

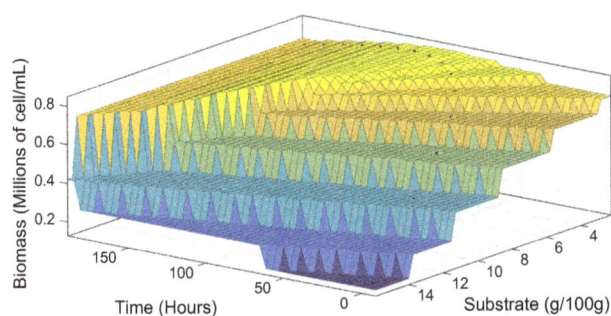

Fig. 16 3D Proximal Interpolant Simulation of Biomass Variation during fermentation using Sudden Growth Stop inhibition kinetic Model

using the models suggested in this work to understand controls variables that can minimize the effect of such inhibitions in batch fermentations.

Acknowledgements The Authors wishes to expresses their sincere appreciations to the Intra-ACP under the project "Strengthening African Higher Education through Academic Mobility", STREAM. Appreciation also goes to some breweries in Ghana for supplying us with industrial data for model validation.

References

1. Alford JS (2006) Bioprocess control: advances and challenges. Comput Chem Eng 30:1464–1475
2. Chen CI, McDonald KA (1990a) Oscillatory behavior of Saccharomyces cerevisiae in continuous culture: I. Effects of pH and nitrogen levels. Biotechnol Bioeng 36:19–27
3. Chen CI, McDonald KA (1990b) Oscillatory behavior of Saccharomyces cerevisiae in continuous culture: II. Analysis of cell synchronization and metabolism. Biotechnol Bioeng 36:28–38
4. Beuse MRB, Kopmann AHD, Thoma M (1998) Effect of dilution rate on the mode of oscillation in continuous culture of Saccharomyces cerevisiae. J Biotechnol 61:15–31
5. Fengwu B (2007) Process Oscillations in Continuous Ethanol Fermentation with Saccharomyces cerevisiae. PhD thesis University of Waterloo, Ontario, Canada
6. Russell AD (2003) Similarities and differences in the responses of microorganisms to biocides. J Antimicrob Chemother 52:750–776
7. Hinshelwood C (1946) NThe chemical kinetics of the bacterial cell. Clarendon Press, Oxford
8. Ghose TK, Tyasi RD (1979) Rapid ethanol fermentation and cellulose hydrolyase. II Product and substrate inhibition and optimization of fermenter design. Biotechnol Bioeng 21(8):1401–1420
9. Aiba S, Shoda M, Nagatani M (1968) Kinetics of product inhibition in alcohol fermentation. Biotechnol Bioeng 10:845–864
10. Ingledew WM (1999) Alcohol production by saccharomyces cerevisiae: a yeast primer. In: the Alcohol Textbook. 3rd Edn, pp 49–87
11. Sutton KB (2011) Fermentation Inhibitors, Novozyme report. 18600-0

3

Removal of Basic Blue 41 dyes using *Persea americana*-activated carbon prepared by phosphoric acid action

Abdelmajid Regti[1] · My Rachid Laamari[1] · Salah-Eddine Stiriba[2,3] ·
Mohammadine El Haddad[1]

Abstract Adsorption study of Basic Blue 41 dye onto activated carbon from *Persea americana* nuts with phosphoric acid activation was achieved. The effect of operating parameters, the effect of pH (2–12), adsorbent amount (5–30 mg/50 mL), dye concentration (25–125 mg/L), contact time (0–200 min) and temperature (298–323 K), on the adsorption capacity was examined. The experimental isotherm data were analyzed using Langmuir and Freundlich models, which showed that the best fit was achieved by the Langmuir model with the maximum monolayer adsorption capacity at 625 mg/g. The adsorption kinetic process followed pseudo-second-order kinetics. Thermodynamic evaluation showed that the process was endothermic ($\Delta H^0 = 144.60$ kJ/mol) and spontaneous (ΔG^0 varied from to -11.64 to -19.50 kJ/mol), while the positive value of entropy ($\Delta S^0 = 524.3$ J/mol K) revealed increased randomness at the adsorbent–adsorbate interface. It was found to be a very efficient adsorbent and a promising alternative for dye removal from aqueous solutions.

Keywords Removal of dye · *Persea americana*-activated carbon · Surface area · Adsorption · Kinetics and thermodynamic studies

Introduction

The textile industry plays a part in the economy of several countries around the world. However, effluents from textile and dyeing have a low biological oxygen demand and strong chemical oxygen demand. Disposal of this colored water into receiving water can be toxic to aquatic life and cause food chain contamination, resulting in deleterious health effect even in very low concentrations. Moreover, most of these dyes can cause allergy, dermatitis, skin irritation and also provoke cancer and mutation in humans [1, 2]. Dyes are usually highly visible, very difficult to biodegrade, and extremely difficult to eliminate in natural aquatic environments [3, 4].

To improve the effluent quality, the addition of physical and/or chemical treatments comprising adsorption [5–9], photocatalytic [10, 11] or electrochemical methods [5] and reverse osmosis [12] are necessary. Adsorption is the most simple and known for the treatment of effluents containing dyes using the new low-cost and environmentally friendly adsorbents in the carbon-based or not activated means [13–18].

The potential properties of activated carbon as adsorbents are due to their highly developed porosity, favorable pore size distribution, large surface area, and high degree of surface reactivity [19]. Chemical activation and physical activation are two methods for the preparation of activated carbon. Chemical activation uses chemical agents for the preparation of activated carbon in a single step method, while physical activation involves carbonization of

✉ Mohammadine El Haddad
elhaddad71@gmail.com

1 Equipe de Chimie Analytique and Environnement, Faculté Poly-disciplinaire, Université Cadi Ayyad, BP 4162, 46000 Safi, Morocco

2 Equipe de Chimie Moléculaire, Matériaux et Modélisation, Faculté Poly-disciplinaire, Université Cadi Ayyad, BP 4162, 46000 Safi, Morocco

3 Instituto de Ciencia Molecular/ICMol, Universidad de Valencia, C/. Catedrático José Beltrán 2, Paterna, 46980 Valencia, Spain

carbonaceous materials followed by activation of the resulting substrate in the presence of dioxide carbon or steam as activating agents [20]. It is recognized that the carbon yields of chemical activation are higher than the physical one. The most common precursors used for the production of activated carbon are organic materials that are rich in carbon.

Several studies to find low-cost carbonaceous materials have been reported. These materials include Jerusalem artichoke [21], waste rice hulls [22], homemade cocoa shell [23], waste tea [24], coir pith [25], orange peels [26], jute sticks [27], walnut [28], palm oil shell [29], Acacia mangium wood [30] and waste tires [31].

In the present study, we examine the feasibility of using activated carbon prepared using *Persea americana* as adsorbent for the removal of Basic Blue 41 dyes from aqueous solutions. The effect of different parameters including solution pH, adsorbent dosage, dye concentration, temperature and contact time were studied to optimize the adsorption process. The isotherm and kinetic and thermodynamic parameters were examined to analyze the experimental data.

Materials and methods

The *Persea americana* nuts were collected, washed with distilled water and dried at ambient temperature for several days. The unmodified *Persea americana* nuts were abbreviated as PAN. The carbonization of PAN was carried out using an appropriate weight of PAN and 25 mL concentrated phosphoric acid with a mass ratio (1:4). A glass beaker of 100 mL was heated to 500 °C for 1 h producing a black carbonaceous residue. The solid material was neutralized with KOH solution until a neutral pH was obtained. The resulting carbonized *Persea americana* nut (C-PAN) was filtered and washed intensively with water. The C-PAN was then dried at 100 °C for 2 h and kept in desiccators for further use.

The characterization of C-PAN was achieved by FT-IR spectroscopy and X-ray powder diffraction measurements. FT-IR spectra (4000–450 cm^{-1} range) were recorded with a Nicolet 5700 FT-IR spectrometer on samples prepared as KBr pellets. The polycrystalline sample of each adsorbent was lightly ground in an agate mortar and pestle and filled into 0.5 mm borosilicate capillary prior to being mounted and aligned on an Empyrean PANalytical powder diffractometer using Cu K$_\alpha$ radiation ($\lambda = 1.54056$ Å). Three repeated measurements were obtained at room temperature in the $10° < 2\theta < 60°$ range with a step size of 0.01°. Scanning electronic microscopy (SEM) images were obtained with HITACHI-S4100 equipment operated at 20 kV.

Adsorption–desorption isotherms of nitrogen at -196 °C were measure with an automatic adsorption instrument (NOVA-1000 Gas Sorption analyzer) to determine the surface areas and total pore volumes. The BET surface area and the total pore volumes of the obtained *Persea americana*-activated carbon were found to be 1593 and 1.053 cm^3/g, respectively.

Deionized water was used throughout the experiments for solution preparation. The adsorption studies for evaluation of the C-PAN adsorbent for the removal of the Blue Basic 41 dye from aqueous solutions were carried out in triplicate using the batch contact adsorption method. Basic Blue 41 dye used in this study, abbreviated as BB41, was purchased from Sigma-Aldrich. The chemical structure of BB41 is shown in Fig. 1.

For the adsorption experiments, fixed amounts of adsorbents (5–30 mg) were placed in a 100 mL glass Erlenmeyer flask containing 50 mL of dye solution at various concentrations (25–125 mg/L), which were stirred for a suitable time (5–200 min) from 293 to 313 K. The pH of the dye solutions ranging from 2 to 10 was adjusted by 0.1 M HCl or 0.1 M NaOH to investigate the effect of pH on the adsorption processes. Subsequently, to separate the adsorbents from the aqueous solutions, the samples were centrifuged at 3600 rpm for 10 min, and aliquots of 1–10 mL of the supernatant were taken. At a predetermined time, the residual dye concentration in the reaction mixture was analyzed by centrifuging the reaction mixture and then measuring the absorbance by UV–visible spectroscopy of the supernatant at the maximum absorbance wavelength of the sample at 606 nm.

The amount of equilibrium adsorption q_e (mg/g) was calculated using the formula:

$$q_e = \frac{C_0 - C_e}{W} V, \tag{1}$$

where C_e (mg/L) is the liquid concentration of the dye at equilibrium, C_0 (mg/L) the initial concentration of the dye in solution, V the volume of the solution (L) and W the mass of the dye biosorbent (g). The BB41 removal percentage (%) can be calculated as follows:

$$\%\text{Removal} = \frac{C_0 - C_e}{C_0} \times 100, \tag{2}$$

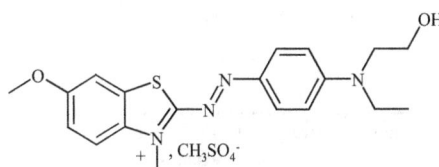

Fig. 1 The chemical structure of Basic Blue 41

where C_0(mg/L) is the initial dye concentration and C_e (mg/L) is the concentration of the dye at equilibrium.

Results and discussion

Characterization of C-PAN adsorbent

To investigate the surface characteristics of the C-PAN adsorbent, FT-IR and XRD spectra were recorded. As shown in the FT-IR spectrum in Fig. 2, the frequencies of the absorption bands of C-PAN are 876, 1074, 1149, 1563, 2916 and 3415 cm^{-1}. The absorption band at 3415 cm^{-1} is attributed to the hydroxyl group (O–H) vibration [32]. The bands at 2916, 1563 and 1381 cm^{-1} correspond, respectively, to unsymmetrical aliphatic C–H stretching, C=C stretching of aromatic rings and aromatic C=C stretching vibration. The band at 1149 cm^{-1} is ascribed to C–O stretching in alcohol or ether or the hydroxyl group [33, 34]. The bands at 1074 and 876 cm^{-1} could result from ionized linkage of P+ O− in acid phosphate esters, to symmetrical vibration in a chain of P–O–P and to P–C phosphorus-containing compound [35, 36]. These functional groups are due to the presence of H$_3$PO$_4$ acid as an activation agent in the preparation of C-PAN.

Figure 3 shows an X-ray powder diffraction pattern of C-PAN. An amorphous peak with the equivalent Bragg angle at $2\theta = 24.6$ was recorded, together with other peaks recorded at $2\theta = 17.5°$, $31.2°$ and $47.5°$. The surface morphology of the C-PAN adsorbent was examined. Figure 4 shows the SEM image indicating that the surface is relatively smooth and contains many pores. The SEM of C-PAN show very distinguished dark spots, which can be taken as a sign for effective adsorption of dye molecules in the cavities and pores of this adsorbent.

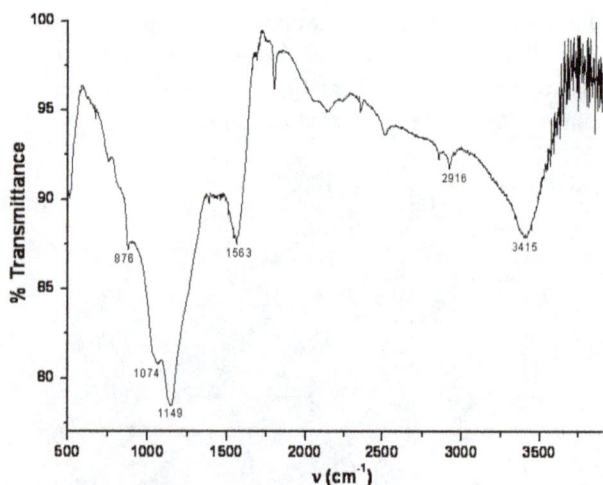

Fig. 2 FT-IR spectrum of C-PAN adsorbent

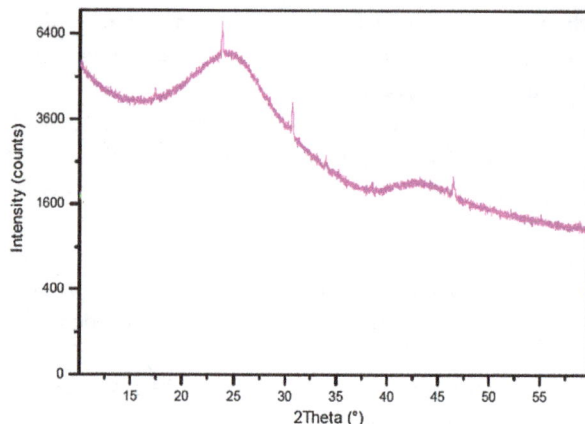

Fig. 3 XRD spectrum of C-PAN adsorbent

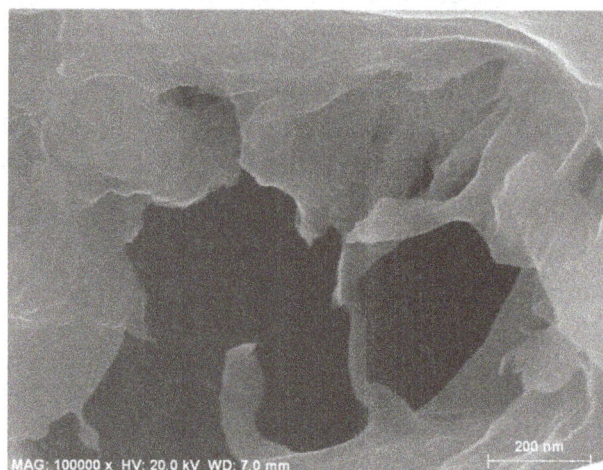

Fig. 4 SEM image of C-PAN adsorbent

Effect of pH

To examine the pH effect of the initial aqueous dye solution, a concentration of BB41 dye at 100 mg/L and 20 mg of C-PAN adsorbent was used, keeping the temperature at 20 °C at different pH values in the range 2–12. Figure 5 shows the variation of dye removal vs. pH. In fact, the amount of dye adsorbed onto C-PAN was found to be constant for all pH values being studied. The experiments carried out at different pH values showed that there was no significant change in the percent removal of dye over the entire pH range. This indicates the strong affinity of the dye to C-PAN and that either H$^+$ or OH$^-$ ions could influence the dye adsorption capacity. The dye removal % uptake comprised between 92 and 97 %. If the adsorption would have occurred through an ion exchange mechanism, there should have been an influence on the dye adsorption while varying the pH. This observation and the high positive ΔH^0 value obtained indicate irreversible adsorption, probably

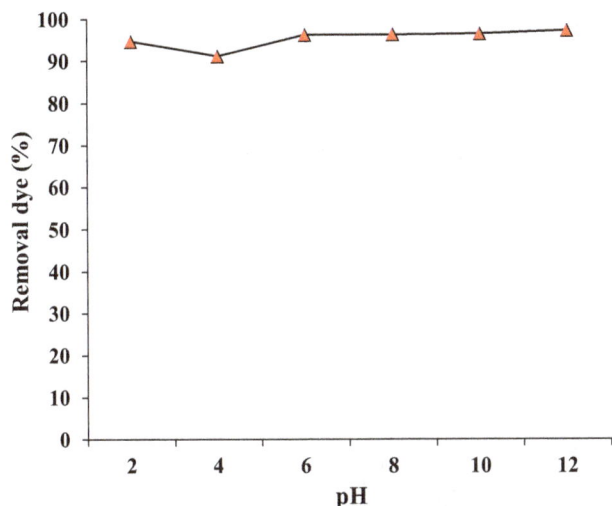

Fig. 5 Effect of pH on the removal of BB41 from aqueous solution onto C-PAN. Initial dye concentration 100 mg/L; adsorbent amount 200 mg; agitation time 80 min; temperature 20 °C

due to nonpolar interactions [37]. Other studies for different dyes were also found to be independent of pH [37–39].

The effect of C-PAN adsorbent dosage

The effect of C-PAN dosage on adsorption was studied at 20 °C with 100 mg/L dye solution. The adsorbent dosage ranged from 5 to 30 mg. The plots of dye removal (%) versus time at different adsorbent dosages are shown in Fig. 6. The results follow the expected pattern, in which the removal (%) of BB41 increased with increase in C-PAN adsorbent dosage. This might be due to an increase in the surface active sites in C-PAN samples. The adsorption

equilibrium was achieved after 120 min of stirring the dye solution with the appropriate amount of C-PAN adsorbent. At this time, the removal of BB41 (%) increased from 45.7 to 99 % for C-PAN adsorbent dosage of 5 mg to 30 mg. However, the amount of BB41 adsorbed onto C-PAN, q (mg/g) was found to decrease from 457.46 to 165.83 mg/g upon increasing the adsorbent dosage. This behavior could be due to the high number of unsaturated sorption sites during the adsorption process.

Effect of BB41 concentration

The initial adsorbate concentration provides an important driving force to overcome all mass transfer resistances of dye between the aqueous and solid phases. The effect of initial concentration of dye on the adsorption was studied at 20 °C with 30 mg of C-PAN adsorbent. The adsorbate concentration ranged from 25 to 125 mg/L. The plots of dye removal (%) vs. time at different initial BB41 concentrations are shown in Fig. 7. The removal dye (%) shows a decreasing trend as the initial dye concentration of the dye is increased. At lower concentrations, all adsorbate ions present in the medium could interact with the binding sites, resulting in higher dye removal (%). At higher concentrations, the dye removal (%) shows a decreasing behavior because of the saturation of the adsorption sites. Similarly, the adsorption equilibrium was achieved after 120 min of stirring the appropriate dye solution with the C-PAN adsorbent. At this time, the removal of BB41 (%) decreased from 99 to 77.5 % for the initial BB41 concentration of 25 to 125 mg/L. However, the amount of BB41 adsorbed onto C-PAN, q_e(mg/g) was found to increase from 124.41 to 484.35 mg/g upon increasing the initial dye concentration

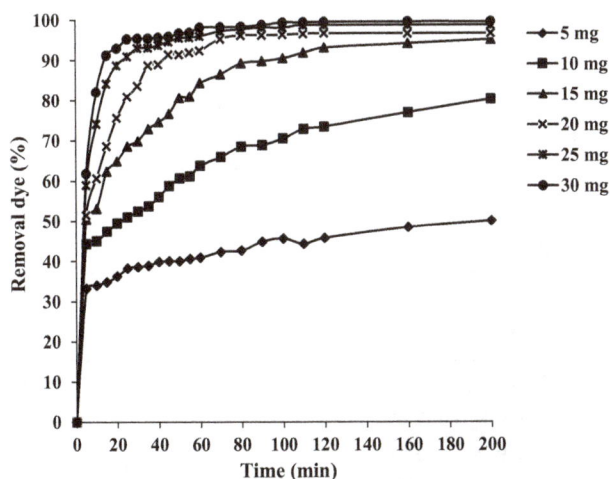

Fig. 6 Effect of C-PAN adsorbent dosage and contact time on the removal of BB41 from aqueous solution. Initial dye concentration 100 mg/L; temperature 20 °C

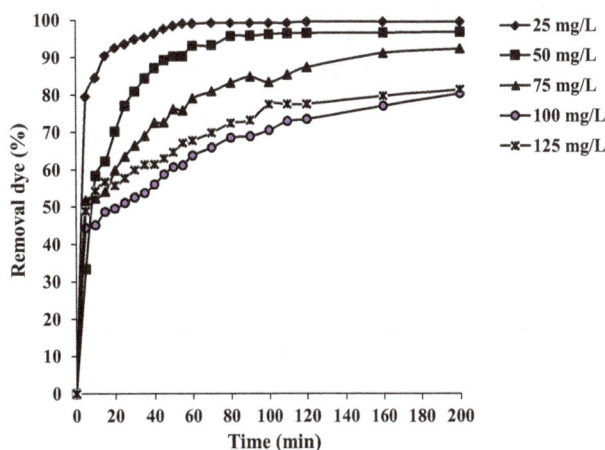

Fig. 7 Effect of initial dye concentration and contact time on the removal of BB41 from aqueous solution. C-PAN 30 mg/L; temperature 20 °C

Adsorption kinetics

To study the sorption process of BB41 onto C-PAN adsorbent, the data obtained from kinetic adsorption experiments were simulated with pseudo-first-order and pseudo-second-order models.

The pseudo-first-order equation is generally represented as follows [40]:

$$\log(q_e - q_t) = \log(q_e) - \frac{k_1}{2.303}t, \tag{3}$$

where q_e is the amount of dye adsorbed at equilibrium (mg/g), q_t the amount of dye adsorbed at time t (mg/g), k_1 the pseudo-first-order rate constant (min^{-1}) and t the time (min).

The values of k_1, q_e calculated from the equation and the correlation coefficient (R^2) values of fitting the first-order rate model at different concentrations are presented in Table 1. The linearity plots of log ($q_e - q_t$) versus time at different initial dye concentrations (Fig. 8) suggested that the process of dye adsorption did not follow the pseudo-first-order rate kinetics. Also from Table 1, it is indicated that the values of the correlation coefficients are not high for the different dye concentrations. Furthermore, the estimated values of q_e calculated from the equation qe differ substantially from those measured experimentally. This gives confirmation that the adsorption process of BB41 onto C-PAN did not obey the pseudo-first-order model.

The pseudo-second-order equation is generally represented as follows [41]:

$$\frac{t}{q_t} = \frac{1}{k_2 q_e^2} + \frac{1}{q_e}t, \tag{4}$$

where k_2 is the pseudo-second-order rate constant (g/mg min).

A plot of t/q_t and t should give a linear relationship if the adsorption follows pseudo-second-order model. To understand the applicability of the model, a linear plot of t/q_t vs. t under different dye concentrations was plotted as in Fig. 9. The constants k_2, q_e and correlation coefficients (R^2)

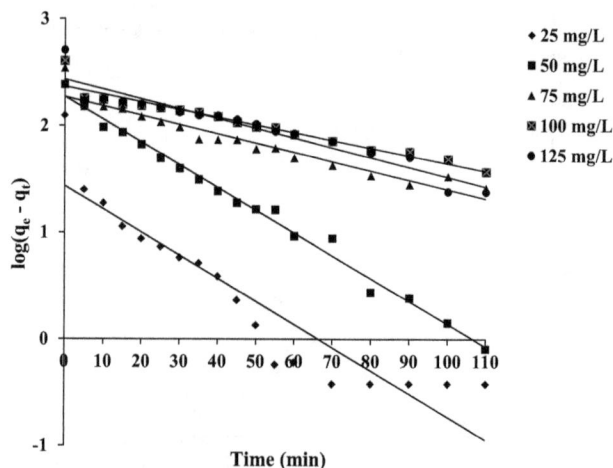

Fig. 8 Pseudo-first-order plots for different initial dye concentrations removal using C-PAN adsorbent. C-PAN adsorbent 30 mg; temperature 20 °C

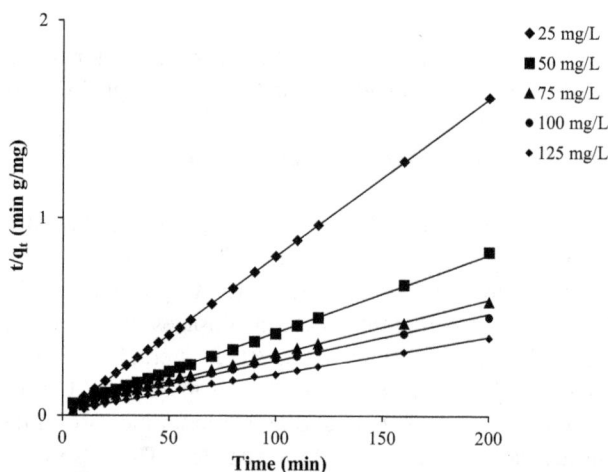

Fig. 9 Pseudo-second-order plots for different initial dye concentration removal using C-PAN adsorbent. C-PAN adsorbent 30 mg; temperature 20 °C

were calculated from the plot and are given in Table 1. The q_e determined from the model along with correlation coefficients indicated that q_e was very close to q_{exp} and the

Table 1 The pseudo-first-order and pseudo-second-order kinetic parameters for BB41 removal using C-PAN

Concentration of BB41 (mg/L)	Pseudo-first order				Pseudo-second order		
	q_e,exp (mg/g)	k_1 (min^{-1})	q_e,cal (mg/g)	r^2	$k_2 \times 10^{-4}$ (g/mg min)	q_e,cal (mg/g)	r^2
25	124.41	0.0493	027.03	0.862	59.65	123.45	0.999
50	241.90	0.0483	186.63	0.989	5.248	256.41	0.999
75	345.84	0.0184	234.96	0.918	2.548	344.82	0.997
100	401.45	0.0161	233.34	0.928	1.689	400.00	0.992
125	508.13	0.0207	271.01	0.906	1.616	502.51	0.995

correlation coefficient was also greater than 0.99. As a matter of consequence, the system BB41–C-PAN could be well described by the pseudo-second-order model. This adequate model shows that the adsorption of BB41 onto C-PAN is controlled by chemisorption mechanism.

Adsorption isotherms

Adsorption isotherms are basic requirements for the design of adsorption systems. It can express the relationship between the amounts of adsorbate by unit mass of adsorbent at a constant temperature. Herein, we analyzed our experimental data by Langmuir and Freundlich isotherms models. The best-fitting model was evaluated using the correlation coefficient.

The Langmuir model [42] permits the evaluation of maximum dye adsorption capacity when all sites have equal affinities and active sites of adsorbent are independent of each other. The linear form of the Langmuir isotherm is expressed as follows:

$$\frac{C_e}{q_e} = \frac{1}{q_m K_L} + \frac{1}{q_m} C_e, \tag{5}$$

where C_e (mg/L) is the equilibrium concentration of the BB41 dye and q_e (mg/g) is the amount of BB41 adsorbed per unit mass of adsorbent. q_m (mg/g) and K_L (L/mg) are the constants related to the maximum adsorption capacity and is the Langmuir constant, respectively.

A straight line with a slope of $1/q_m$ and intercept of $1/q_m K_L$ is obtained when C_e/q_e is plotted against C_e. Table 2 shows the values of these parameters.

The essential characteristics of the Langmuir equation can be expressed in terms of the dimensionless separation factor, R_L, defined as:

$$R_L = \frac{1}{1 + K_L C_0}, \tag{6}$$

where C_0 is the initial concentration of the BB41 dye; the R_L value implies whether the adsorption is unfavorable: $R_L > 1$, linear: $R_L = 1$, favorable: $0 < R_L < 1$, or irreversible: $R_L = 0$.

Figure 10 depicts the plot of the calculated R_L values versus the initial dye concentration at 298, 303 and 313 K. It was observed that all the R_L values obtained were between 0 and 1, showing that the adsorption of BB41 onto C-PAN was favorable. The R_L values decrease upon increasing the initial dye concentration, which indicates that the adsorption is more favorable at higher BB41 concentrations. The adsorption capacity increases on increasing the temperature, attaining a higher value of adsorption capacity of 625 mg/g at 313 K. The values of K_L increased on an increase in temperature. Therefore, by increasing the temperature, higher adsorption capacities were achieved.

Freundlich isotherm [43] is observed if the sites with stronger binding affinities are occupied by the dye molecules first and the binding strength decreases with increase in the degree of site occupation. The linear form of the Freundlich isotherm is expressed as follows:

$$\log(q_e) = \log(K_f) + \frac{1}{n}\log(C_e), \tag{7}$$

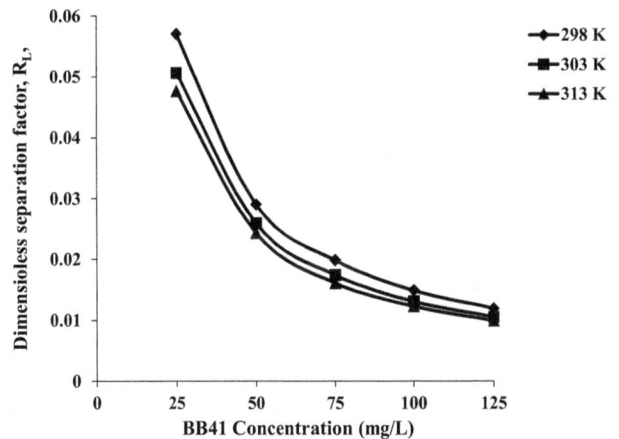

Fig. 10 Dimensionless separation factor, R_L, versus concentration dye

Table 2 Adsorption isotherm constants for removal of BB41 onto the C-PAN adsorbent

	Temperature		
	298 K	303 K	313 K
Langmuir isotherm			
q_m (mg/g)	500.00	555.55	625.00
K_L (L/mg)	0.66	0.75	0.80
r^2	0.997	0.995	0.991
Freundlich isotherm			
K_F (mg/g) (L/g)	213.80	269.15	359.74
n	4.03	3.97	4.10
r^2	0.982	0.984	0.979

Table 3 Thermodynamic data for the adsorption of BB41 onto C-PAN

T (K)	K_C	ΔG^0 (kJ/mol)	ΔH^0 (kJ/mol)	ΔS^0 (J/mol K)
298	106.51	−11.64	144.6	524.3
303	287.90	−14.26		
313	624.08	−16.88		
323	1354.26	−19.50		

Table 4 Comparison of the maximum monolayer adsorption capacities of C-PAN with those of various AC adsorbents

AC adsorbent	Dye	Isotherm	q_m (mg/g)	Reference
Pomegranate peel	Remazol brilliant	Freundlich	370.86	[44]
Homemade cocoa shell	Reactive violet 5	Liu	603.3	[45]
Rambutan peel	Acid yellow 17	Langmuir	215.05	[46]
Rice husks	Methylene blue	Langmuir	578	[47]
Pomelo skin	Methylene blue	Langmuir	501.1	[48]
Stricta algae based	Safranin O	Langmuir	526	[49]
Persea americana	Basic Blue 41	Langmuir	625	This study

where K_f (mg/g) (L/g) and n are Freundlich constants related to the adsorption capacity and adsorption intensity, respectively.

The Freundlich constants n and K_f were obtained from the plot of log (q_e) versus log (C_e) that should give a straight line with a slope of ($1/n$) and intercept of log (K_f) (as shown in Table 2). In this study, the values found for n were superior to 1, which indicates that the adsorption of BB41 onto C-PAN is favorable.

Thermodynamic study

The thermodynamic data reflect the feasibility and favorability of the adsorption. The parameters such as free energy change (ΔG^0), enthalpy change (ΔH^0) and entropy change (ΔS^0) can be estimated by the change of equilibrium constants with temperature. The free energy change of the sorption reaction is given by:

$$\Delta G^0 = -RT \ln K_C, \qquad (8)$$

where ΔG^0 is the free energy change (kJ/mol), R the universal gas constant (8.314 J/mol K), T the absolute temperature (K) and K_C the equilibrium constants (q_e/C_e). The values of ΔH^0 and ΔS^0 can be calculated from the Van't Hoff equation:

$$\ln K_C = -\frac{\Delta H^0}{RT} + \frac{\Delta S^0}{R}, \qquad (9)$$

where $\ln K_C$ is plotted against $1/T$, and a straight line with the slope ($-\Delta H^0/R$) and intercept ($\Delta S^0/R$) are found. The calculated thermodynamic parameters are depicted in Table 3.

In the study, ΔG^0 values were determined at different temperatures and decrease from -11.64 to -19.50 kJ/mol when the temperature increases from 298 to 323 K. The negative values of ΔG^0 suggest that the adsorption of BB41 onto C-PAN is a highly favorable process. The values of ΔH^0 and ΔS^0 were obtained as 144.6 and 524.3 J/mol K, respectively. The positive value of ΔH^0 shows that the adsorption is an endothermic process, while a positive value of ΔS^0 reflects the increase of randomness state at the solid/solution interface during the adsorption.

Comparison of adsorption capacities of C-PAN with those of various AC adsorbents

Table 4 depicts the values of maximum monolayer adsorption capacities of many AC adsorbents. It appeared that C-PAN can be classed belong the best ACs adsorbents with an adsorption capacity of 625 mg/g.

Conclusion

Based on the results obtained in this study, it appears that the activated carbon prepared from *Persea American* nut constitutes a good adsorbent for removing a dye from aqueous solutions:

- The percent (%) removal of BB41 was observed to increase with increasing initial dye concentration and increasing adsorbent dose.
- The Langmuir isotherm best described the equilibrium data with acceptable R^2, which signifies that a homogeneous adsorption takes place between the BB41 dye and C-PAN.
- The pseudo-second-order equation best describes the kinetics of the C-PAN adsorption system due to its high R^2. In addition, the theoretical q_e generated by the pseudo-second-order equation is in good agreement with the experimental q_e value. This implies that the rate-limiting step is a chemisorption process.
- Thermodynamic studies indicated that the adsorption process is endothermic and spontaneous.

References

1. Yagub MT, Sen TK, Afroze S, Ang HM (2014) Dye and its removal from aqueous solution by adsorption: a review. Adv Colloid Interface Sci 209:172–184
2. Banat IM, Nigam P, Singh D, Marchant R (1996) Microbial decolorization of textile-dye containing effluents: a review. Bioresour Technol 58:217–227
3. Rafatullah M, Sulaiman O, Hashim R, Ahmad A (2010) Adsorption of methylene blue on low-cost adsorbents: a review. J Hazard Mater 177:70–80

4. Vakili M, Rafatullah M, Salamatinia B, Abdullah AZ, Ibrahim MH, Tan KB, Gholami Z, Amouzgar P (2014) Application of chitosan and its derivatives as adsorbents for dye removal from water and wastewater: a review. Carbohyd Polym 113:115–130

5. Gupta VK, Jain R, Nayak A, Agarwal S, Shrivastava M (2011) Removal of the hazardous dye-Tartrazine by photodegradation on titanium dioxide surface. J Hazard Mater 31:1062–1067

6. Ahmad T, Danish M, Rafatullah M, Ghazali A, Sulaiman O, Hashim R, Ibrahim MN (2012) The use of date palm as a potential adsorbent for wastewater treatment: a review. Environ Sci Technol 19:1464–1484

7. Low LW, Teng T, Rafatullah M, Morad N, Azahari B (2014) adsorption studies of methylene blue and malachite green from aqueous solutions by pretreated lignocellulosic materials. Separ Sci Technol 48:1688–1698

8. Mittal A, Mittal J, Malviya A, Kaur D, Gupta VK (2010) Decoloration treatment of a hazardous triarylmethane dye, light green SF (Yellowish) by waste material adsorbents. J Colloid Interface Sci 342:518–527

9. Gupta VK, Nayak A (2012) Cadmium removal and recovery from aqueous solutions by novel adsorbents prepared from orange peel and Fe_2O_3 nanoparticles. Chem Eng J 180:81–90

10. Gupta VK, Jain R, Mittal A, Tawfik A, Saleh A, Naya A, Agarwal S, Sikarwa S (2012) Photo-catalytic degradation of toxic dye amaranth on TiO_2/UV in aqueous suspensions. Mater Sci Eng C 32:12–17

11. Saleh TA, Gupta VK (2012) Photo-catalyzed degradation of hazardous dye methyl orange by use of a composite catalyst consisting of multi-walled carbon nanotubes and titanium dioxide. J Colloid Interface Sci 371:101–106

12. Gupta VK, Ali I, Saleh TA, Nayak A, Agarwal S (2012) Chemical treatment technologies for waste-water recycling-an overview. RSC Adv 2:6380–6388

13. Gupta VK, Srivastava SK, Mohan D, Sharma S (1997) Design parameters for fixed bed reactors of activated carbon developed from fertilizer waste for the removal of some heavy metal ions. Waste Manage 17:517–522

14. Mittal A, Kaur A, Malviya A, Mittal J, Gupta VK (2009) Adsorption studies on the removal of coloring agent phenol red from wastewater using waste materials as adsorbents. J Colloid Interface Sci 337:345–354

15. Mittal A, Mittal J, Malviya A, Gupta VK (2009) Adsorptive removal of hazardous anionic dye "Congo red" from wastewater using waste materials and recovery by desorption. J Colloid Interface Sci 340:16–26

16. Gupta VK, Kumar R, Nayak A, Saleh TA, Barakat MA (2013) Adsorptive removal of dyes from aqueous solution onto carbon nanotubes: a review. Adv Colloid Interface Sci 193–194:24–34

17. Gupta VK, Nayak A, Bhushan B, Agarwal S (2015) A critical analysis on the efficiency of activated carbons from low-cost precursors for heavy metals remediation. Crit Rev Environ Sci Technol 45:613–668

18. Mittal A, Mittal J, Malviya A, Gupta VK (2010) Removal and recovery of Chrysoidine Y from aqueous solutions by waste materials. J Colloid Interface Sci 344:497–507

19. Ncibi MC, Sillanpää M (2015) Mesoporous carbonaceous materials for single and simultaneous removal of organic pollutants: activated carbons vs. carbon nanotubes. J Mol Liq 207:237–247

20. Ahmad AA, Hameed BH, Ahmad AL (2009) Removal of disperse dye from aqueous solution using waste-derived activated carbon: optimization study. J Hazard Mater 170:612–619

21. Yu L, Luo Y (2014) The adsorption mechanism of anionic and cationic dyes by Jerusalem artichoke. J Environ Chem Eng 2:220–229

22. Luna MD, Flores ED, Genuino DA, Futalan CM, Wan MW (2013) Adsorption of Eriochrome Black T (EBT) dye using activated carbon prepared from waste rice hulls-optimization, isotherm and kinetic studies. J Taiwan Ins Chem Eng 44:646–653

23. Ribas MC, Adebayo MA, Prola LD, Lima EC, Cataluña R, Feris LA, Puchana-Rosero MJ, Machado FM, Pavan FA, Calvete T (2014) Comparison of a homemade cocoa shell activated carbon with commercial activated carbon for the removal of reactive violet 5 dye from aqueous solutions. Chem Eng J 248:315–326

24. Auta M, Hameed BH (2011) Optimized waste tea activated carbon for adsorption of Methylene Blue and Acid Blue 29 dyes using response surface methodology. Chem Eng J 175:233–243

25. Santhy K, Selvapathy P (2006) Removal of reactive dyes from wastewater by adsorption on coir pith activated carbon. Bioresour Technol 97:1329–1336

26. Fernandez ME, Nunell GV, Bonelli PR, Cukierman AL (2014) Activated carbon developed from orange peels: batch and dynamic competitive adsorption of basic dyes. Ind Crops Products 62:437–445

27. Asadullah M, Asaduzzaman M, Kabir MS, Mostofa MG, Miyazawa T (2010) Chemical and structural evaluation of activated carbon prepared from jute sticks for Brilliant Green dye removal from aqueous solution. J Hazard Mater 174:437–443

28. Heibati B, Rodriguez-Couto S, Amrane A, Rafatullah M, Hawari A, Al-Ghouti MA (2014) Uptake of Reactive Black 5 by pumice and walnut activated carbon: chemistry and adsorption mechanisms. J Ind Eng Chem 20(5):2939–2947

29. Rafatullah M, Ahmad T, Ghazali A, Sulaiman O, Danish M, Hashim R (2013) Oil palm biomass as a precursor of activated carbons: a review. Crit Rev Environ Sci Technol 43:1117–1161

30. Danish M, Hashim R, Mohamad Ibrahim MN, Rafatullah M, Ahmad T, Sulaiman O (2011) Characterization of acacia mangium wood based activated carbons prepared in the presence of basic activating agents. Bioresour Technol 6(3):3019–3033

31. Saleh TA, Gupta VK (2014) Processing methods: characteristics and adsorption behavior of tire derived carbons: a review. Adv Colloid Interf Sci 211:92–100

32. Liang S, Guo X, Tian Q (2011) Adsorption of Pb^{2+} and Zn^{2+} from aqueous solutions by sulfured orange peel. Desalination 275:212–216

33. He J, Ma Y, He J, Zhao J, Yu JC (2002) Photooxidation of azo dye in aqueous dispersions of $H2O2$/α-FeOOH. Appl Catal B Environ 39:211–220

34. Pavia DL, Lampman, GM, Kaiz GS (1987) Introduction to spectroscopy: a guide for students of organic chemistry. W B Saunders Company

35. Liu H, Zhang J, Bao N, Cheng C, Ren L, Zhang C (2012) Textural properties and surface chemistry of lotus stalk-derived activated carbons prepared using different phosphorus oxyacids: adsorption of trimethoprim. J Hazard Mater 235:236367–236375

36. Wang Z, Nie E, Li J, Zhao Y, Luo X, Zheng Z (2011) Carbons prepared from Spartina alterniflora and its anaerobically digested residue by $H3PO4$ activation: characterization and adsorption of cadmium from aqueous solutions. J Hazard Mater 188:29–36

37. Syed Shabudeen PS, Venckatsh R, Pattabhi S (2006) Preparation and utilization of kapok hull carbon for the removal of rhodamine B from aqueous solution. J Chem 3:83–96

38. Mohan D, Singth KP, Sinha S, Gosh D (2001) Removal of pyridine from aqueous solution using low cost activated carbons derived from agricultural waste materials. Carbon 42:2409–2421

39. Hu Z, Chen H, Ji K, Yuan S (2010) Removal of Congo red from aqueous solution by cattail root. J Hazard Mater 173:292–297

40. Ozcan A, Omeroglu C, Erdogan Y, Ozcan AS (2007) Modification of bentonite with a cationic surfactant: an adsorption study of textile dye Reactive Blue 19. J Hazard Mater 140:173–179

41. Moussavi G, Mahmoudi M (2009) Removal of azo and anthra-

quinone reactive dyes from industrial wastewaters using MgO nanoparticles. J Hazard Mater 168:806–812

42. Langmuir I (1916) The constitution and fundamental properties of solids and liquids. J Am Chem Soc 39:2221–2295

43. Freundlich MF (1906) Over the adsorption in solution. J Phys Chem 57:385–470

44. Mohd AA, Nur AA, Olugbenga SB (2014) Kinetic, equilibrium and thermodynamic studies of synthetic dye removal using pomegranate peel activated carbon prepared by microwave-induced KOH activation. Water Resour Ind 6:18–35

45. Marielen CR, Matthew AA, Lizie TP, Eder CL, Renato C, Liliana AF, Puchana-Rosero MJ, Fernando MM, Flávio P, Calvete T (2014) Comparison of a homemade cocoa shell activated carbon with commercial activated carbon for the removal of reactive violet 5 dye from aqueous solutions. Chem Eng J 248:315–326

46. Njoku VO, Foo KY, Asif M, Hameed BH (2014) Preparation of activated carbons from rambutan(*Nephelium lappaceum*) peel by microwave-induced KOH activation for acid yellow 17 dye adsorption. Chem Eng J 250:198–204

47. Chen Y, Zhai SR, Liu N, Song Y, An QD, Song XW (2013) Dye removal of activated carbons prepared from NaOH-pretreated rice husks by low-temperature solution-processed carbonization and H_3PO_4 activation. Bioresour Technol 144:401–409

48. Foo KY, Hameed BH (2011) Microwave assisted preparation of activated carbon from pomelo skin for the removal of anionic and cationic dyes. Chem Eng J 173:385–390

49. Attouti S, Bestani B, Benderdouche N, Laurent D (2013) Application of Ulva lactuca and Systoceira stricta algae-based activated carbons to hazardous cationic dyes removal from industrial effluents. Water Res 47:3375–3388

4

Flower-shaped gold nanoparticles synthesized using *Kedrostis foetidissima* and their antiproliferative activity against bone cancer cell lines

M. Jannathul Firdhouse[1] · P. Lalitha[1]

Abstract Three different methods were employed for the synthesis of biogenic gold nanoparticles using the aqueous extracts of *Kedrostis foetidissima*. The interaction of gold nanoparticles with the phytoconstituents was investigated by FTIR. The complete reduction of chloroaurate ions to gold nanoparticles was monitored using UV–visible spectroscopy under the different plant extract concentration and conditions. The formation of gold nanoparticles was confirmed using XRD, SEM and TEM analysis. The anisotropic and flower-shaped gold nanoparticles of size below 25 nm was confirmed by TEM analysis. Cucurbitacins, the chief constituents of *K. foetidissima* probably might have interacted with the chloroaurate ions, facilitating the formation of gold nanoparticles. KFL-mediated AuNPs showed 88 % cell viability against bone cancer cell lines at 200 µg/ml concentration using MTT assay. This may be due to the formation of flower-shaped nature of KFL-mediated AuNPs. The novelty of the work lies in the plant-mediated synthesis of biocompatible gold nanoparticles of size less than 25 nm showing 88 % of cell viability against bone cancer cell lines.

Keywords *Kedrostis foetidissima* · Gold nanoparticles · UV–Visible · SEM · XRD · TEM

Introduction

In modern science, one of the most dynamic areas of research is found to be in nanotechnology field which creates a force towards human life [1]. Nanochemistry mainly focuses on the synthesis of various metal nanoparticles like gold, silver, zinc, platinum [2–5], etc., due to its unique properties, which opens a new venue for the production of nanodevices, therapeutics, drugs, etc., in the field of nanomedicine and nanobiotechnology [6–12]. Gold nanoparticles have diverse applications in cancer treatment, electrochemical sensors, biosensors, gene and drug delivery, catalysis, etc., [13, 14].

The superior properties of the metal nanoparticles depend upon the size, shape and dispersed nature which is different from that of the bulk materials. Nanoparticles can be prepared by physical and chemical methods. Vapour deposition and exfoliation are the physical method of synthesis, whereas hazardous chemicals are used as starting materials in most of the chemical methods of synthesis. Hence, the eco-friendly synthetic procedures for nanoparticles have been adopted by researchers [15, 16].

In the biosynthetic method, plant extracts, bacteria, fungi, etc., are used as reductants for the synthesis of metal nanoparticles. The nanoparticles synthesized using plant extracts are more stable and the rate of synthesis is also rapid than the microorganisms. Plants have been exploited for the large scale synthesis of nanoparticles due to its accessibility and loaded source of secondary metabolites which serve as reducing and capping agents. Gold nanoparticles synthesized using plant extracts have received much attention due to its enhanced conductivity, ability to transfer electrons, catalytic and antimicrobial activities [17–20].

In the past few years, non-spherical gold nanoparticles such as rods, wires, cubes, nanocages, triangular prisms and

✉ P. Lalitha
goldenlalitha@gmail.com

[1] Department of Chemistry, Avinashilingam Institute for Home Science and Higher Education for Women University, Coimbatore 641043, Tamil Nadu, India

other interesting structures such as hollow tubes, capsules, even branched nanocrystals have garnered significant attention in research. These unsymmetrical nanoparticles strongly differ from those of symmetric, spherical gold nanoparticles due to their unique and fine-tuned properties. Their improved mechanical, optical and electronic properties and specific surface-enhanced spectroscopies make them ultimate structures for budding applications in electronics, photonics, optical sensing and imaging, biomedical labeling and sensing, catalysis and electronic devices among others [21].

Spherical, oval and other polyhedral gold nanoparticles have been synthesized using three different plant extracts Angelica, Hypericum, and Hamamelis [22]. An excellent electrocatalytic activity towards the oxidation of methanol and the reduction of oxygen was achieved by gold nanoflowers (GNFs) synthesized using N-2-hydroxyethylpiperazine-N-2-ethanesulphonic acid (HEPES) in one-pot synthesis [23]. Exploration of the optical properties of those triangular particles has shown that the surface plasmon bands frequency is strongly sensitive to the corner sharpness. Simulations by Discrete Dipolar Approximation (DDA) approach qualitatively indicate that the electric field is mainly associated with the particle corners [24].

Osteosarcoma, the most common primary malignant bone tumour, usually arises in the metaphysis of long bones such as the distal femur, proximal tibia and proximal humerus during the second decade (60–80 years) of life. Overall, osteosarcoma has an incidence rate of 10–26 per million new cases worldwide each year [25]. Various side effects such as severe nausea, vomiting, nephrotoxicity, myelosuppression and neurotoxicity were noted after cisplatin administration, whereas gastrointestinal disturbances, acute nausea, vomiting, stomatitis, alopecia baldness, neurologic disturbances, bone marrow, aplasia, cumulative cardio toxicity and bone marrow depressant effects due to doxorubicin administration [26]. Because of these shortcomings, effective approaches with less or no side effects should be considered. Glycogenic AuNPs have been found to be active against human osteosarcoma cell line (Saos2) with an IC_{50} of 0.187 mM [27].

In developing countries, traditional plants have served huge source of medicine, due to their observed antibacterial and antiproliferative properties. *Kedrostis foetidissima*, belongs to Cucurbitaceae family and its characteristic constituents are cucurbitacins, tetracyclic triterpenoids, with varied pharmacological activities. Miro [28] and Rios et al. [29] reported the presence of cucurbitacins B, D, E and I in *K. foetidissima*. A significant level of cytotoxicity was also observed by *K. foetidissima* on both MCF-7 and YMB-1 cell lines [30]. Biosynthesis of silver nanoparticles using *K. foetidissima* [31] and its antimicrobial activities [32] are reported. The presence of metabolites like terpenoids in the above extracts as noted from phytochemical

colour tests, prompted us to utilize these extracts for this study.

In our present study, gold nanoparticles was synthesized using the aqueous extracts of *K. foetidissima* by three different conditions viz. room temperature, higher temperature and sonication. The synthesized KFL-mediated AuNPs were tested for their antiproliferative activity against bone cancer (MG-63) cell lines.

Materials and methods

Chloroauric acid ($HAuCl_4$) was purchased from Loba chemicals, India. Fresh plant (*K. foetidissima*) was collected from a local area in Coimbatore district, Tamil Nadu, India. The voucher specimen was submitted to Botanical Survey of India, Southern Regional Centre, Tamil Nadu Agricultural University, Coimbatore, Tamil Nadu and the Specimen was identified as *K. foetidissima* (jacq.) cogn, Cucurbitaceae family and authenticated (BSI/SRC/5/23/2010-11/Tech-1309).

Extraction of plant

The fresh leaves and stem parts of *K. foetidissima* were labelled as KFL and KFS, respectivley, (20 g) were weighed, washed thoroughly thrice with distilled water and cut into fine pieces, then 100 ml of Millipore water was added and boiled for 5 min in a 500 ml Erlenmeyer flask. The prepared solutions were filtered using Whatmann filter paper No. 42 to remove the plant debris, sonicated for 15 min and refrigerated at −4 °C for further studies. This filtrate obtained (designated as aqueous extract) was used within a week for the synthesis of nanoparticles to avoid the annihilation of phytoconstituents.

Phytochemical screening test

The phytochemical screening of leaf and stem extract of *K. foetidissima* was done according to standard procedures [33, 34].

Synthesis of gold nanoparticles

Different concentrations of aqueous extracts (KFL and KFS) in the ratio 1:1, 2:1, 3:1, 4:1 and 5:1 were mixed with 3 mM $HAuCl_4$ solution at room temperature (28–30 °C). The aqueous extracts (KFL and KFS) were kept constant (1 ml) and treated with different concentrations (1, 2, 3, 4 and 5 ml) of gold chloride solutions and kept at room temperature (RT). The colour change from pale green to violet colour characteristic of gold nanoparticles was indicative of its formation. The λ_{max} value was recorded using a spectrophotometer. Similar experiment was repeated by sonicating the mixture using an ultrasonic bath [PCI Ultrasonics 1.5 L (H)] (SN). In another set of experiments, the mixture was heated to higher temperature (75 °C) (HT) and the time of formation of gold nanoparticles was recorded, as evidenced from the violet colour formation.

Separation of gold nanoparticles

The violet coloured gold nanoparticles solution was centrifuged at 13,000 rpm for 15 min in a centrifuge (Spectrofuge 7M) twice, to remove the plant cell debris. The supernatant solution obtained after centrifugation was analyzed by XRD, SEM and TEM.

Characterization techniques

UV–Visible absorption spectra were measured using Double beam spectrophotometer 2202-(SYSTRONICS). The crystallite size of synthesized gold nanoparticles was examined by X-ray diffraction analysis (X' pert powder PANalytical) with a Cu Kα radiation monochromatic filter in the range 10°–80°. The crystallite size of gold nanoparticles was calculated using Debye–Scherrer's equation:

$$D = k\,\lambda/\beta \cos \theta$$

where, k is dimensionless constant; λ is wavelength of X-ray (Å); β is angular FWHM of the XRD peak at the diffraction angle (radian); and θ is diffraction angle (degrees).

The topography and morphology of gold nanoparticles were investigated by scanning electron microscope using TESCAN instrument provided with Vega TC software. The FTIR spectrum (FTIR-8400S, Shimadzu) was also recorded for the synthesized gold nanoparticles. Transmission electron microscopy (TEM) offers high resolution of particle size, shape and also dispersity. The nanodimension of the synthesized AuNPs was established through recording TEM images using FEI's Tecnai G2 Transmission Electron Microscope.

Antiproliferative activity of biosynthesized AuNPs against bone cancer cell lines—MTT assay

The cell lines MG-63 (Human Osteosarcoma) were obtained from National Centre for Cell Science (NCCS), Pune, India and grown-up in Eagles Minimum Essential Medium.

Cancer cell treatment procedure

The monolayer cells were separated with trypsin-ethylene diamine tetraacetic acid to formulate single cell suspensions and feasible cells were counted using a hemocytometer and diluted with medium enclosing 5 % FBS to provide final density of 1×10^5 cells/ml. The cell suspensions (100 µl/well) were seeded into 96-well plates at plating density of 10,000 cells/well and incubated to allow for cell attachment at 37 °C, 5 % CO_2, 95 % air and 100 % humidity. After 24 h the cells were treated with serial concentrations of the test samples (KFL-mediated AuNPs). The tested samples were initially dissolved in DMSO and twofold series dilutions were made to get the desired test concentration (12.5, 25, 50, 100 and 200 µg/ml) in serum free medium for KFL-mediated AuNPs. Aliquots of different sample dilutions were added to suitable wells already containing 100 µl medium. Following sample addition, the plates were incubated at 37 °C for 48 h at 100 % relative humidity, 95 % air and 5 % CO_2. The medium without samples was taken as control. The study was run in triplicate to ensure accuracy of the results.

MTT assay

3-[4, 5-Dimethylthiazol-2-yl] 2,5-diphenyltetrazolium bromide (MTT), a yellow water soluble tetrazolium salt gets converted to an insoluble purple formazan by the mitochondrial enzyme in existing cells, succinate-dehydrogenase. The quantity of formazan produced is unswervingly proportional to the number of viable cells. After forty-eight hours of incubation, 15 µl of MTT (5 mg/ml) in PBS (phosphate buffered saline) was added to each well and incubated at 37 °C for 4 h. The medium with MTT was then flicked off and the formed formazan crystals were solubilized in 100 µl DMSO and absorbance was measured at 570 nm using a micro plate reader.

Calculation

The percentage cell inhibition was determined using the formula:

$$\% \text{ Cell inhibition} = 100 - \text{ABS (sample)}/\text{ABS (control)} \times 100.$$

A nonlinear regression graph was plotted between percentage cell inhibition and Log_{10} concentration. The IC_{50} value was determined using GraphPad Prism software.

Results and discussion

The phytochemical screening of KFL and KFS extracts revealed the presence of secondary metabolites like alkaloids, flavonoids, saponins, steroids and primary metabolites like proteins and glycosides. The aqueous extracts of

Table 1 Comparison of time of formation of AuNPs synthesized using KFL extracts at different conditions

S. no	Concentration of KFL extracts + HAuCl$_4$ (ml)	Time of formation (min)			Concentration of KFL extracts + HAuCl$_4$ (ml)	Time of formation (min)		
		RT	HT	SN		RT	HT	SN
1.	1 + 1	40	5	5	1 + 1	40	5	5
2.	1 + 2	60	15	10	2 + 1	30	18	15
3.	1 + 3	90	20	15	3 + 1	25	25	20
4.	1 + 4	120	20	15	4 + 1	10	30	25
5.	1 + 5	180	25	20	5 + 1	15	35	25

Table 2 Comparison of time of formation of AuNPs synthesized using KFS extracts at different conditions

S. no	Concentration of KFS extracts + HAuCl$_4$ (ml)	Time of formation (min)			Concentration of KFS extracts + HAuCl$_4$ (ml)	Time of formation (min)		
		RT	HT	SN		RT	HT	SN
1.	1 + 1	120	15	15	1 + 1	150	25	35
2.	1 + 2	90	25	20	2 + 1	120	25	30
3.	1 + 3	60	25	25	3 + 1	90	20	20
4.	1 + 4	45	35	30	4 + 1	60	15	15
5.	1 + 5	45	35	30	5 + 1	60	10	9

KFL and KFS when treated with gold chloride solution produced a visible colour change from pale green to violet colour after 60 min. The colour change was observed rapidly in the highest concentration (5 ml) of KFL and KLS extracts (Tables 1, 2). The varied concentrations (1–5 ml) of KFL extracts with 1 ml of gold chloride solution produce AuNPs within 20–30 min (Table 1). It was noted that the rate of synthesis of gold nanoparticles increases as the concentration of KFS extracts increases as in sonication, room and higher temperature methods (Table 2).

Comparing the variation in the concentration of KFL and KFS extracts and gold chloride solution reveals the increase in the concentration of aqueous extracts to decrease the time taken for the formation of AuNPs. This may be due to the presence of more constituents and less metal ions in the mixture. Hence, rapid synthesis of AuNPs was obtained with different concentrations of gold chloride solution compared to variation of the plant extract concentration. The plant *K. foetidissima* as a whole is a rich source of cucurbitacins as reported earlier and the phytochemical screening also revealed the presence of terpenoids [28, 29]. This revealed that the phytoconstituents present in the extracts are responsible for the rapid reduction of chloroaurate ions to gold nanoparticles.

Plant-mediated synthesis of AuNPs reveal increase in concentration of plant extract cause agglomeration in AuNPs. As the temperature is increased, more and more of gold chloride is reduced, leading to an increase in the formation of gold nanoparticles using both KFL and KFS extracts. This is due to the enhanced activation of chloroaurate ions which quickens the formation of gold nanoparticles. The time variation in the interaction of chloroaurate ions with the metabolites present in the extracts can be overcome by the sonication method compared to the other methods employed in the present synthesis.

The rate of reduction of chloroaurate ions to be managed using KFL and KFS extracts under different conditions is in the order:

Sonication > Higher temperature > Room temperature

This may be due to the acoustic cavitations produced by ultrasound irradiation which enhances the reduction of metal ions with addendum from increase in temperature and hence the rate of synthesis than the conventional method. It was observed that the ultrasound technique aids the reaction rate and simplifies the synthetic methods of novel materials. Hence, the dispersion produced by the cavitational motion may facilitate the interaction of metabolites present in the aqueous extracts with the chloroaurate ions aiding gold nanoparticles formation.

UV–Visible spectroscopy

UV–Visible spectroscopy is a good analytical tool which authenticates the complete formation and stability of gold nanoparticles. The UV-visible spectra (Fig. 1) of gold

Fig. 1 UV-visible spectra of gold nanoparticles synthesized using the solvent extracts (KFL and KFS) of *Kedrostis foetidissima* under sonication

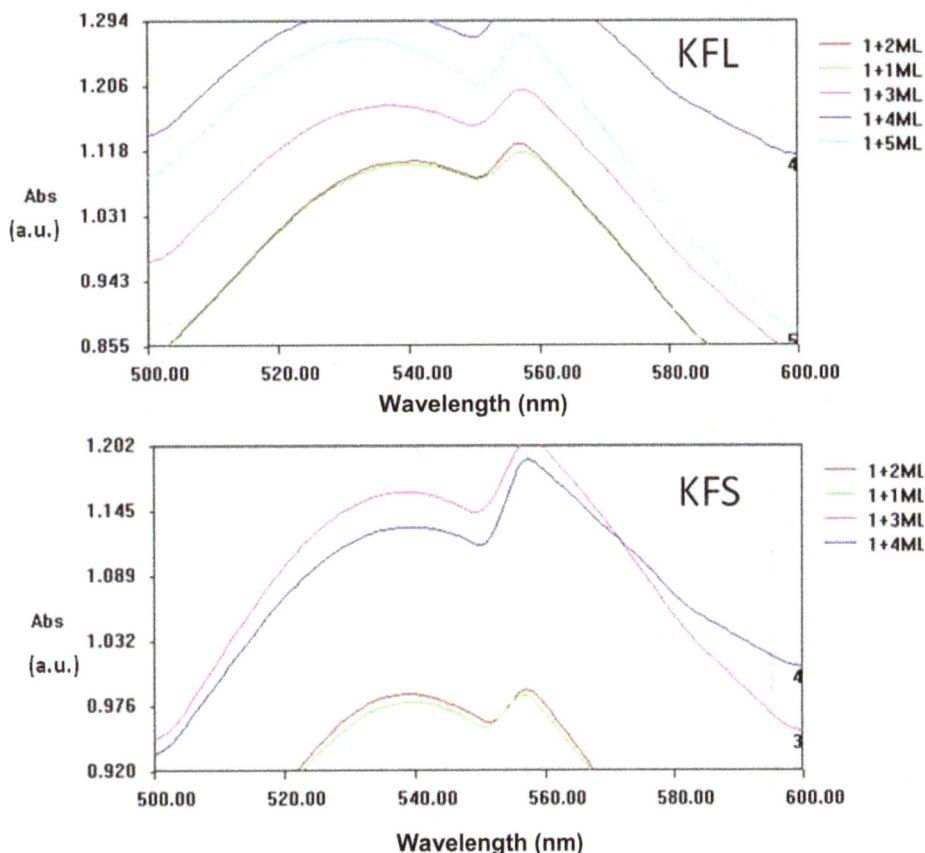

nanoparticles shows a broad absorption band at 510–600 nm. The appearance of a broad SPR band from 510 to 550 nm, with a sharp band at 570 nm signifies the presence of large and polyshaped gold nanoparticles. The intensity of the bands increases with extract concentration. At the highest concentration (5:1) of plant extracts (KFL and KFS), the intensity of the band starts to decrease, implying the dominant nature of the phytoconstituents. The shape and position of the bands remain the same under different conditions, which may be due to the adsorption of phytoconstituents onto the surface of chloroaurate ions through Vander Waal's forces. The synthesized gold nanoparticles were stable withstanding the colour of the solution for 30 days, prompting us to explore its application as chemotherapeutic drugs.

XRD analysis

The crystalline nature of the synthesized gold nanoparticles was analyzed by X-ray diffractometer. Figure 2 represents the XRD patterns of the synthesized gold nanoparticles coated on a glass substrate. The diffraction peaks at 38°, 44°, 64°, 77° which were indexed to the face-centered cubic lattice was observed for the extracts (KFL and KFS) under room temperature (KFLRT and KFSRT) and

sonication (KFLSN and KFSSN), respectively. There is little difference in the diffraction peaks and full-width half maximum values, but the crystallite size 10.6 nm was observed in both extracts under room temperature and sonication, respectively. The additional peak at 28° may be due to the presence of phytoconstituents capped onto the synthesized AuNPs. The determination of crystallite size of the gold nanoparticles using Debye–Scherrer's equation is given in Table 3.

SEM analysis

To determine the shape, size and distribution of the nanoparticles, SEM Images were also recorded. The SEM analysis confirmed the synthesized gold nanoparticles to be in nanometer size. They were spherical in shape and the size of the gold nanoparticles found to be uniformly distributed (Fig. 3). The size of the gold nanoparticles varies in SEM compared to that of XRD analysis. This may be due to the capping nature of the phytoconstituents as the time increases from the day of synthesis. It was noted that in room temperature condition the crystallite size is 90 nm, whereas it reduced to 50 nm in sonication due to the fine dispersion produced by the ultrasonic irradiation.

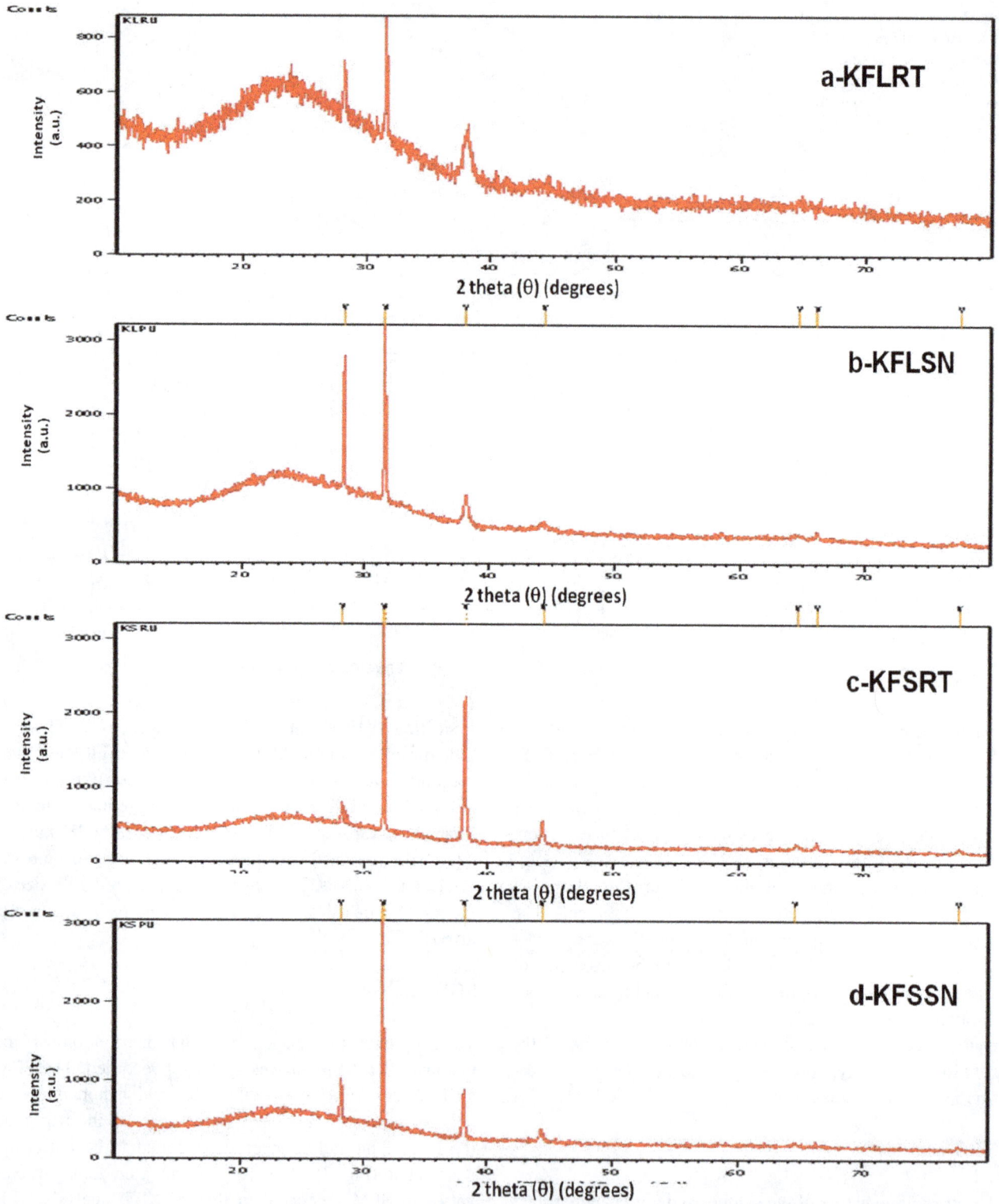

Fig. 2 XRD patterns of the synthesized gold nanoparticles using the KF extracts under room temperature (KFLRT and KFSRT) and sonication (KFLSN and KFSSN)

Table 3 Determination of crystallite size of gold nanoparticles synthesized using extracts (KFL and KFS) under room temperature and sonication

S. no	Sample	2θ (degrees)	$\cos\theta$	$\beta = \pi*\text{FWHM}/180$ (radians)	$D = k\,\lambda/\beta \cdot \cos\theta$ (nm)	Average size (nm)
1.	KFLRT	38.1224	0.94517	0.01648	8.90	10.6
		44.4012	0.92600	0.16149	9.27	
		64.6879	0.84507	0.01473	11.14	
		77.7243	0.77933	0.01359	13.09	
2.	KFLSN	38.1216	0.94517	0.01648	8.90	10.6
		44.3707	0.92600	0.16149	9.27	
		64.6438	0.84507	0.01473	11.14	
		77.6067	0.77933	0.01359	13.09	
3.	KFSRT	38.1616	0.94506	0.01648	8.90	10.6
		44.3498	0.92606	0.01615	9.27	
		64.6068	0.84526	0.01474	11.13	
		77.5398	0.77977	0.01359	13.08	
4.	KFSSN	38.1686	0.94506	0.01648	8.90	10.6
		44.3738	0.92593	0.01614	9.28	
		64.5678	0.84544	0.01474	11.13	
		77.5923	0.77933	0.01359	13.09	

TEM analysis

The AuNPs synthesized using *K. foetidissima* leaf extract was found to be less than 50 nm as obtained from SEM analysis and hence it was further analyzed for TEM. *Kedrostis foetidissima* (1 ml) sonicated with chloroaurate ions (3 ml) for 15 min results in purple gold nanoparticles were chosen for TEM analysis. The TEM micrographs shows the formation of anisotropic (triangular plate, spherical, hexagon) gold nanoparticles of size less than 25 nm (Fig. 4a, b). Figure 4c, d showed the formation of flower-shaped gold nanoparticles. The arrow in Fig. 4d indicates the encapsulation of AuNPs with the protein moiety present in *K. foetidissima* leaf extract. Hence the proteins in the leaf extract of *K. foetidissima* act as a capping and stabilizing agent. The flower-shape gold nanoparticles are synthesized without the use of toxic chemicals and stabilizers.

Literature survey revealed synthesis of biological triangular gold nanoprisms by the reduction of aqueous chloroaurate ions by extracts of *Cymbopogon flexuosus* in high yield at room temperature [35]. Tamarind leaf extract can also be used as the reducing agent for making gold nanotriangles [36]. Diverse shapes of nanoparticles such as hexagon, truncated triangle and triangle can also be synthesized using the extract of seaweed, *Sargassum* sp., at room temperature. The growth of nanoparticles may be assisted by the capping agents which limits the size and controls the shape. The non-specific nature of binding on all exposed surfaces of gold may lead to the anisotropic growth [37].

Irregular surfaces and round edges of non-spherical nanoparticles possess more surface area than spherical particles with smooth surfaces. Hence the binding capacity of these particles to cancerous cells and tissues will be enhanced due to their surface area and extended residence times in vivo [38]. The rearrangement and aggregation of the smaller particles results in anisotropic triangular structures. The fabrication of anisotropic nanoparticles can be facilitated by the low rate of reduction, room temperature and sometimes with slight modifications in the temperature. This nanostructure remains no longer as triangular nanoprisms, when significant rounding occurs and gets transformed into nanodisks or hexagonal nanoprisms. Due to their aforesaid reasons, different shapes of gold nanoparticles using *K. foetidissima* extracts obtained in this study is justified.

FTIR analysis

Figure 5a, b represents the FTIR spectra of the plant extracts (KFL and KFS) and synthesized gold nanoparticles (KFL-mediated AuNPs and KFS-mediated AuNPs). In KFL extract, the band at 1638 cm^{-1} may be due to the presence of a carbonyl group. The two bands at 2926 cm^{-1} corresponds to the stretching vibrations of the –C–H bond and 3345 cm^{-1} may be assigned to –OH group (Fig. 5a). In KFS extract, the bands at 1639 and 3336 cm^{-1} corresponds to the carbonyl and hydroxyl group, respectively. An another band at 2931 cm^{-1} may be due to the –C–H bond stretching frequencies (Fig. 5b). The absence and shift of the aforesaid bands indicate that the metabolites like triterpenoids might have been responsible for the reduction of gold to gold nanoparticles

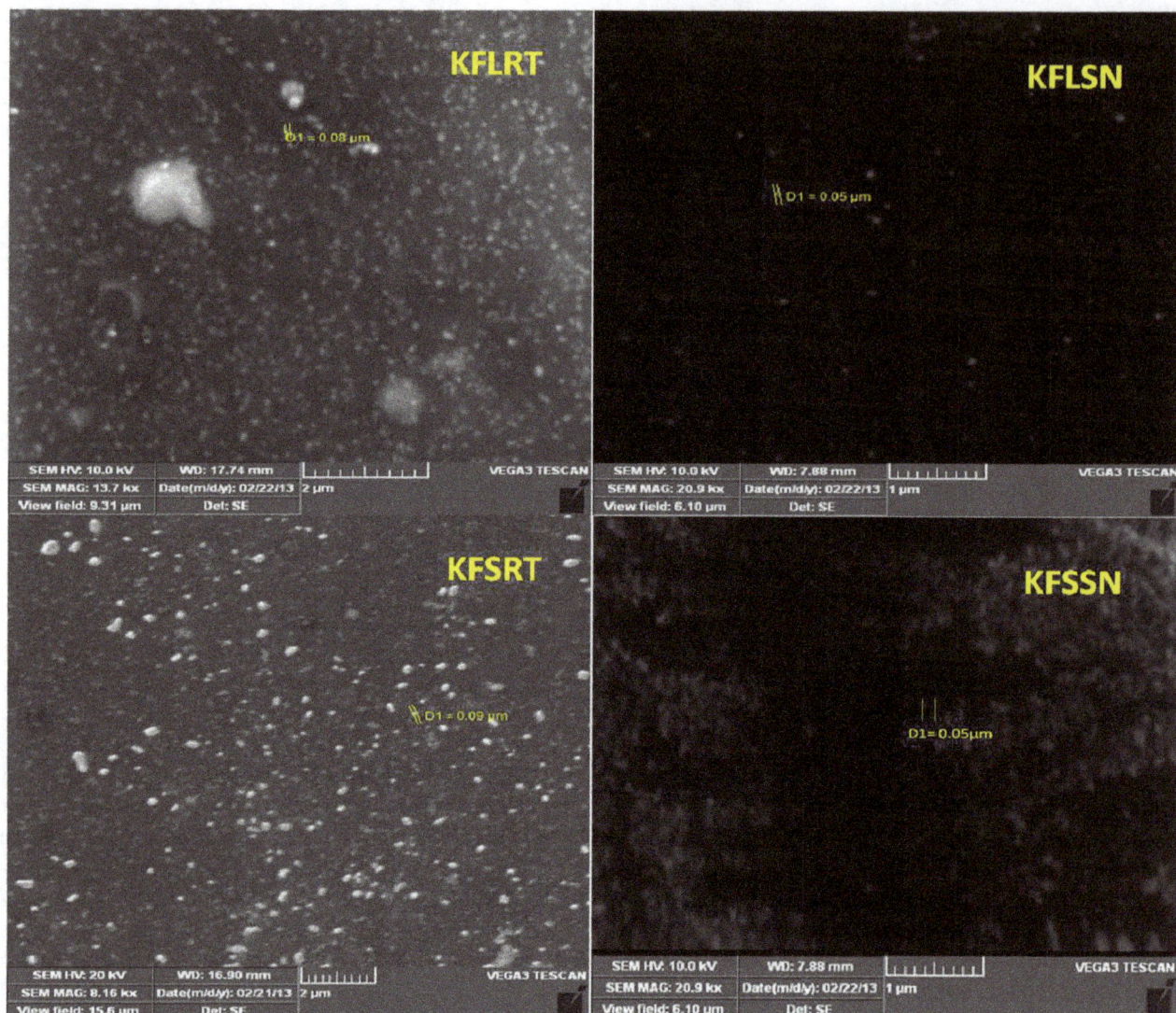

Fig. 3 SEM micrographs of the synthesized gold nanoparticles under room temperature and sonication using the extracts (KFL and KFS)

in Fig. 5 (a—KFL-mediated AuNPs and b—KFS-mediated AuNPs).

Antiproliferative study—MTT assay

The antiproliferative activity of KFL-mediated AuNPs under in vitro conditions was examined against human osteosarcoma cell lines (MG-63) by MTT assay. The histogram plot of MTT assay results for cell viability studies of MG-63 cell lines after exposing to varying concentrations of KFL-mediated AuNPs was shown in Fig. 6. The red bars correspond to the % cell viability at the given concentration of AuNPs. The plots clearly show 88 % cell viability for the bone cancer cells that were treated with 200 µg/ml concentration of KFL-mediated AuNPs for 24 h (Fig. 6). The borohydride-reduced gold nanoparticles show nearly 80 % cell viability at the exposure to 100 µM

concentration for 24 h. Gold nanotriangles synthesized using the leaf extract of lemon grass plant do not show any cellular toxicity to non-phagocytotic cancerous cells while the gold salt precursor showed acute toxicity to the cells. Thus, it inferred that biogenic gold nanotriangles are more biocompatible than chemically synthesized gold nanoparticles [39]. In the present work, the MG-63 cells also show almost 100 % viability after exposure to 25 µg/ml concentration of flower-shaped gold nanoparticles for 24 h. It can be concluded that the KFL-mediated AuNPs do not show acute toxicity even at high concentrations taken in the present study revealing the biocompatibility.

As concentration increases from 12.5 to 200 µg/ml, the cell viability gets decreased indicates the anitproliferative activity of KFL-mediated AuNPs on MG-63 cell lines. The results revealed that the flower-shaped KFL-mediated AuNPs may be the reason for the less toxic nature against

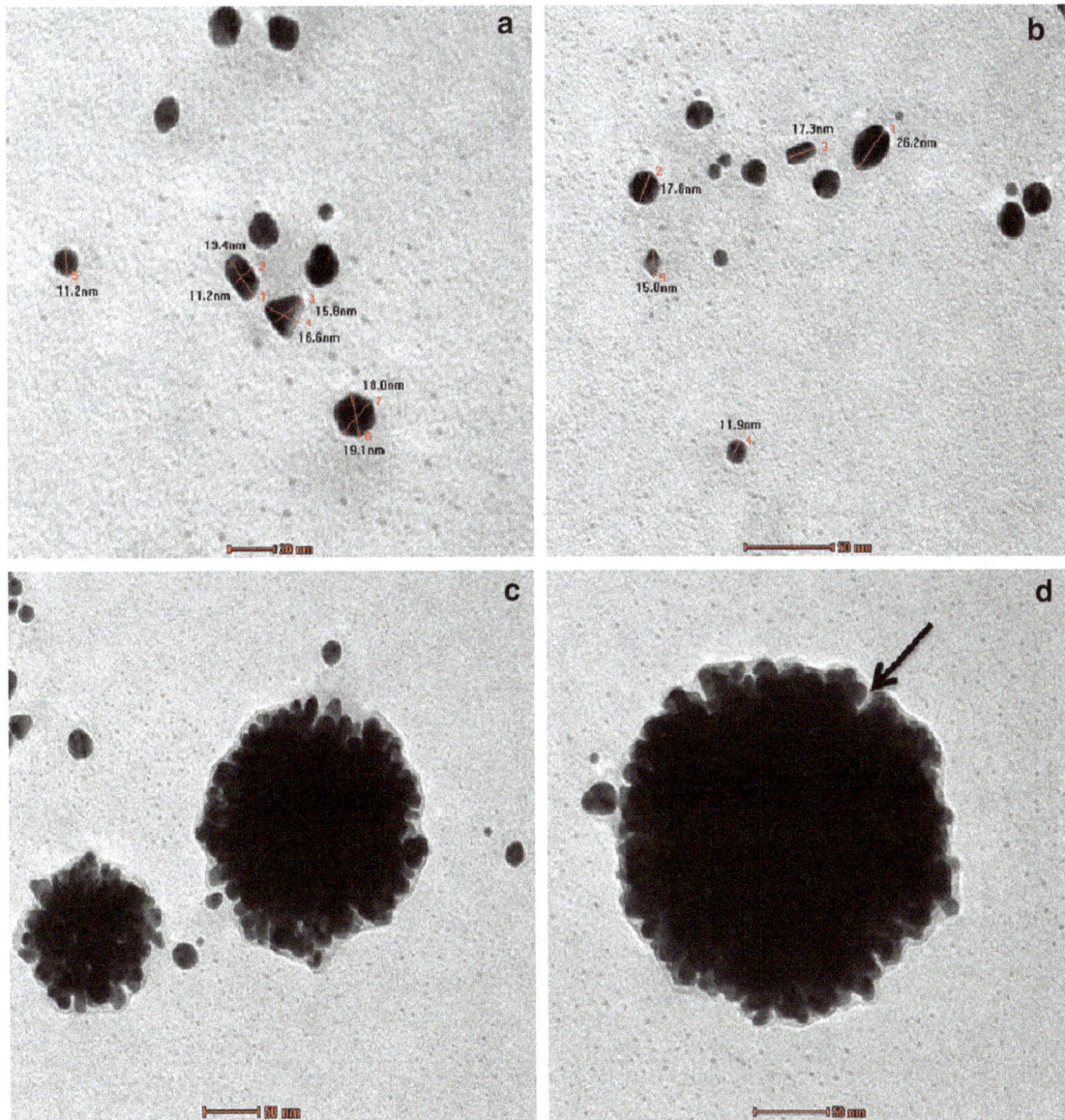

Fig. 4 TEM micrographs of the anisotropic gold nanoparticles viz. (**a**) *triangular plate* and *hexagon*, *spherical* and *rod*, (**b**) and *flower-shaped* AuNPs (**c**, **d**) synthesized under sonication using KFL extract

MG-63 cell lines. This present results suggest that the gold nanoparticles do not show any cellular toxicity to cancerous as well as non-cancerous cells at concentration 50 μg/ml. Cytotoxicity also depends on the type of cells used. Pioneering studies on nanoparticles show that non-spherical shapes show great promise as cancer drug delivery vectors. Filamentous or worm-like micelles together with other rare morphologies such as needles or disks may become the norm for next-generation drug carriers [40]. Resv-Dox mixtures and Dox-GNPs complexes mediate the anticancer activity of Dox at very low concentration (0.1 μg/ml) against HeLa and CaSki cells [41].

Figure 7 shows the slight changes in the morphology of cells and a decrease in the number at 200 μg/ml concentration of KFL-mediated AuNPs on MG-63 cell lines. Further tests such as apoptosis assays can aid in the elucidation of the cellular mechanism of cell death. The results suggest the antiproliferative activity of the KFL-mediated AuNPs may be due to the presence of anticancer molecules like cucurbitacins present in these extracts. Meat-ball like AuNPs developed using green tea extract under microwave irradiation are reported to be biocompatible up to 500 μg/mL against MCF-7 and HeLa cell lines [42]. The cytotoxic activity of the synthesized nanosilver using *A. sessilis* studied against PC-3 [43] and MCF-7 [44] cell lines by MTT assay was found to show significant activity, but *A. sessilis*-mediated AuNPs was non-toxic against MCF-7 cell lines [45]. Thus, the previous

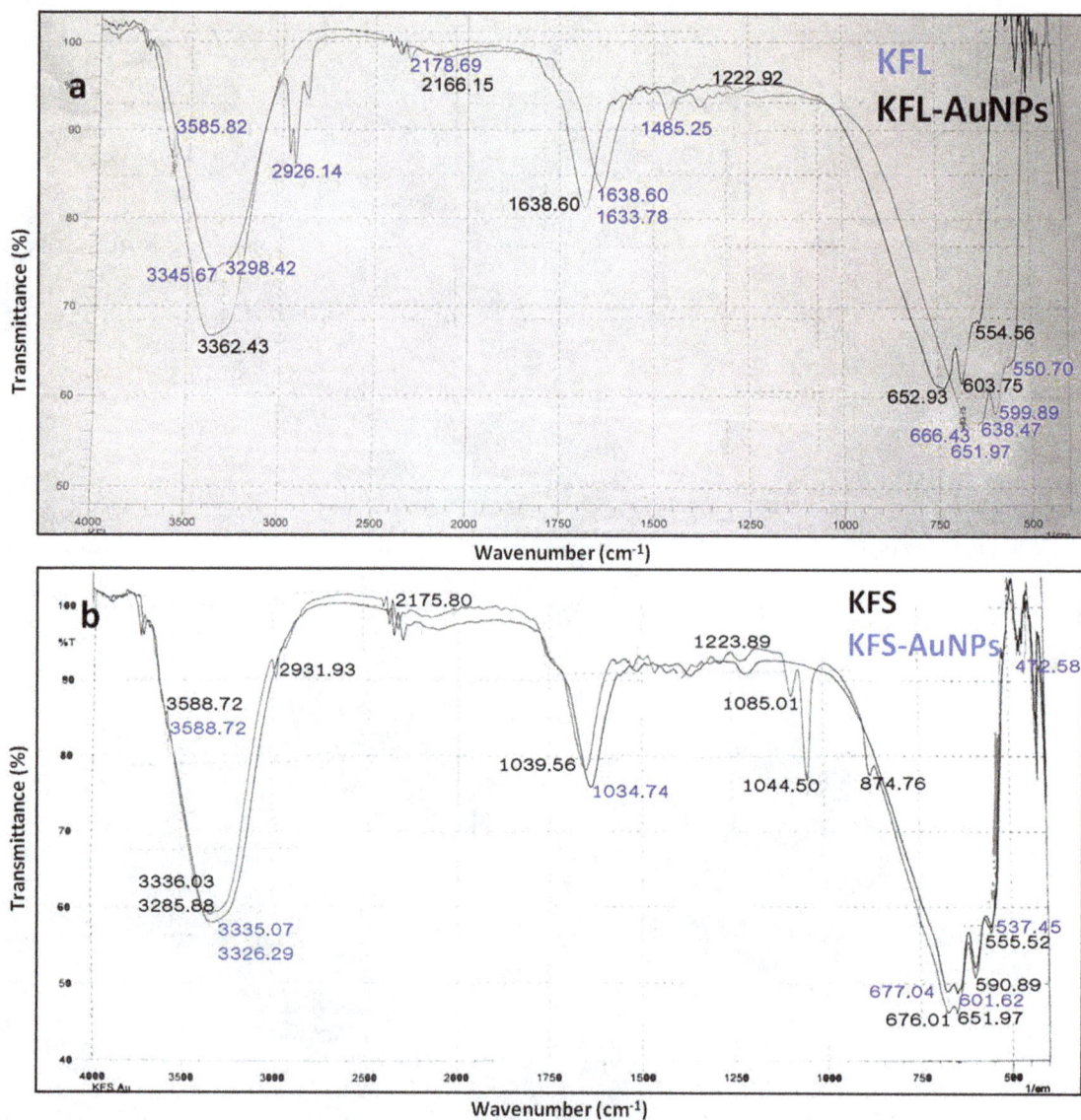

Fig. 5 FTIR spectra of the plant extracts (KFL and KFS) and synthesized gold nanoparticles (KFL-mediated AuNPs and KFS-mediated AuNPs)

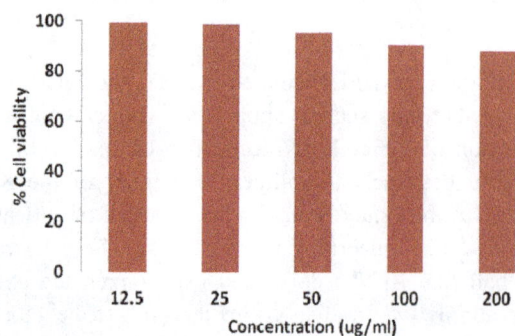

Fig. 6 Histogram of cell viability studies of MG-63 cell lines at various concentrations of KFL-mediated AuNPs

literature revealed that the cytotoxicity depends upon the nature of the capping agents and also the metal. The non-toxic nature of the biogenic gold nanoparticles provides new opportunities for the safe application in molecular imaging and therapy.

Conclusion

In this study, the aqueous extracts of KFL and KFS was tested for the reduction of chloroaurate ions to gold nanoparticles. The formation of the stable gold nanoparticles was confirmed by SPR band obtained in the UV–visible spectroscopy. The crystalline nature of the gold nanoparticles was revealed by the presence of diffraction peaks in the XRD patterns. The gold nanoparticles were spherical in shape and uniformly distributed as confirmed by SEM analysis. Gold nanoparticles below 50 nm size were anisotropic and flower shaped as confirmed from

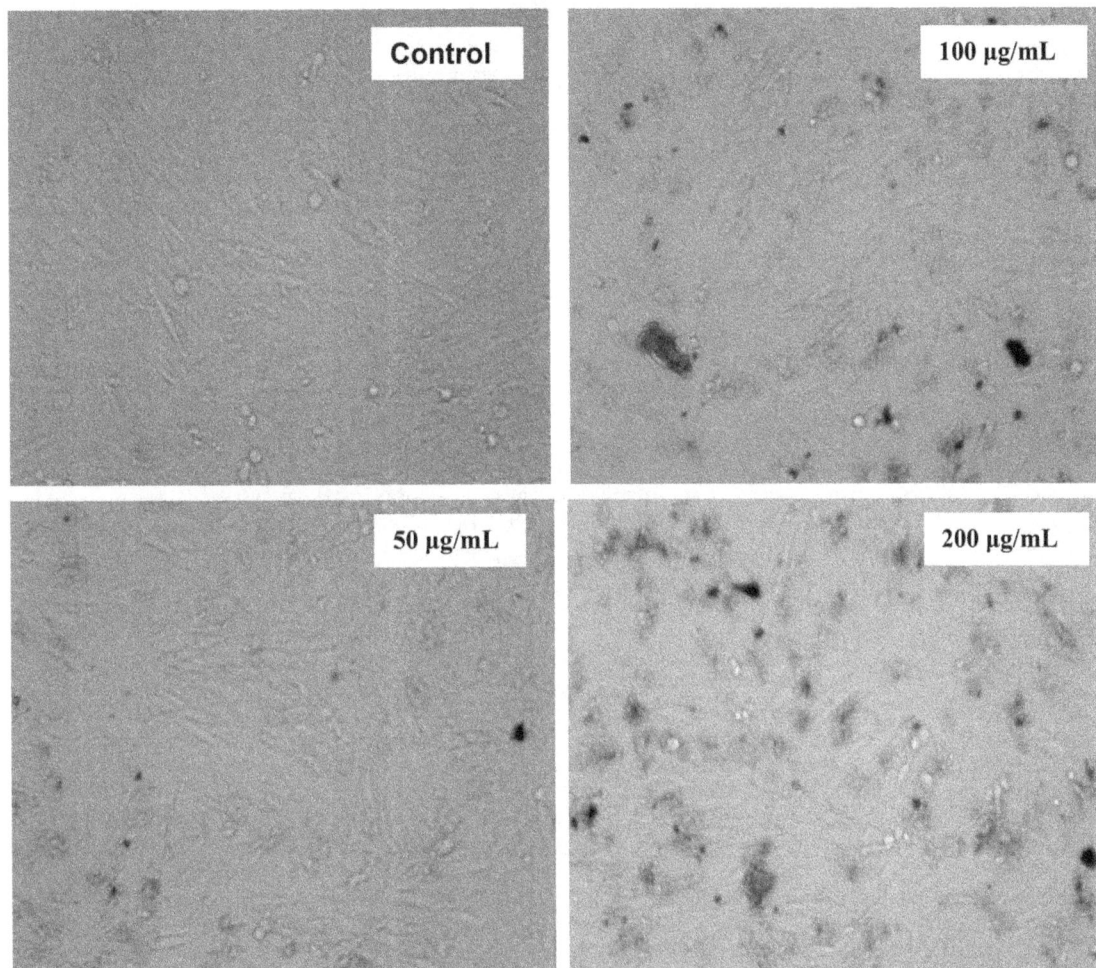

Fig. 7 Cytomorphological changes and growth inhibition of AuNPs synthesized using KFL extracts at different concentrations on MG-63 cell lines

TEM analysis. FTIR measurements revealed that the phytoconstituents can serve as an excellent reducing and capping agents. The results of the cell viability studies of KFL-mediated AuNPs revealed the antiproliferative activity of AuNPs to depend on the size, shape and the nature of capping agents. Thus, the biogenic gold nanoparticles may find applications in the treatment of cancer.

Acknowledgments The authors sincerely thank the Avinashilingam Institute for Home Science and Higher Education for Women University, Coimbatore, Tamil Nadu, for providing research facilities, Department of Physics, Avinashilingam University for Women, for recording XRD, Periyar Maniammai University for recording SEM and TNAU, Coimbatore for recording TEM and KMCH College of Pharmacy for certifying anticancer activity of test samples.

References

1. Bhattacharya R, Mukherjee P (2008) Biological properties of "naked" metal nanoparticles. Adv Drug Deliv Rev 60:1289–1306. doi:10.1016/j.addr.2008.03.013

2. Ghule K, Ghule AV, Liu JY, Ling YC (2006) Microscale size triangular gold prisms synthesized using Bengal gram beans (*Cicer arietinum* L.) extract and $HAuCl_4 \times 3H_2O$: a green biogenic approach. J Nanosci Nanotechnol 6:3746–3751

3. Li S, Shen Y, Xie A, Yu X, Qiu L, Zhang L (2007) Green synthesis of silver nanoparticles using *Capsicum annuum* L. extracts. Green Chem 9:852–858. doi:10.1039/B615357G

4. Song JY, Eun YK, Beom SK (2009) Biological synthesis of platinum nanoparticles using *Diopyros kaki* leaf extract. Bioproc Biosyst Eng 3:159–164. doi:10.1007/s00449-009-0373-2

5. Yan S, Wen H, Caiyun S (2009) The biomimetic synthesis of zinc phosphate nanoparticles. Dyes Pigm 80:254–258. doi:10.1016/j.dyepig.2008.06.010

6. Weller H (1993) Colloidal semiconductor Q-particles: chemistry in the transition region between solid state and molecules. Angew Chem Int Ed 32:41–53. doi:10.1002/anie.199300411

7. Henglein A (1993) Physicochemical properties of small metal particles in solution: "microelectrode" reactions, chemisorption, composite metal particles and the atom-to-metal transition. J Phys Chem 97:5457–5471. doi:10.1021/j100123a004

8. Alivisatos AP (1996) Semiconductor clusters, nanocrystals and quantum dots. Science 271:933–937

9. Colvin VL, Schlamp MC, Alivisatos AP (1994) Light-emitting diodes made from cadmium selenide nanocrystals and a semi-

conducting polymer. Nature 370:354–357. doi:10.1126/science. 271.5251.933

10. Sastry M, Ahmad A, Khan MI, Kumar R (2003) Biosynthesis of metal nanoparticles using fungi and *actinomycete*. Curr Sci 85:162–170. Accessed 25 Jul 2003

11. Ahmad T, Wania IA, Manzoor N, Ahmed J, Asiri AM (2013) Biosynthesis, structural characterization and antimicrobial activity of gold and silver nanoparticles. Colloids Surf B 107:227–234. doi:10.1016/j.colsurfb.2013.02.004

12. Mocanu A, Horovitz O, Racz P, Tomoaia-Cotisel M (2015) Green synthesis and characterization of gold and silver nanoparticles. Rev Roum Chim 60(7–8):721–726

13. Schrinner M, Polzer F, Mei Y, Lu Y, Haupt B, Ballauff M (2007) Mechanism of the formation of amorphous gold nanoparticles within spherical polyelectrolyte brushes. Macromol Chem Phys 208:1542–1547. doi:10.1002/macp.200700161

14. Doria G, Conde J, Veigas B, Giestas, L, Almedia C, Assuncao M, Rosa J, Baptista P (2012) Noble metal nanoparticles for biosensing applications. J Drug Deliv 1657–1687. doi:10.3390/s120201657

15. Parsons JG, Peralta-Videa JR, Gardea-Torresdey JL (2007) Use of plants in biotechnology: Synthesis of metal nanoparticles by inactivated plant tissues, plant extracts, and living plants. In: Sarkar D, Datta R, Hannigan R (eds) Developments in environmental science, vol. 5, Chapter 21. Elsevier, Minessota, pp 436–485

16. Iravani S (2011) Green synthesis of metal nanoparticles using plants. Green Chem 13:2638–2650. doi:10.1039/c1gc15386b

17. Gan PP, Li SFY (2012) Potential of plant as a biological factory to synthesize gold and silver nanoparticles and their applications. Rev Environ Sci Biotechnol 11:169–206. doi:10.1007/s11157-012-9278-7

18. Goyal RN, Gupta VK, Oyama M, Bachheti N (2007) Gold nanoparticles modified indium tin oxide electrode for the simultaneous determination of dopamine and serotonin: application in pharmaceutical formulations and biological fluids. Talanta 72:976–983. doi:10.1016/j.talanta.2006.12.029

19. Inbakandan D, Venkatesan R, Ajmal Khan S (2010) Biosynthesis of gold nanoparticles utilizing marine sponge *Acanthella elongata* (Dendy, 1905). Colloids Surf B 81:634–639. doi:10.1016/j.colsurfb.2010.08.016

20. Yan N, Xiao C, Kou Y (2010) Transition metal nanoparticles catalysis in green solvents. Coord Chem Rev 254:1179–1218. doi:10.1016/j.ccr.2010.02.015

21. Karuppaiya P, Satheeshkumar E, Chao WT, Kao LY, Chen EC, Tsay HS (2013) Anti-metastatic activity of biologically synthesized gold nanoparticles on human fibrosarcoma cell line HT-1080. Colloids Surf B 110:163–170. doi:10.1016/j.colsurfb.2013.04.037

22. Pasca RD, Mocanu A, Cobzac SC, Petean I, Horovitz O, Tomoaia-Cotisel M (2014) Biogenic syntheses of gold nanoparticles using plant extracts. Particul. Sci. Technol. 32(2):131–137

23. Jena BJ, Raj CR (2007) Synthesis of flower-like gold nanoparticles and their electrocatalytic activity towards the oxidation of methanol and the reduction of oxygen. Langmuir 23(7):4064–4070. doi:10.1021/la063243z

24. Kelly KL, Coronado E, Zhao LL, Schatz GC (2003) The optical properties of metal nanoparticles: the influence of size, shape, and dielectric environment. J Phys Chem B 107:668–677

25. Ando K, Heymann MF, Stresing V, Mori K, Rédini F, Heymann D (2013) Current therapeutic strategies and novel approaches in osteosarcoma. Cancers 5:591–616. doi:10.3390/cancers5020591

26. Khan AK, Rashid R, Murtaza G, Zahra A (2014) Gold nanoparticles: synthesis and applications in drug delivery. Trop J Pharma Res 13(7):1169–1177

27. Rahim M, Iram S, Khan MS, Khan MS, Shukla AR, Srivastava AK, Ahmad S (2014) Glycation-assisted synthesized gold nanoparticles inhibit growth of bone cancer cells. Colloids Surf B Biointerfaces 1(117):473–479. doi:10.1016/j.colsurfb.2013.12.008

28. Miro M (1995) Cucurbitacins and their pharmacological effects. Phytother Res 9:159–168. doi:10.1002/ptr.2650090302

29. Rios JL, Escandell JM, Recio MC (2005) New insights into the bioactivity of cucurbitacins. Stud Nat Prod Chem 32:429–469. doi:10.1016/S1572-5995(05)80062-6

30. Choene M, Motadi LR (2012) Anti-Proliferative effects of the methanolic extract of *Kedrostis foetidissima* in Breast cancer cell lines. Mol Biol 1(2):1–5. doi:10.4172/2168-9547.1000107

31. Amutha M, Firdhouse MJ, Lalitha P (2014) Biosynthesis of silver nanoparticles using *Kedrostis foetidissima* (Jacq.) Cogn J Nanotechnol 860875:5. doi:10.1155/2014/860875

32. Firdhouse MJ, Lalitha P (2014) Biocidal potential of biosynthesized silver nanoparticles against fungal threats. J Nanostruc Chem 5:25–33. doi:10.1007/s40097-014-0126-x-

33. Harborne JB (1973) Phytochemical methods: a guide to modern techniques of plant analysis. Chapman and Hall, New York, p 279

34. Raaman N (2006) Phytochemical techniques. New Indian Publishing Agencies, New Delhi, p 19

35. Shankar SS, Rai A, Ahmad A, Sastry M (2005) Controlling the optical properties of lemongrass extract synthesized gold nanotriangles and potential application in infrared absorbing optical coatings. Chem Mater 17:566–572

36. Ankamwar B, Chaudhary M, Sastry M (2005) Gold nanotriangles biologically synthesized using tamarind leaf extract and potential application in vapor sensing. Synth React Inorg Metals 35:19–26

37. Liu B, Xie J, Lee JY, Ting YP, Chen JP (2005) Optimization of high yield biological synthesis of single crystalline gold nanoplates. J Phys Chem B 109:15256–15263

38. Xue X, Wanga F, Liu X (2011) Emerging functional nanomaterials for therapeutics. J Mater Chem 21:13107–13127. doi:10.1039/c1jm11401h

39 Singh A (2011) Metal nanoparticles in therapeutic and sensor applications. PhD thesis, University of Pune

40 Truong NP, Whittaker MR, Mak CW, Davis TP (2015) The importance of nanoparticle shape in cancer drug delivery. Expert Opin Drug Deliv 12(1):129–142. doi:10.1517/17425247.2014.950564

41 Tomoaia G, Horovitz O, Mocanu A, Nita A, Avram A, Racz CP, Soritau O, Cenariu M, Tomoaia-Cotisel M (2015) Effects of doxorubicin mediated by gold nanoparticles and resveratrol in two human cervical tumor cell lines. Colloids Surf B 135:726–734

42 Wu S, Zhou X, Yang X, Hou Z, Shi Y, Zhong L, Jiang Q, Zhang Q (2014) A rapid green strategy for the synthesis of Au ''meatball''-like nanoparticles using green tea for SERS applications. J Nanopart Res 16(2325):1–13

43 Firdhouse MJ, Lalitha P (2013) Biosynthesis of silver nanoparticles using the extract of *Alternanthera sessilis*—antiproliferative effect against prostate cancer cells. Cancer Nano 4:137–143. doi:10.1007/s12645-013-0045-4

44 Firdhouse MJ, Lalitha P (2015) Apoptotic efficacy of biogenic silver nanoparticles on human breast cancer MCF-7 cell lines. Prog Biomater 4:113–121. doi:10.1007/s40204-015-0042-2

45 Firdhouse MJ, Lalitha P (2014) Cell viability studies of cubic gold nanoparticles synthesized using the extract of *Alternanthera sessilis*. World J Pharma Res 3(2):2868–2879

5

Estimation of excess molar volumes and theoretical viscosities of binary mixtures of benzene + *n*-alkanes at 298.15 K

Omer El-Amin Ahmed Adam[1,3] · Akl M. Awwad[2]

Abstract Excess molar volumes, (V_m^E), have been derived from the literature viscosity data for the binary mixtures of benzene with *n*-hexane, *n*-octane, *n*-decane, *n*-dodecane, *n*-tetradecane, and *n*-hexadecane as a function of composition at 298.15 K and atmospheric pressure conditions. The V_m^E values were found to be positive over the entire composition range for all mixtures. Concentration dependence of V_m^E were fitted with Redlich–Kister polynomial equation to estimate the binary coefficients and standard errors. From density data, the partial molar volumes (V_m), partial molar volumes at infinite dilution ($\overline{V}_m^0$), excess partial molar volumes at infinite dilution ($\overline{V}_m^{0,E}$), and apparent molar volumes (V_ϕ), were calculated over the whole composition range as were the limiting apparent molar volumes at infinite dilution ($\overline{V}_\phi^0$) and excess apparent molar volumes at infinite dilution ($\overline{V}_\phi^{0,E}$). Viscosity of the binary mixtures of benzene with n-alkanes were estimated using Kendall-Monroe, Frenkel, Hind et al., Katti-Chaudhri, Grunberg-Nissan, Wilke and Herráez et al. equations. The agreement between experimental and predicted values for all systems was found to be quite reasonable as evidenced from

computed standard deviation and average percentage deviation (APD). Wilke relation gives maximum deviations for all the systems in comparison to other methods employed. Other relations give comparatively good results.

Keywords Density · Viscosity · Binary mixture · Excess molar volume · Molecular interactions · Viscosity deviation

Introduction

Excess thermodynamic properties and deviations of non-thermodynamic ones from ideal behavior of binary liquid mixtures are fundamental for the design of industrial equipment and for the interpretation of the liquid state, particularly when polar components are involved [1]. These quantities have the advantage of illustrating the sign and magnitude of the nonideality [2].

Volumetric properties of binary mixtures are complex properties because they depend not only on solute–solute, solvent–solvent and solute–solvent interactions, but also on the structural effects arising from interstitial accommodation due to the difference in molar volume and free volume between components present in the solution [3].

Partial molar properties are useful in providing information about solute–solvent interactions. This is because at infinite dilution, solute–solute interactions disappear. Of course this information is of great interest because it is composition independent.

Alkanes are important series of homologous, nonpolar, and organic solvents. They have often been used in the study of solute dynamics because their physicochemical properties as a function of chain length are well-known [4]. They are also employed in a large range of chemical processes [5].

✉ Omer El-Amin Ahmed Adam
omaramin1967@gmail.com

[1] Chemistry Department, University of Kassala,
P.O. Box 266, Kassala 31111, Sudan

[2] Royal Scientific Society, P.O. Box 1438, Al-Jubaiha,
Amman 11941, Jordan

[3] Present Address: Chemistry Department, Faculty of Science
and Arts in Baljurashi, Al baha University,
P.O. Box 1988, Baljurashi 65635, Saudi Arabia

Properties such as viscosity or surface tension are required in many empirical equations for different operations such as mass and heat transfer processes. Determination of equations that modelize the mass transfer process requires knowledge of the density, viscosity, and surface tension of the liquid phase [6].

The measurement of viscosity reveals information about the molecular packing, molecular motion, and various types of intermolecular interactions as related to size, shape, and chemical nature of the component molecules [7].

In recent years, there has been considerable interest in theoretical and experimental investigations of the excess thermodynamic properties of binary mixtures [8, 9].

Generally, V_m^E can be considered as a result of three types of interactions between component molecules of liquid mixtures [10, 11].

(1) Physical interactions consisting mainly of dispersion forces or weak dipole–dipole interaction making a positive contribution, (2) chemical or specific interactions, which include charge transfer, H-bonding and other complex formation interactions, resulting in a negative contribution, and (3) structural contribution due to differences in size and shape of the component molecules of the mixtures, due to fitting of component molecules into each other's structure, hereby reducing the volume and compressibility of the mixtures, resulting in a negative contribution.

In a previous work by Akl et al. [12], results of density and viscosity measurement were determined for binary mixtures of benzene + n-alkanes at 298.15 K and atmospheric pressure. Excess molar volumes (V_m^E, excess molar viscosities ($\Delta \ln\eta$), and excess molar activation energies, (ΔG^{*E}) were calculated. The effect of orientational order of n-alkane on solution molar volumes and viscosities is investigated as well as the adequacy of the absolute rate and free volume theories to predict solution viscosities. For longer n-alkane ΔG^{*E} and $\Delta \ln\eta$ are positive and associated with the orientational order.

In the present work, the data of density and viscosity reported in the literature [12] have been used to evaluate the excess molar volume (V_m^E) along with other derived parameters, such as partial molar volumes at infinite dilution ($\overline{V}_{m,1}^0$ and $\overline{V}_{m,2}^0$), excess partial molar volumes at infinite dilution ($\overline{V}_{m,1}^{0,E}$ and $\overline{V}_{m,2}^{0,E}$), apparent molar volumes ($V_{\phi,1}$ and $V_{\phi,2}$), limiting apparent molar volumes at infinite dilution ($\overline{V}_{\phi,1}^0$ and $\overline{V}_{\phi,2}^0$) and excess apparent molar volumes at infinite dilution ($\overline{V}_{\phi,1}^{0,E}$ and $\overline{V}_{\phi,2}^{0,E}$).

These parameters have been interpreted in terms of molecular interactions and structural effects. The work provides a test of various empirical equations to correlate viscosity data of binary mixtures in terms of pure component viscosities.

Theoretical analysis

Excess molar volume

Excess molar volumes (V_m^E), were calculated for the binary mixtures of benzene with n-hexane, n-octane, n-decane, n-dodecane, n-tetradecane, and n-hexadecane using viscosity data by a correlation proposed by Singh [13]. According to the relation, the deviations in viscosity, $\Delta\eta$, and excess molar volumes, V_m^E, are related to each other as:

$$\Delta\eta = K \times V_m^E \tag{1}$$

where, K is a fitting parameter. The values of K for the investigated mixtures were evaluated using the experimentally reported $\Delta\eta$ and V_m^E data [12] (Table 1).

From the experimental $\Delta\eta$ data, V_m^E values at the whole mole fraction range were calculated at 298.15 K and results were presented in Table 2.

Figures 1, 2, 3 show that V_m^E values calculated from viscosity data are positive over the entire composition range for all investigated mixtures and follow the sequence: n-hexane < n-octane < n-decane < n-dodecane < n-tetradecane < n-hexadecane. V_m^E values increase systematically (0.43031, 0.74732, 0.89911, 1.01524, 1.06634 and 1.19575 cm^3 mol^{-1}) at benzene mole fraction (x_1) range (0.39880–0.43415) as chain length increases, indicating volume expansion upon mixing of higher n-alkanes with benzene. With increasing size of n-alkanes, volume expansion will be larger, indicating dispersion type of interactions between the component liquids [14].

The volumetric behavior of this class of mixtures could be explained by random mixing model [15], and the interaction between benzene and n-alkane depends strongly on the size of the n-alkane. For each system studied it was observed that the excess molar volume is slightly skewed towards the benzene-rich region of the mole fraction. Calculated values of V_m^E compare well with those reported by Peña and Delgado [16] for all mixtures, Calvar et al. [17] for benzene + n-hexane, +n-octane, and Letcher and Perkins [18] for benzene + n-dodecane, +n-hexadecane.

The excess molar volumes (V_m^E) were fitted by the Redlich–Kister [19] polynomial equation:

$$V_m^E = x_2(1 - x_2) \sum_{i=0}^{i=n} A_i(1 - 2x_2)^i \tag{2}$$

where x_2 is the mole fraction of n-alkane.

The appropriate degree (n) of Eq. (2) was determined by standard deviation (σ) being calculated as:

$$\sigma = \left[\sum (V_{exp}^E - V_{cal}^E)^2 / (N - (n + 1)) \right]^{1/2} \tag{3}$$

where N is the total number of experimental points; $(n + 1)$ is the number of coefficients (A_i) in Eq. (2).

Table 1 Values of fitting parameter (K) of binary mixtures, and molar volumes of pure components (V_m^*) at 298.15

System	K/m Pa.s.cm^{-3}.mol	$V_m^*/cm^3.mol^{-1}$	
Benzene + n-hexane	−0.2145	Benzene	89.458
Benzene + n-octane	−0.1071	n-Hexane	131.540
Benzene + n-decane	−0.0719	n-Octane	163.531
Benzene + n-dodecane	−0.0931	n-Decane	195.913
Benzene + n-tetradecane	−0.1389	n-Dodecane	228.242
Benzene + n-hexadecane	−0.2116	n-Tetradecane	261.287
		n-Hexadecane	294.066

Table 2 Values of excess molar volumes, V_m^E, calculated using Eq. (1) for binary mixtures for (x) n-alkane + $(1-x)$ benzene at 298.15 K

$V_m^E/ cm^3.mol^{-1}$

(x) n-Hexane + $(1-x)$ benzene		(x) n-Octane + $(1-x)$ benzene		(x) n-Decane + $(1-x)$ benzene	
0.05393	0.13220	0.04472	0.22234	0.03584	0.25984
0.11224	0.25405	0.09184	0.40061	0.10038	0.48841
0.20795	0.35508	0.17152	0.55951	0.20129	0.77266
0.30955	0.41102	0.29540	0.72549	0.29022	0.89092
0.41965	0.43031	0.40878	0.74732	0.39880	0.89911
0.50041	0.41434	0.49074	0.70538	0.49997	0.83648
0.59341	0.36720	0.58551	0.60583	0.59432	0.71743
0.68570	0.30241	0.73898	0.47019	0.70210	0.55340
0.79132	0.21284	0.92796	0.12684	0.83527	0.32117
0.88344	0.11426			0.91651	0.16049

(x) n-Dodecane + $(1-x)$ Benzene		(x) n-Tetradecane + $(1-x)$ Benzene		(x) n-Hexadecane + $(1-x)$ Benzene	
0.04733	0.22407	0.04444	0.19194	0.05350	0.29540
0.10840	0.48939	0.09068	0.46134	0.09944	0.47192
0.20364	0.79727	0.21254	0.78494	0.21994	0.83828
0.29778	0.94472	0.41693	1.06634	0.32338	1.16970
0.41395	1.01524	0.51212	1.05153	0.43415	1.19575
0.51924	0.96690	0.62706	0.81988	0.51942	1.15131
0.61546	0.86817	0.71387	0.62522	0.64097	1.04078
0.72914	0.67885	0.80062	0.55735	0.69281	1.00728
0.80113	0.55101	0.88223	0.51832	0.83610	0.85819
0.92306	0.22493			0.94161	0.22225

Calculated values of A_i and σ for all systems studied are given in Table 3.

Partial molar volume, $V_{m,1}$ and $V_{m,2}$, of benzene (1) and n-alkanes (2), respectively, are given by:

$$V_{m,1} = V_m^E + V_{m,1}^* + (1 - x_1)(\partial V^E/\partial x_1)_{P,T} \qquad (4)$$

$$V_{m,2} = V_m^E + V_{m,2}^* - x_1(\partial V^E/\partial x_1)_{P,T} \qquad (5)$$

where $V_{m,1}^*$ and $V_{m,2}^*$ are the molar volumes of the pure liquids (1) and (2), respectively. Combination of Eqs. (2), (4) and (5) leads to Eqs. (6) and (7) for $V_{m,1}$ and $V_{m,2}$:

$$V_{m,1} = V_{m,1}^* + x_2^2 \sum_{i=0}^{i=n} A_i(1 - 2x_1)^i$$
$$- 2x_1x_2^2 \sum_{i=1}^{i=n} A_i(1 - 2x_1)^{i-1} \qquad (6)$$

$$V_{m,2} = V_{m,2}^* + x_1^2 \sum_{i=0}^{i=n} A_i(1 - 2x_1)^i$$
$$+ 2x_1^2x_2 \sum_{i=1}^{i=n} A_i(1 - 2x_1)^{i-1} \qquad (7)$$

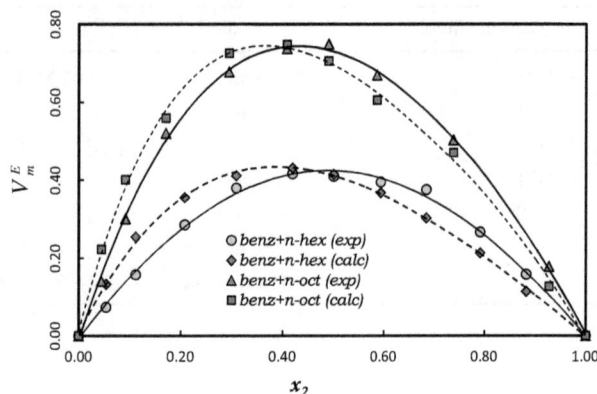

Fig. 1 Excess molar volume V_m^E versus x_2 for benzene(1) with n-hexane(2) and n-octane(2) at 298 K

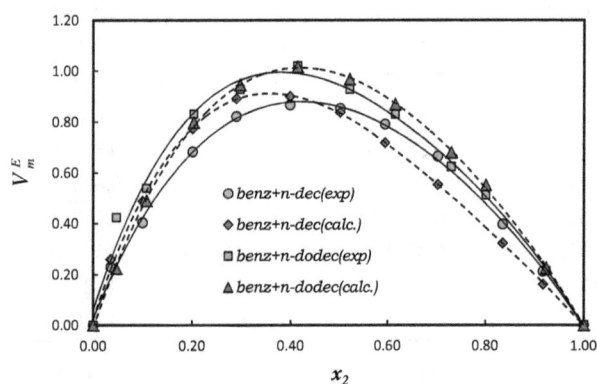

Fig. 2 Excess molar volume V_m^E versus x_2 for benzene(1) with n-decane(2) and n-dodecane(2) at 298 K

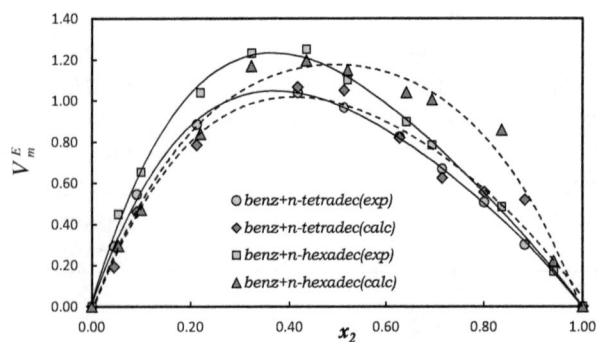

Fig. 3 Excess molar volume V_m^E versus x_2 for benzene(1) with n-tetradecane(2) and n-hexadecane(2) at 298 K

Values of the partial molar volumes at infinite dilution, $\overline{V}_{m,1}^0$ and $\overline{V}_{m,2}^0$, were obtained by the linear extrapolation of corresponding partial molar volumes using Eqs. (6) and (7). Extrapolation of $V_{m,1}$ to $x_1 \rightarrow 0$ results in $\overline{V}_{m,1}^0$, and consequently, extrapolation of $V_{m,2}$ to $x_2 \rightarrow 0$ results in $\overline{V}_{m,2}^0$. The excess partial molar volumes at infinite dilution, $\overline{V}_{m,1}^{0,E}$ and $\overline{V}_{m,2}^{0,E}$, were calculated using the following relations [20]:

$$\overline{V}_{m,1}^{0,E} = \overline{V}_{m,1}^0 - V_{m,1}^* \tag{8}$$

$$\overline{V}_{m,2}^{0,E} = \overline{V}_{m,2}^0 - V_{m,2}^* \tag{9}$$

Apparent molar volumes of benzene $V_{\phi,1}$ and apparent molar volume of n-alkanes $V_{\phi,2}$ were calculated from experimental data using the following relations [21]:

$$V_{\phi,1} = \frac{(V_m - x_2 V_2^0)}{x_1} = \frac{(\rho_2 - \rho)x_2 M_2}{x_1 \rho \rho_2} + \frac{M_1}{\rho} \tag{10}$$

$$V_{\phi,2} = \frac{(V_m - x_1 V_1^0)}{x_2} = \frac{(\rho_1 - \rho)x_1 M_1}{x_2 \rho \rho_1} + \frac{M_2}{\rho} \tag{11}$$

Extrapolation of $V_{\phi,1}$ to $x_1 \rightarrow 0$ and $V_{\phi,2}$ to $x_2 \rightarrow 0$ give the values of the limiting apparent molar volumes $\overline{V}_{\phi,1}^0$ and $\overline{V}_{\phi,2}^0$ at infinite dilution, represented earlier as $\overline{V}_{m,1}^0$ and $\overline{V}_{m,2}^0$. The excess apparent molar volumes at infinite dilution, $\overline{V}_{\phi,1}^{0,E}$ and $\overline{V}_{\phi,2}^{0,E}$ were also calculated by equations similar to Eqs. (8) and (9). The values $\overline{V}_{m,1}^0$, $\overline{V}_{\phi,1}^0$, $V_{m,1}^*$, $\overline{V}_{m,1}^{0,E}$, $\overline{V}_{\phi,1}^{0,E}$, $\overline{V}_{m,2}^0$, $\overline{V}_{\phi,2}^0$, $V_{m,2}^*$, $\overline{V}_{m,2}^{0,E}$ and $\overline{V}_{\phi,2}^{0,E}$ for all six binary mixtures at 298.15 K are listed in Table 4. An examination of data in Table 4 reveals that the values of $\overline{V}_{m,1}^0$ and $\overline{V}_{\phi,1}^0$ are nearly of the same magnitude, and the values of $\overline{V}_{\phi,2}^0$ are slightly greater than $\overline{V}_{m,2}^0$ for all binary mixtures. Values of $\overline{V}_{m,1}^{0,E}$, $\overline{V}_{\phi,1}^{0,E}$, $\overline{V}_{m,2}^{0,E}$ and $\overline{V}_{\phi,2}^{0,E}$ are positive over the entire composition range for all systems. This indicates that molar volumes and apparent molar volumes of both components in the mixture are larger than their respective values in the pure state, which indicates the presence of

Table 3 Coefficients A_i of Eq. 2 and standard deviation, σ, at 298.15 K

Binary mixture	A_0	A_1	A_2	A_3	A_4	σ
Benzene + n-hexane	0.0051	2.6233	−5.1111	3.4665	−0.9871	0.008
Benzene + n-octane	−0.0123	4.0913	−7.0177	4.2741	−1.3307	0.062
Benzene + n-decane	0.0234	5.9634	−12.8940	9.8982	−2.9941	0.018
Benzene + n-dodecane	−0.0114	5.7372	−10.2400	6.5010	−1.9869	0.011
Benzene + n-tetradecane	−0.0279	6.2820	−12.5210	9.7973	−3.4848	0.080
Benzene + n-hexadecane	−0.0194	6.4531	−12.9770	12.9030	−6.3815	0.080

Table 4 The values $\overline{V}_{m,1}^{0}$, $\overline{V}_{m,2}^{0}$, $\overline{V}_{\phi,1}^{0}$, $\overline{V}_{\phi,2}^{0}$, $V_{m,1}^{*}$, $V_{m,2}^{*}$, $\overline{V}_{m,1}^{0,E}$, $\overline{V}_{m,2}^{0,E}$, $\overline{V}_{\phi,1}^{0,E}$ and $\overline{V}_{\phi,2}^{0,E}$ for all six mixtures at 298.15 K

Benzene (1)+	$\overline{V}_{m,1}^{0}$	$\overline{V}_{\phi,1}^{0}$	$V_{m,1}^{*}$	$\overline{V}_{m,1}^{0,E}$	$\overline{V}_{\phi,1}^{0,E}$	$\overline{V}_{m,2}^{0}$	$\overline{V}_{\phi,2}^{0}$	$V_{m,2}^{*}$	$\overline{V}_{m,2}^{0,E}$	$\overline{V}_{\phi,2}^{0,E}$
n-Hexane(2)	90.111	91.082	89.458	0.653	1.620	131.754	133.143	131.540	0.214	1.603
n-Octane(2)	90.255	92.159	89.458	0.797	2.701	164.070	166.932	163.531	0.539	3.401
n-Decane(2)	91.397	92.423	89.458	1.939	2.965	196.460	200.061	195.913	0.547	4.148
n-Dodecane(2)	91.623	92.630	89.458	2.165	3.172	228.661	233.215	228.242	0.419	4.973
n-Tetradecane(2)	92.181	92.646	89.458	2.723	3.188	261.598	267.354	261.287	0.311	6.067
n-Hexadecane(2)	92.938	92.989	89.458	3.480	3.531	294.715	301.119	294.066	0.649	7.053

All quantities have the unit: $cm^3\ mol^{-1}$

significant solute–solute and solvent–solvent interactions between like molecules in the mixture [22].

Correlation equations for viscosity

Viscosity data are given in Table 5 as a function of the mole fraction of the n-alkane at 298.15 K. The viscosity deviation was calculated according to the following equation:

$$\Delta\eta = \eta - (x_1\eta_1 + x_2\eta_2) \tag{12}$$

where η_1, η_2 and η are the viscosities of component 1(benzene), component 2 (n-alkane) and the mixture, respectively. All mixtures deviate from ideality with a negative deviation, which indicate that dispersion forces are predominant between benzene and n-alkane [23].

Various equations are used in the literature to calculate the viscosity of mixtures in terms of pure component data. To carry out a comparative study, Kendall–Munroe, Frenkel, Hind et al., Katti-Chaudhri, Grunberg-Nissan, Wilke and Herráez et al. equations have been employed to estimate the viscosity of binary liquid mixtures of benzene + n-alkanes.

1. Kendall-Monroe [24] derived Eq. (13) for analyzing the viscosity of binary mixtures based on zero adjustable parameter. This equation calculates the mixture viscosity as the cubic-root average of the component viscosities:

$$\eta = (x_1\eta_1^{1/3} + x_2\eta_2^{1/3})^3 \tag{13}$$

2. Frenkel with the help of Eyring's model [25, 26] took into consideration the interaction between molecules and developed the following logarithmic relation for nonideal binary mixtures:

$$\ln\eta = x_1^2 \ln\eta_1 + x_2^2 \ln\eta_2 + 2x_1x_2 \ln\eta_{12} \tag{14}$$

where, η_{12} is a constant attributed to unlike pair interactions. Its value is obtained from the following equation:

$$\eta_{12} = 0.5\eta_1 + 0.5\eta_2 \tag{15}$$

3. Hind et al. [27] have suggested the following equation for the viscosity of binary liquid mixtures:

$$\ln\eta = x_1^2\eta_1 + x_2^2\eta_2 + 2x_1x_2H_{12} \tag{16}$$

where, H_{12} is the Hind interaction parameter and is attributed to unlike pair interactions, and other terms have their usual meaning.

4. Katti-Chaudhri equation is expressed as [28]:

$$\ln(\eta V) = x_1 \ln(\eta_1 V_1) + x_2 \ln(\eta_2 V_2) + x_1x_2 (W/RT) \tag{17}$$

where, W is the interaction energy parameter, V is the volume of the mixture, V_1 and V_2 are the volumes of component 1 and component 2, respectively.

5. Grunberg and Nissan [29] have formulated equation to assess the molecular interactions leading to viscosity changes:

$$\ln\eta = x_1 \ln\eta_1 + x_2 \ln\eta_2 + x_1x_2G_{12} \tag{18}$$

where, G_{12} is a constant, proportional to interchange energy, η is the dynamic viscosity and the subscripts 1, 2 and 12 stands for the pure components, benzene, n-alkanes and mixtures, respectively.

6. Wilke [30] proposed the viscosity equation:

$$\eta = \frac{x_1\eta_1}{x_1 + x_2\phi_{12}} + \frac{x_2\eta_2}{x_2 + \phi_{21}} \tag{19}$$

where, ϕ_{12} and ϕ_{21} are calculated by the following equations:

$$\phi_{12} = \frac{\left[1 + \left(\frac{\eta_1}{\eta_2}\right)^{1/2} (M_2/M_1)^{1/4}\right]}{\{8 + [1 + (M_1/M_2)]\}^{1/2}} \tag{20}$$

$$\phi_{21} = \phi_{12} \frac{\eta_2}{\eta_1} \frac{M_1}{M_2} \tag{21}$$

where, ϕ and M are the volume fraction and molar mass, and the subscripts 1 and 2 stands for pure components benzene and n-alkane, respectively.

Table 5 Viscosity deviation ($\eta_{exp} - \eta_{calc.}$) for the binary mixtures at 298.15 K

Viscosity deviation ($\eta_{exp.} - \eta_{calc.}$)

x_2	$\eta_{exp.}$ (mPa.s)	K. & M.	Frenkel	Hind	Katti	G. & N.	Wilke	Herráez et al.
Benzene + n-hexane								
0.05393	0.5662	−0.0248	0.0186	−0.0100	−0.0039	−0.0284	−0.0366	0.0134
0.11224	0.5225	−0.0476	0.0372	−0.0186	−0.0089	−0.0545	−0.0695	−0.0006
0.20795	0.4720	−0.0648	0.0754	−0.0168	−0.0060	−0.0762	−0.0983	−0.0146
0.30955	0.4294	−0.0735	0.1086	−0.0111	−0.0034	−0.0882	−0.1138	−0.0260
0.41965	0.3921	−0.0757	0.1319	−0.0045	−0.0024	−0.0923	−0.1182	−0.0342
0.50041	0.3712	−0.0720	0.1412	0.0012	−0.0009	−0.0889	−0.1133	−0.0358
0.59341	0.3533	−0.0626	0.1433	0.0082	0.0019	−0.0788	−0.1002	−0.0331
0.68570	0.3394	−0.0505	0.1335	0.0128	0.0039	−0.0649	−0.0823	−0.0280
0.79132	0.3268	−0.0347	0.1063	0.0139	0.0045	−0.0457	−0.0577	−0.0201
0.88344	0.3202	−0.0177	0.0703	0.0126	0.0056	−0.0245	−0.0314	−0.0097
Benzene + n-octane								
0.04472	0.5829	−0.0236	0.0245	−0.0102	−0.0073	−0.0238	−0.0407	0.0194
0.09184	0.5595	−0.0425	0.0514	−0.0162	−0.0115	−0.0429	−0.0748	0.0038
0.17152	0.5352	−0.0592	0.1007	−0.0145	−0.0088	−0.0599	−0.1115	−0.0127
0.29540	0.5061	−0.0767	0.1575	−0.0112	−0.0077	−0.0777	−0.1476	−0.0341
0.40878	0.4934	−0.0788	0.1931	−0.0028	−0.0025	−0.0800	−0.1561	−0.0418
0.49074	0.4904	−0.0743	0.2069	0.0043	0.0025	−0.0755	−0.1508	−0.0418
0.58551	0.4924	−0.0637	0.2094	0.0127	0.0088	−0.0649	−0.1345	−0.0368
0.73898	0.4929	−0.0494	0.1676	0.0113	0.0064	−0.0504	−0.1018	−0.0322
0.92796	0.5124	−0.0133	0.0620	0.0078	0.0058	−0.0136	−0.0300	−0.0084
Benzene + n-decane								
0.03584	0.6003	−0.0179	0.0314	−0.0100	−0.0079	−0.0187	−0.0408	−0.0105
0.10038	0.5986	−0.0330	0.0958	−0.0125	−0.0091	−0.0351	−0.0912	−0.0125
0.20129	0.6012	−0.0517	0.1775	−0.0152	−0.0114	−0.0556	−0.1512	−0.0118
0.29022	0.6130	−0.0591	0.2346	−0.0124	−0.0095	−0.0641	−0.1824	−0.0041
0.39880	0.6372	−0.0589	0.2830	−0.0045	−0.0033	−0.0646	−0.1971	0.0105
0.49997	0.6648	−0.0541	0.3023	0.0025	0.0023	−0.0601	−0.1937	0.0234
0.59432	0.6949	−0.0457	0.2980	0.0089	0.0078	−0.0516	−0.1767	0.0336
0.70210	0.7313	−0.0347	0.2635	0.0126	0.0111	−0.0398	−0.1451	0.0386
0.83527	0.7784	−0.0197	0.1764	0.0114	0.0100	−0.0231	−0.0901	0.0321
0.91651	0.8085	−0.0097	0.0994	0.0076	0.0068	−0.0115	−0.0481	0.0204
Benzene + n-dodecane								
0.04733	0.6259	−0.0121	0.0685	−0.0040	0.0003	−0.0209	−0.0607	0.0023
0.10840	0.6476	−0.0267	0.1459	−0.0095	−0.0036	−0.0456	−0.1297	−0.0020
0.20364	0.6913	−0.0423	0.2471	−0.0137	−0.0066	−0.0742	−0.2121	−0.0098
0.29778	0.7491	−0.0464	0.3264	−0.0099	−0.0041	−0.0880	−0.2622	−0.0119
0.41395	0.8308	−0.0458	0.3862	−0.0040	−0.0007	−0.0945	−0.2926	−0.0134
0.51924	0.9153	−0.0394	0.4046	0.0031	0.0031	−0.0900	−0.2908	−0.0116
0.61546	0.9976	−0.0325	0.3881	0.0075	0.0047	−0.0808	−0.2691	−0.0100
0.72914	1.1016	−0.0224	0.3281	0.0105	0.0054	−0.0632	−0.2190	−0.0071
0.80113	1.1682	−0.0182	0.2644	0.0082	0.0027	−0.0513	−0.1765	−0.0073
0.92306	1.2912	−0.0060	0.1198	0.0056	0.0020	−0.0209	−0.0767	−0.0022
Benzene + n-tetradecane								
0.04444	0.6498	−0.0032	0.0880	−0.0031	0.0060	−0.0267	−0.0719	0.0062
0.09068	0.6807	−0.0182	0.1585	−0.0183	−0.0021	−0.0641	−0.1513	−0.0085
0.21254	0.8158	−0.0143	0.3427	−0.0162	0.0087	−0.1090	−0.2837	−0.0173

Table 5 continued

x_2	$\eta_{exp.}$ (mPa.s)	K. & M.	Frenkel	Hind	Katti	G. & N.	Wilke	Herráez et al.
0.41693	1.0787	−0.0065	0.5080	−0.0132	0.0023	−0.1481	−0.3987	−0.0389
0.51212	1.2214	0.0014	0.5283	−0.0074	−0.0032	−0.1461	−0.4035	−0.0412
0.62706	1.4234	0.0262	0.5173	0.0159	0.0050	−0.1139	−0.3558	−0.0223
0.71387	1.5787	0.0370	0.4645	0.0265	0.0059	−0.0868	−0.2996	−0.0105
0.80062	1.7163	0.0204	0.3534	0.0112	−0,0142	−0.0774	−0.2454	−0.0202
0.88223	1.8423	−0.0076	0.2084	−0.0143	−0.0371	−0.0720	−0.1828	−0.0361
Benzene + n-hexadecane								
0.05350	0.6803	−0.0033	0.1243	−0.0099	0.0082	−0.0625	−0.1211	0.0046
0.09944	0.7563	0.0058	0.2305	−0.0069	0.0223	−0.0999	−0.2034	0.0053
0.21994	0.9761	0.0297	0.4555	0.0007	0.0432	−0.1774	−0.3761	−0.0106
0.32338	1.1612	0.0217	0.5597	−0.0204	0.0161	−0.2475	−0.5023	−0.0564
0.43415	1.4290	0.0553	0.6532	0.0020	0.0155	−0.2530	−0.5420	−0.0568
0.51942	1.6488	0.0744	0.6772	0.0155	0.0024	−0.2436	−0.5407	−0.0554
0.64097	1.9721	0.0792	0.6287	0.0186	−0.0349	−0.2202	−0.5004	−0.0585
0.69281	2.1071	0.0662	0.5720	0.0078	−0.0603	−0.2131	−0.4755	−0.0676
0.83610	2.4922	0.0026	0.3239	−0.0393	−0.1200	−0.1816	−0.3581	−0.0940
0.94161	2.8871	0.0282	0.1558	0.0100	−0.0347	−0.0470	−0.1210	−0.0134

Viscosity deviation ($\eta_{exp.} - \eta_{calc.}$)

7. Herráez et al. [31] proposed a new correlation equation based on the linear behavior of binary mixtures:

$$\eta = \eta_1 + (\eta_2 - \eta_1) x_2 \qquad (22)$$

They introduce an exponential function of the mole fraction, x_2, in Eq. 22 above, to yield:

$$\eta = \eta_1 + (\eta_2 - \eta_1) x_2^{(\sum_{i=0}^{n} (Bi.x_2^i))} \qquad (23)$$

which for $n = 0$ would be:

$$\eta = \eta_1 + (\eta_2 - \eta_1) x_2^{[Bo]} \qquad (24)$$

where, B_0 is the universal exponent constant.

The predicted values of viscosities of the binary mixtures, using Eqs. (13), (14), (16), (17), (18), (19) and (24), including standard deviation, at 298.15 K were compared with the experimentally measured values, and results are presented in terms of viscosity deviations ($\Delta\eta$) (Table 5).

Validity of aforementioned relations has been checked by calculating the viscosity deviations. The average percentage deviation (APD) between calculated and experimental viscosity values is calculated by [32]:

$$APD = \left(\frac{100}{N}\right) \sum \left[\frac{(|\eta_{exp} - \eta_{cal}|)}{\eta_{exp}}\right] \qquad (25)$$

where N is the number of data points in each set.

The interaction parameters of Eqs. (13), (14), (16), (17), (18), (19) and (24), along with APD and σ values for all binary mixtures are presented in Table 6. A careful perusal

of Table 6 reveals that maximum deviations are obtained using Wilke relation for the prediction of viscosity of binary liquid mixtures under the present study while other relations give comparatively good results. The best results obtained by using Katti–Chaudhri relation. The trend of validity of the presently used relations is as follows: Katti-Chaudhri > Hind et al. > Herráez et al. > Grunberg and Nissan > Kendall-Munroe > Frenkel > Wilke.

The values of interaction parameters η_{12}, H_{12}, W/RT, G_{12} and B_0 are presented in Table 6. η_{12} and H_{12} values are positive for all binary mixtures and increases with increasing alkyl chain length of n-alkanes. Correlation parameter, W/RT has negative values for the first two binary mixtures and has positive values for the rest of binary mixtures. The negatives values of W/RT suggest weak interactions, and positive values indicate strong interactions between the unlike molecules [33].

G_{12} values are negative for all binary mixtures except for the two last ones (n-tetradecane + benzene and n-hexadecane + benzene). The negative values of G_{12} indicates the dominance of dispersion forces [34, 35], while the positive values are attributed to the presence of strong specific interactions between the mixture components [36, 37]. B_0 values are positive for all binaries mixtures, $B_0 > 1$ for mixtures where the sub index 1 represent the component of least viscosity (benzene + n-decane, +n-dodecane, +n-tetradecane, and +n-hexadecane), whereas $B_0 < 1$ for mixtures where the sub index 2 represent the component of least viscosity (benzene + n-hexane, +n-octane).

Table 6 Adjustable parameters of Eqs. (13), (14), (16), (17), (18), (19) and (24), APD values and standard deviations of binary mixture viscosities at 298.15 K

	Benzene + n-hexane	Benzene + n-octane	Benzene + n-decane	Benzene + n-dodecane	Benzene + n-tetradecane	Benzene + n-hexadecane
Kendall and Monroe						
σ	0.056	0.058	0.042	0.032	0.019	0.046
APD	10.98	8.59	4.93	2.96	1.04	1.90
Frenkel						
η_{12}	0.4602	0.5651	0.7245	0.9907	1.3496	1.8445
σ	0.059	0.0581	0.0432	0.040	0.023	0.024
APD	11.60	8.63	5.05	3.69	1.49	0.64
Hind et al.						
H_{12}	0.2800	0.4053	0.5996	0.8041	1.0721	1.3255
σ	0.072	0.062	0.084	0.071	0.106	0.199
APD	13.55	8.73	1.24	0.76	1.02	0.71
Katti and Chaudhri						
W/RT	−0.5332	−0.3768	0.0132	0.3674	0.8304	1.2899
σ	0.005	0.008	0.009	0.005	0.015	0.051
APD	0.84	1.06	0.99	0.32	0.58	1.82
Grunberg and Nissan						
G_{12}	−0.6226	−0.5840	−0.3170	−0.0891	0.2626	0.5781
σ	0.005	0.009	0.014	0.014	0.009	0.023
APD	0.91	1.31	1.52	1.20	0.92	0.66
Wilke						
ϕ_{12}	0.53315	0.56841	0.61068	0.66773	0.72554	0.78865
ϕ_{21}	1.16051	0.97753	0.80972	0.64889	0.53899	0.45366
σ	0.088	0.109	0.143	0.215	0.272	0.407
APD	17.32	16.96	16.65	19.19	18.75	21.98
Herráez et al.						
B_0	0.5643	0.2123	2.8978	1.3382	1.2230	1.2425
σ	0.024	0.029	0.023	0.009	0.026	0.052
APD	4.733	4.161	2.365	0.751	1.516	2.089

Conclusions

Excess molar volume have been calculated from the experimental viscosity data at 298.15 K for benzene + n-hexane, or +n-octane, or +n-decane, or +n-dodecane, or +n-tetradecane, or + n-hexadecane binary mixtures.

The values of V_m^E, $\overline{V}_{m,1}^0$, $\overline{V}_{m,2}^0$, $\overline{V}_{\phi,1}^0$, $\overline{V}_{\phi,2}^0$, $V_{m,1}^*$, $V_{m,2}^*$, $\overline{V}_{m,1}^{0,E}$, $\overline{V}_{m,2}^{0,E}$, $\overline{V}_{\phi,1}^{0,E}$ and $\overline{V}_{\phi,2}^{0,E}$ were calculated for all six mixtures at 298.15 K. The V_m^E values were found to be positive over the whole composition range for all mixtures. Molar volumes and apparent molar volumes of both components in the mixture are larger than their respective values in the pure state. The order of interaction between benzene and n-alkanes follows the sequence: n-hexane < n-octane < n-decane < n-dodecane < n-tetradecane < n-hexadecane, i.e., the interaction increase with increasing chain length of the n-alkane.

Moreover, an attempt has been made to check the suitability of empirical and semiempirical relations for experimental viscosities data of n-alkanes + benzene fits by taking into account a number of empirical adjustment coefficients. The predicted viscosities show good accuracy in comparison with the experimental viscosities. The trend of validity of the presently used relations is as follows: Katti-Chaudhri > Hind et al. > Herráez et al. > Grunberg and Nissan > Kendall-Munroe > Frenkel > Wilke.

References

1. Tôrres RB, Francesconi AZ, Volpe PLO (2003) Experimental study and modeling using the ERAS-Model of the excess molar volume of acetonitrile–alkanol mixtures at different temperatures and atmospheric pressure. Fluid Phase Equilib 210:287–306
2. Desnoyers JE, Perron G (1997) Treatment of excess thermodynamic quantities for liquid mixtures. J Solut Chem 26:749–755
3. Tôrres RB, Francesconi AZ, Volpe PLO (2007) Volumetric properties of binary mixtures of acetonitrile and alcohols at different temperatures and atmospheric pressure. J Mol Liq 131(132):139–144
4. Zhang Y, Venable RM, Pastor RW (1996) Molecular dynamics simulations of neat alkanes: the viscosity dependence of rotational relaxation. J Phys Chem 100:2652–2660
5. Aminabhavi TM, Aralaguppi MI, Gopalakrishna B, Khinnavar RS (1994) Densities, shear viscosities, refractive indices, and speeds of sound of bis(2-methoxyethyl) ether with hexane, heptane, octane, and 2,2,4-trimethylpentane in the temperature interval 298.15–318.15 K. J Chem Eng Data 39:522–528
6. Gómez-Díaz D, Mejuto JC, Navaza JM, Rodríguez-Alvarez A (2002) Viscosities, densities, surface tensions, and refractive indexes of 2,2,4-trimethylpentane + cyclohexane + decane ternary liquid systems at 298.15K. J Chem Eng Data 47:872–875
7. Oswal RL, Phalak RP (1992) Viscosities of nonelectrolyte liquid mixtures. I. Binary mixtures containing p-dioxane. Int J Thermophys 13:251–267
8. Baragi JG, Aralaguppi MI, Kariduraganavar MY, Kulkarni SS, Kittur AS, Aminabhavi TM (2006) Excess properties of the binary mixtures of methylcyclohexane alkanes (C_6 to C_{12}) at $T = 298.15$ K to $T = 308.15$ K. J Chem Thermodyn 38:75–83
9. Garcia B, Alcalde R, Aparicio S, Leal JM (2002) Volumetric properties, viscosities and refractive indices of binary mixed solvents containing methyl benzoate. Phys Chem Chem Phys 4:5833–5840
10. Ali A, Nain AK, Sharma VK, Ahmad S (2004) Molecular interactions in binary mixtures of tetrahydrofuran with alkanols (C6, C8, C10): an ultrasonic and volumetric study. Indian J Pure Appl Phys 42:666–673
11. Iloukhani H, Rezaei-Sameti M, Basiri-Parsa J (2006) Excess molar volumes and dynamic viscosities for binary mixtures of toluene + n-alkanes (C5–C10) at $T = 298.15$ K, comparison with Prigogine-Flory-Patterson theory. J Chem Thermodyn 38:975–982
12. Awwad AM, Al-azzawi SF, Salman MA (1986) Volumes and viscosities of benzene + n-alkane mixtures. Fluid Phase Equilib 31:171–182
13. Singh PP (1988) Topological investigations of the viscous behavior of binary mixtures of non-electrolyte. Ind J Chem Sect A 27:469–473
14. Yadava SS, Yadav N (2011) Excess molar volumes and refractive indices of binary mixtures of isopropylethanoate and symmetrical hydrocarbons at 308.15 K. Can J Chem Eng 89:576–581
15. Kehiaian HV (1985) Thermodynamics of binary liquid organic mixtures. Pure Appl Chem 57(1):15–30
16. Peña MD, Delgado JN (1975) Excess volumes at 323.15 K of binary mixtures of benzene with n-alkanes. J Chem Thermodyn 7:201–204
17. Calvar N, Gómez E, González B, Domínguez Á (2009) Experimental densities, refractive indices, and speeds of sound of 12 binary mixtures containing alkanes and aromatic compounds at $T = 313.15$ K. J Chem Thermodyn 41:939–944
18. Letcher TM, Perkins DM (1984) Application of the Flory theory of liquid mixtures to excess volumes and enthalpies of benzene + cycloalkane and +n-alkane mixtures. Thermochim Acta 77:267–274
19. Redlich O, Kister AT (1948) Thermodynamics of nonelectrolyte solutions-x-y-t relations in a binary system. Ind Eng Chem 40:341–345
20. Sinha B (2010) Excess molar volumes, viscosity deviations and speeds of sound for some alkoxyethanols and amines in cyclohexanone at 298.15 K. Phys Chem Liq 48:183–198
21. Egorov GI, Makarov DM (2012) Volumetric properties of the binary mixture of ethylene glycol + tert-butanol at $T = (278.15, 288.15, 298.15, 308.15, 323.15, 333.15, 348.15)$ K under atmospheric pressure. J Mol Liq 171:29–36
22. Nain AK (2013) Densities and volumetric properties of butyl acrylate + 1-butanol, or +2-butanol, or +2-methyl-1- propanol, or +2-methyl-2-propanol binary mixtures at temperatures from 288.15 to 318.15 K. J Solution Chem 42:1404–1422
23. Ouerfelli N, Bouaziz M, Herráez JV (2013) Treatment of Herráez equation correlating viscosity in binary liquid mixtures exhibiting strictly monotonous distribution. Phys Chem Liq 51(1):55–74
24. Babu CP, Kumar GP, Samatha K (2015) Comparison of experimental viscosities by theoretically for 1-bromopropane in chlorobenzene mixture at (303.15, 308.15, 313.15 and 318.15) K. Int J Adv Sci Tech 76:27–34
25. Medvedevskikh Y, Khavunko O (2012) Phenomenological coefficients of the viscosity for low-molecular elementary liquids and solutions. Ch Ch T 6(4):363–370
26. Dikio ED, Nelana SM, Isabirye DA, Ebenso EE (2012) Density, dynamic viscosity and derived properties of binary mixtures of methanol, ethanol, n-propanol, and n-butanol with pyridine at $T = (293.15, 303.15, 313.15$ and $323.15)$ K. Int J Electrochem Sci 7:11101–11122
27. Hind RK, McLaughlin E, Ubbelohde AR (1960) Structure and viscosity of liquids camphor + pyrene mixtures. Trans Faraday Soc 56:328–330
28. Sanz LF, Gonzalez JA, de la Fuente IG, Cobos JC (2015) Thermodynamics of mixtures with strongly negative deviations from Raoult's law. XII. Densities, viscosities and refractive indices at $T = (293.15$ to $303.15)$ K for (1-heptanol, or 1-decanol + cyclohexylamine) systems. Application of the ERAS model to (1-alkanol + cyclohexylamine) mixture". J Chem Thermodyn 80:161–171
29. Hernández-Galván MA, García-Sánchez F, Macías-Salinas R (2007) Liquid viscosities of benzene, n-tetradecane, and benzene + n-tetradecane from 313 to 393 K and pressures up to 60 MPa: experiment and modeling. Fluid Phase Equilib 262:51–60
30. Wilke CR (1950) A viscosity equation for gas mixtures. J Chem Phys 18:517–520
31. Herráez JV, Belda R, Díez O, Herráez M (2008) An equation for the correlation of viscosities of binary mixtures. J Solut Chem 37:233–248
32. Mahajan AR, Mirgane SR (2013) Excess molar volumes and viscosities for the binary mixtures of n-octane, n-decane, n-dodecane, and n-tetradecane with octan-2-ol at 298.15 K. J Thermodyn 2013:1–11
33. Venkatalakshmi V, Chowdappa A, Venkateswarlu P, Reddy KS (2014) Volumetric, speed of sound data and viscosity for the binary mixtures of 2-methylaniline with aliphatic ketones and cyclic ketones at different temperatures. Int J Innov Res Sci Eng Tech 3(11):17556–17566
34. Sharma S, Thakkar K, Patel P, Makavana M (2013) Volumetric, viscometric and excess properties of binary mixtures of 1-iodobutane with benzene, toluene, o-xylene, m-xylene, p-xylene, and mesitylene at temperatures from 303.15 to 313.15 K. Adv Phys Chem 2013:1–12
35. Agarwal D, Singh M (2004) Viscometric studies of molecular interactions in binary liquid mixtures of nitromethane with some

polar and non-polar solvents at 298.15K. J Indian Chem Soc 81:850–859

36. Dubey GP, Kumar K (2011) Thermodynamic properties of binary liquid mixtures of diethylenetriamine with alcohols at different temperatures. Thermochim Acta 524:7–17

37. Chand GP, Sankar MG, Ramachandran D, Rambabu C (2016) Densities, viscosities and speeds of sound of binary mixtures of 2-chloroaniline with o-chlorotoluene, m-chlorotoluene and p-chlorotoluene at different temperatures. J Solut Chem 45:153–187

6

Phase transfer catalyst aided radical polymerization of *n*-butyl acrylate in two-phase system: a kinetic study

Vajjiravel Murugesan[1] · M. J. Umapathy[2]

Abstract The kinetics of radical polymerization of butyl acrylate initiated by potassium peroxydisulphate and cetyltrimethylammonium bromide as phase transfer catalyst (PTC) was carried out under inert and unstirred conditions at constant temperature of 60 ± 2 °C in ethyl acetate–water biphase media. The polymerization reactions were relatively fast in the two-phase systems with phase transfer agent whereas extremely sluggish in the system without PTC. Use of PTC accelerates the reaction effectively if the reactants located in two phase. The effects of rate of polymerization (R_p) on various experimental conditions such as different concentrations of monomer, initiator, phase transfer catalyst, temperature, and different ionic strength of the medium were explored. The order with respect to monomer, initiator and phase transfer catalyst was found to be unity. The R_p is independent of ionic strength and pH of the medium. However, an increase in the polarity of solvent has slightly increased the R_p value. Based on the results obtained, a plausible mechanism has been proposed for the polymerization reaction. The obtained polymer was confirmed by FT-IR analysis.

Keywords Kinetics · Phase transfer catalyst · Rate of polymerization · Radical polymerization · Two-phase system

Introduction

The growth and use of phase transfer catalyst (PTC) in the field of chemistry such as organic chemistry [1], inorganic chemistry [2], analytical applications [3], electrochemistry [4–8], photochemistry [9, 10] and in polymer chemistry [11–15] has become increasingly popular within industrial and academic arenas, because it is a potent and versatile technology which offers (1) less dependence on organic solvents, (2) excellent scalability and inherent compatibility with moisture, (3) enhancement of reactivity, which permits shortened reaction times and increased yields, (4) ability to substitute inconvenient reagents [like lithium diisopropylamide (LDA)] and (5) to control enantioselective variants and eco-friendliness. The efficient source of PTC technology in synthesis of polymers offers important technical rewards compared to other conventional polymerization methods [16]. PTC technique make easy of the reactions that are heterogeneously located in an immiscible phases by operating through the transfer of an anionic species from an aqueous (or solid) phase to the organic phase, thus polymerization and organic reactions will take place.

The acrylic esters such as *n*-butyl acrylate (*n*-BA) are commercially attractive and important functional monomers for the synthesis of acrylic resins. Because of their optical clarity, mechanical properties, adhesion and chemical stability, acrylic resins have many applications in paints, adhesives and coatings [17, 18]. Radical polymerization is one of the most widely facilitated commercial processes for the synthesis of polymers with high molecular weight. The rewards of radical polymerization are obvious: it can be applied to almost all vinyl monomers under mild reaction conditions over a wide temperature range, it is water tolerant and its cost is relatively low. The

✉ Vajjiravel Murugesan
chemvel@rediffmail.com

1 Department of Chemistry, B. S. Abdur Rahman University, Vandalur, Chennai 600 048, India

2 Department of Chemistry, College of Engineering Guindy, Anna University, Chennai 600 025, India

development of a new kinetic model for the polymerization of acrylic monomers, particularly n-butyl acrylate (n-BA), using efficient catalyst at moderate (low) temperature is one of the best approaches to an industrial perspective [19].

The growth of novel catalysts and efficient kinetic methodologies for the synthesis of polymers is an important target of research. Recently, phase transfer-catalyzed polymerization of vinyl monomers was gaining remarkable interest [20–34]. Inspired by the versatile application of PTC, herein, we report the systematic kinetic study of radical polymerization of n-butyl acrylate using cetyltrimethylammonium bromide (CTMAB) as phase transfer catalyst and potassium peroxydisulphate (PDS) as water-soluble initiator in an ethyl acetate–water two-phase system. The various kinetic parameters on the rate of polymerization were ascertained and based on the experimental observation plausible mechanism has been derived and its significance was discussed. Moreover, the resultant kinetics was evaluated with the reported radical polymerization of n-BA using different catalyst [35].

Experimental

Materials and solvents

Butyl acrylate (n-BA, Lancaster, India) was purified by washing three times with a 10 % sodium hydroxide solution then washing three times with de-ionized water, drying over calcium carbonate and finally distilling under reduced pressure, no more than 24 h prior to use. The initiator potassium peroxydisulphate (PDS, Merck, India) was purified twice by recrystallization in cold water. The cetyltrimethylammonium bromide (CTMAB, SRL, India) was used without further purification. The solvents, benzene, ethyl acetate and acetone (SRL, India) was used as received. The double-distilled water was used to make an aqueous phase.

Polymerization procedure

A polymerization experiment was carried out in annular glass ampoules with dimensions of 30 and 26 mm for outer and inner diameter, respectively, and 120 mm height. These ampoules have a surface area/volume ratio large enough for the heat transfer necessary to maintain the isothermal conditions during the polymerization. The polymerization ampoule consists of equal volumes of aqueous and organic phase (10 mL each). The monomer (n-BA) in ethyl acetate was the organic phase and the catalyst, sodium bisulfate (for adjusting the ionic strength [μ]) and sulphuric acid (maintaining the [H⁺]), was in the aqueous phase. The ampoule was degassed using nitrogen

gas continuously about 15 min after which it was sealed. Polymerizations were performed by placing the ampoules in a constant water bath at 60 ± 2 °C and the ampoules were removed from the water bath after a recorded time interval. The polymer were precipitated into a tenfold excess of methanol, filtered and dried at high vacuum until a constant weight was reached.

The rate of polymerization (R_p) was calculated from the gravimetric determination of the polymer formed in a given time of polymerization. The R_p was calculated from the weight of polymer obtained using the formula: $R_p = 1000 \, W/V \times t \times M$; where W is the weight of the polymer in gram; V is the volume of the reaction mixture in mL; t is the reaction time in seconds; M is the molecular weight of the monomer in g/mol. The kinetic experiment was carried out by changing the concentration of monomer, initiator, catalyst, temperature, etc., by adopting above-stated polymerization procedure.

Viscosity measurements

The viscosity average molecular weight of the polymer was determined in acetone at 30 ± 1 °C using Ostwald viscometer with the values of Mark–Houwink constant [36] using $[\eta] = K \, (M_v)^\alpha$. From the molecular weight of the polymer, the degrees of polymerization (P_n) values were calculated.

Kinetic model

The incompatible reactants dissolve in the different phases, that is, the ionic reactant [QX (phase transfer catalyst) and KY (initiator)] in the aqueous phase and the organic compound (monomer) in the organic phase and the reaction can take place at the interface between the two phases. Usually, this kind of situation in two-phase approach, because of a small interface, generally causes low reaction rate. This situation can be dramatically improved with the help of phase transfer catalysis (PTC). Phase transfer agent (PTA), a quaternary ammonium or phosphonium salt (QX salt) or a crown ether is added to the two-phase system. These PTAs have the ability to carry the ionic reactant into the organic phase. Once in the organic phase, the reactant becomes highly reactive because the degree of solvation is low; thus, the anion behaves like a 'naked' ion. Therefore, the reaction in the aqueous phase produces QY at the interface between the two phases where it was decomposed and produced the radical ions which initiate the polymerization at 60 ± 2 °C. The simple reaction kinetic model is shown in Schemes 1 and 2. A similar type of kinetic model was reported for phase transfer-catalyzed reactions [37–39].

Scheme 1 Phase transfer-catalyzed radical polymerization of *n*-butylacrylate (*n*-BA) in two-phase system

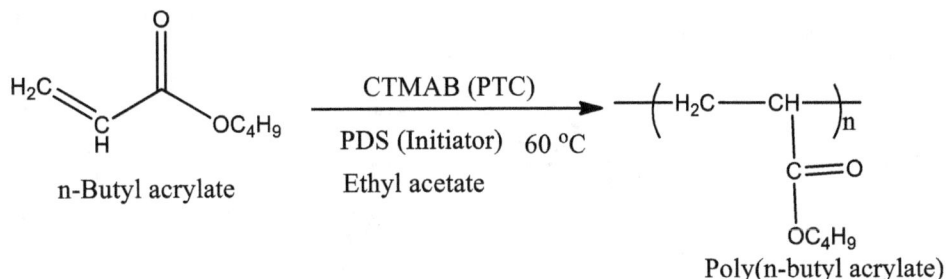

n-Butyl acrylate

CTMAB (PTC)

PDS (Initiator)   60 °C

Ethyl acetate

Poly(n-butyl acrylate)

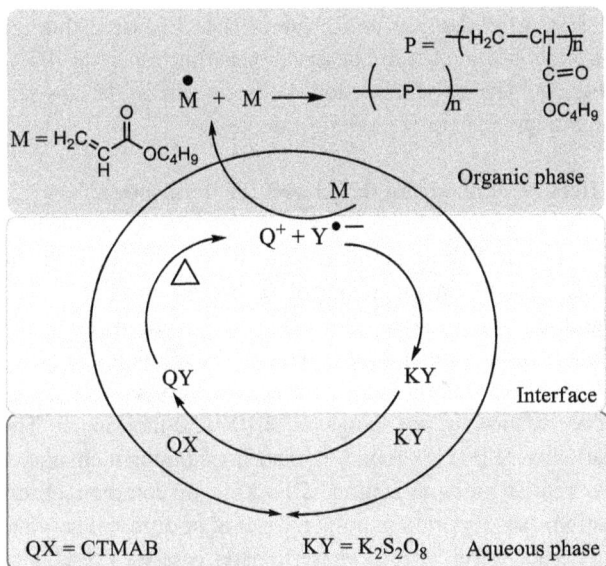

Scheme 2 Simple kinetic model of phase transfer catalyst aided polymerization of *n*-BA

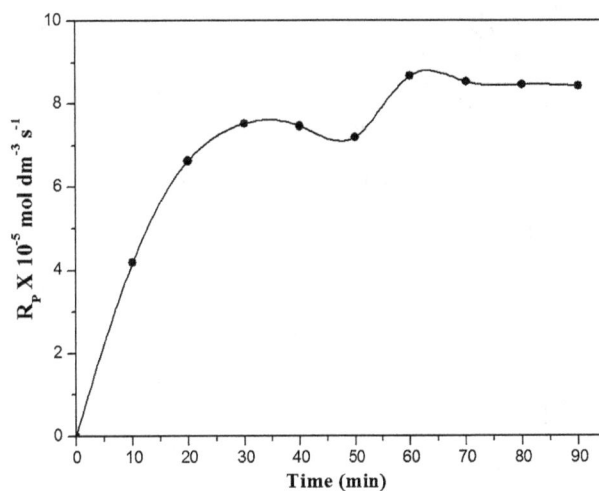

Fig. 1 Steady-state rate of polymerization. Polymerization condition: [BA]: 2.0 mol dm^{-3}; [K$_2$S$_2$O$_8$]: 2.0 × 10^{-2} mol dm^{-3}; [CTMAB]: 2.0 × 10^{-2} mol dm^{-3}; [H$^+$]: 0.5 mol dm^{-3}; [μ]: 0.2 mol dm^{-3}; temp: 60 ± 2 °C

Results and discussion

The radical polymerization of *n*-butyl acrylate initiated by PDS–PTC in an ethyl acetate–water two-phase system was studied under different experimental conditions to evaluate the various parameters, which influence the rate of polymerization reaction.

Steady-state rate of polymerization

The steady-state rate of polymerization for the *n*-butyl acrylate was studied by determining R_p at different intervals of time. A plot of R_p versus time shows an increase to some extent and then reaches constant. The steady-state rate of polymerization of the *n*-butyl acrylate was fixed at 40 min (Fig. 1).

Effect of [BA] on the rate of polymerization (R_p)

The effect of *n*-butyl acrylate [*n*-BA] on the rate of polymerization (R_p) has been carried out at 60 ± 2 °C using

fixed concentration of PTC, PDS, ionic strength and pH of aqueous phase, the monomer concentrations ranging from 1.0 to 2.1 mol dm^{-3}. The R_p increase with an increase in the concentration of the monomer was noticed. The reaction order with respect to monomer concentration was determined from the slope of 6 + log R_p versus 3 + log [BA] and the reaction order with respect to the monomer concentration was found to be 0.91, it is approximately equal to order of unity. Also the plot of R_p versus [BA] passing through the origin confirms the above observations with respect to [BA] (Fig. 2). The monomer reaction order of unity has been reported for the phase transfer catalyst and PDS-aided polymerization of various vinyl monomers [26, 28, 30, 34].

Effect of [PDS] on the rate of polymerization (R_p)

At fixed concentrations of monomer, catalyst and volume ratio of aqueous to organic phase, the effect of concentration of PDS on the rate of polymerization of *n*-butyl acrylate was studied by varying the concentrations in the range of 1.5–6.5 mol dm^{-3}. The R_p was found to an

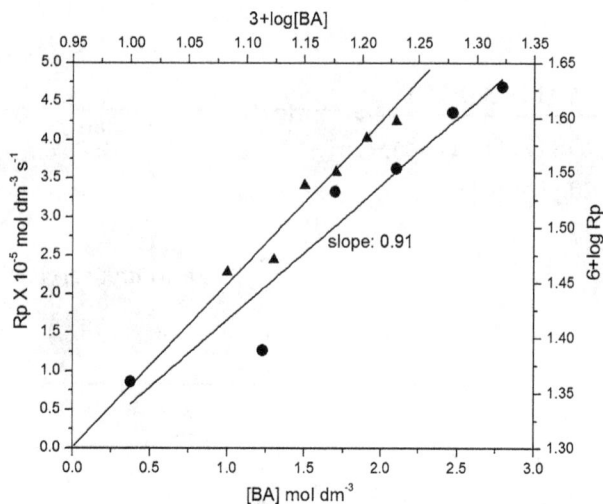

Fig. 2 Effect of [BA] on the R_p. Polymerization condition: [PDS]: 2.0×10^{-2} mol dm^{-3}; [CTMAB]: 2.0×10^{-2} mol dm^{-3}; [H$^+$]: 0.5 mol dm^{-3}; [μ]: 0.2 mol dm^{-3}; temp: 60 ± 2 °C; time: 40 min

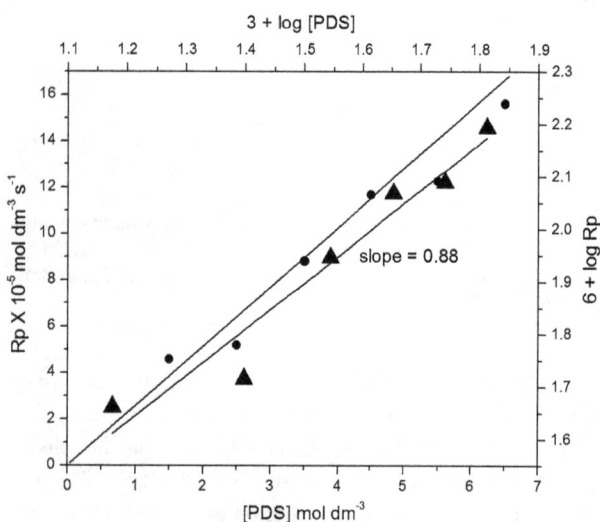

Fig. 3 Effect of [PDS] on the R_p. Polymerization condition: [BA]:2.0 mol dm^{-3}; [CTMAB]: 2.0×10^{-2} mol dm^{-3}; [H^{+-}]:0.5 mol dm^{-3}; [μ]: 0.2 mol dm^{-3}; temp: 60 ± 2 °C; time: 40 min

increase with an increasing concentration of PDS in this system. From the plot of $6 + \log R_p$ versus $3 + \log$ [PDS], the slope (0.88) is almost equal to unity. As expected, a plot of R_p versus [PDS] is linear passing through the origin supporting the above deduction (Fig. 3). Similar order of reaction with respect to initiator concentration has been reported for the polymerization of other vinyl monomers [29, 31, 32, 40] using different phase transfer catalyst initiated by PDS. The higher order of polymerization in the case of initiator variation may be either due to gel effect or diffusion-controlled termination rate constant [40].

Effect of [PTC] on the rate of polymerization (R_p)

The effect of concentration of phase transfer catalyst (PTC), cetyltrimethylammonium bromide (CTMAB), on the rate of polymerization was studied by varying its concentration in the range of 1.5–6.5 mol dm^{-3} at definite concentrations of other parameters. The rate of polymerization (R_p) was increased with an increasing concentration of PTC for *n*-butyl acrylate system. The order with respect to the concentration of the PTC was found to be around unity (0.97) from the plot of $6 + \log R_p$ versus $3 + \log$ [PTC]. Also the plot of R_p versus [PTC] passing through origin confirms that the observations with respect to [PTC] (Fig. 4). The polymerization did not occur in the absence of catalyst even after several minutes.

Effect of ionic strength (μ) and [H$^+$] variation

The effect of ionic strength was observed by varying the ionic strength in the range of 0.40–0.60 mol dm^{-3}. The variation of ionic strength was found to exert no significant change in the rate of polymerization. In addition, the effect of [H$^+$] variation was observed by varying the acidic concentration in the range of 0.10–0.30 mol dm^{-3}. The variation of [H$^+$] is found to exert no significant change in the rate of polymerization. This kind of common observations was reported in polymerization of different acrylate monomer using various phase transfer catalyst [20–23].

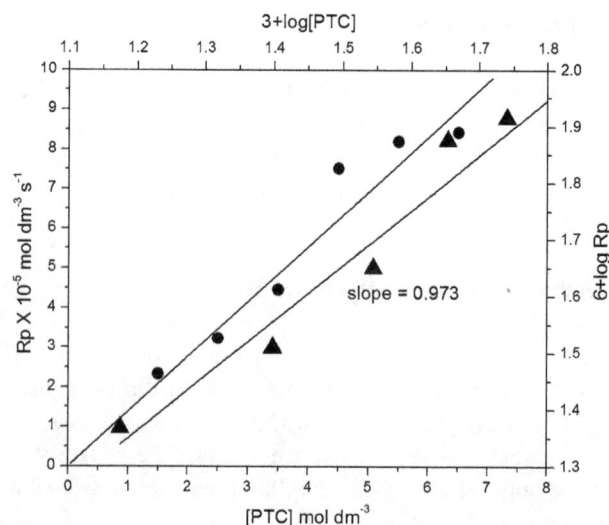

Fig. 4 Effect of PTC concentration on the R_p. Polymerization condition: [BA]: 2.0 mol dm^{-3}; [PDS]: 2.0×10^{-2} mol dm^{-3}; [H$^+$]: 0.5 mol dm^{-3}; [μ]: 0.2 mol dm^{-3}; temp: 60 ± 2 °C; time: 40 min

Effect of temperature on R_p

The effect of variation of temperature from 45 to 60 °C on the rate of polymerization was studied by keeping the variables such as n-BA, PDS, PTC, ionic strength and pH constant. The rate of polymerization increases with an increase in temperature. The overall activation energy of polymerization (E_a) obtained from plot of log R_p versus $1/T$ is 66.36 kJ/mol for n-butyl acrylate system (Fig. 5). The higher E_a (91.35 kJ/mol) value was reported for the polymerization of n-BA using di-site PTC [ref]. From the activation energy value we can conclude that current study has more efficiency than reported [35]. The thermodynamic parameters such as entropy of activation ($\Delta S^{\#}$), enthalpy of activation ($\Delta H^{\#}$) and free energy of activation ($\Delta G^{\#}$) have been calculated and reported in Table 1.

Effect of organic solvent polarity on R_p

The effect of organic solvents on R_p was examined by carrying out the polymerization of n-butylacrylate in three solvents, cyclohexane, ethylacetate and cyclohexanone having the dielectric constants 2.02, 6.02 and 18.03, respectively. It was found that the R_p decreased in the following order:

Cyclohexanone > Ethylacetate > Cyclohexane

An increase in the rate of polymerization may be due to an increase in polarity of the medium which makes easy transfer of more species of peroxydisulfate to the organic phase [41] (Table 2).

Radical mechanism of n-BA–PTC–PDS in two-phase system

Scheme 3 represents the reactions characterizing the polymerization of n-butyl acrylate (M) initiated by PDS/PTC in ethyl acetate/water two-phase systems. It is assumed that dissociation of QX and $K_2S_2O_8$, formation of QS_2O_8 in aqueous phase, and initiation of monomer in

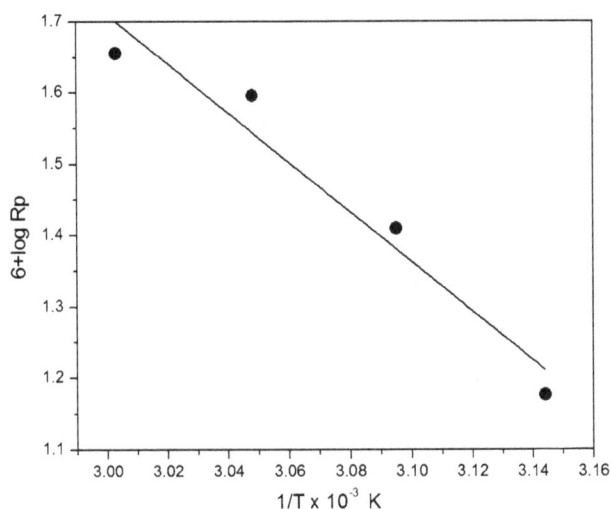

Fig. 5 Effect of temperature on the R_p. Polymerization condition: [BA]: 2.0 mol dm^{-3}; [PDS]: 2.0 × 10^{-2} mol dm^{-3}; [CTMAB]: 2.0 × 10^{-2} mol dm^{-3}; [H$^+$]: 0.5 mol dm^{-3}; [μ]: 0.2 mol dm^{-3}; time: 40 min

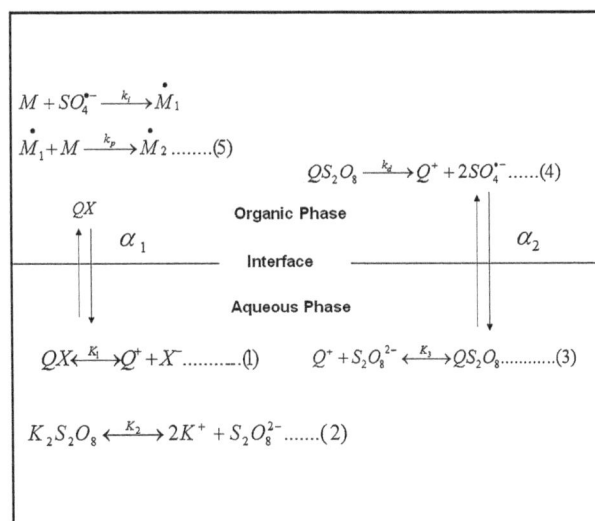

Scheme 3 Polymerization pathways of n-BA–PTC–PDS in an aqueous organic two-phase system

Table 1 Thermodynamic parameters

E_a (kJ/mol)	$\Delta G^{\#}$ (kJ/mol)	$\Delta H^{\#}$ (kJ/mol)	$\Delta S^{\#}$ (EU)
66.36	55.24	176.72	−95.75

Table 2 Effect of solvents

Experimental conditions	R_p × 10^{-5} mol dm^{-3} s^{-1}		
	Cyclohexanone (18.3)	Ethylacetate (3.91)	Cyclohexane (1.13)
[BA]: 2.0 mol dm^{-3}	2.50	1.65	1.08
[PDS]: 2.0 × 10^{-2} mol dm^{-3}			
[PTC]: 2.0 × 10^{-2} mol dm^{-3}			
[H$^+$]: 0.5 mol dm^{-3}			
[μ]: 0.2 mol dm^{-3}; temp: 60 ± 2 °C			

organic phase occur along the reactions shown in Eqs. (1)–(5). The equilibrium constants (K_1 and K_2) in the reactions in Eqs. (1)–(3) and distribution constants (α_1 and α_2) of QX and QS_2O_8 are defined as follows, respectively,

$$K_1 = \frac{[Q^+]_w[X^-]_w}{[QX]_w} \tag{6}$$

$$K_2 = \frac{[K^+]_w^2[S_2O_8^{2-}]_w}{[K_2S_2O_8]_w} \tag{7}$$

$$K_3 = \frac{[QS_2O_8]_w}{[Q^+]_w[S_2O_8^{2-}]_w} \tag{8}$$

$$\alpha_1 = \frac{[Q^+X^-]_w}{[QX]_o} \tag{9}$$

$$\alpha_2 = \frac{[Q^+S_2O_8^{2-}]_w}{[QS_2O_8]_o}. \tag{10}$$

The initiation rate (R_i) of radical SO_4^- in Eq. (4) may be represented as follows, f is initiator efficiency

$$R_i = \frac{d[SO_4^{0-}]}{dt} = 2K_dfK_3[Q^+]_w[S_2O_8^{2-}]_w. \tag{11}$$

The growth of polymer chain is according to the reaction in Eq. (5), the propagation step is represented as follows

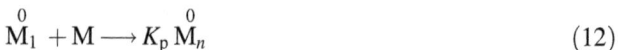

$$\overset{0}{M_1} + M \longrightarrow K_p \overset{0}{M_n} \tag{12}$$

$$\cdots \quad \cdots \quad \cdots$$
$$\cdots \quad \cdots \quad \cdots$$

$$\overset{0}{M_{n-1}} + M \xrightarrow{K_p} \overset{0}{M_n} \tag{13}$$

The rate of propagation (R_p) step in the reaction in Eq. (12) is given as

$$R_p = k_p\overset{0}{[M]}[M] \tag{14}$$

$$\overset{0}{[M]} = \frac{R_p}{k_p[M]} \tag{15}$$

The termination occurs by the combination of two growing polymer chain radicals, it can be represented as

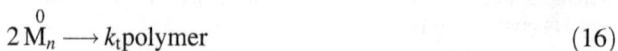

$$2\overset{0}{M_n} \longrightarrow k_t\text{polymer} \tag{16}$$

The rate equation of termination (R_t) process according to the Eq. (16)

$$R_t = 2k_t\overset{0}{[M]}^2 \tag{17}$$

The steady state prevails, the rate of initiation equals to rate of termination, i.e.,

$$R_i = R_t \tag{18}$$

$$2K_dfK_3[Q^+]_w[S_2O_8^{2-}]_w = 2k_t[M^0]^2 \tag{19}$$

$$[\overset{0}{M}]^2 = \frac{K_dfK_3[Q^+]_w[S_2O_8^{2-}]_w}{k_t} \tag{20}$$

$$[\overset{0}{M}] = \left[\frac{K_dfK_3[Q^+]_w[S_2O_8^{2-}]_w}{k_t}\right]^{1/2} \tag{21}$$

Using Eqs. (15) and (21), the rate of polymerization is represented as follows

$$R_p = k_p\left[\frac{k_dK_3f}{k_t}\right]^{1/2}[Q^+]_w^1[S_2O_8^{2-}]_w^1[M]^1 \tag{22}$$

The above equation satisfactorily explains all the experimental observations. The expression for the degree of polymerisation is

$$\bar{P}_n = \frac{R_p}{R_t} \tag{23}$$

$$\bar{P}_n = \left[\frac{K_p[M]}{2(K_3k_tk_df)^{1/2}[Q^+]_w^1[S_2O_8]_w^1}\right] \tag{24}$$

This Eq. (24) for the degree of polymerization $\bar{P}_n$ is directionally proportional to $[M]^1$. It is found that a plot of $\bar{P}_n$ versus $[M]^1$ gives straight line passing through the origin for n-butyl acrylate system (Fig. 6). This observation supports the proposed mechanism.

Characterization of polymer: FT-IR analysis

The FT-IR spectrum of poly(n-butyl acrylate) was recorded with a Perkin-Elmer RXI spectrophotometer in the spectral region from 4500 to 500 cm^{-1}. The pellets of about 25 mg of polymer in KBr powder containing grained powder of sample were made before recording. The FT-IR

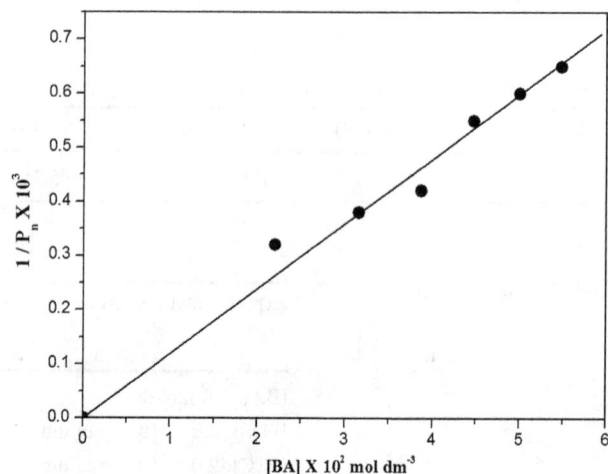

Fig. 6 n-BA–PTC–PDS system dependence of P_n on [n-BA]

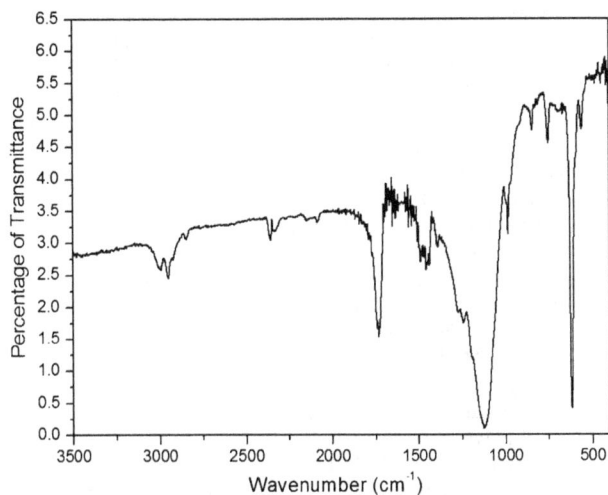

Fig. 7 FT-IR spectral analysis of poly(n-butylacrylate)

spectroscopy confirms a band of 1732 cm^{-1} of ester group of poly(butylacrylate) (Fig. 7). The following bands were observed in the spectra 1125–1260 cm^{-1} (C–O–C stretching band), 1455 cm^{-1} (C–H deformation), 2950 cm^{-1} (C–H stretching band).

Conclusions

The present work shows that the radical polymerization of n-butylacrylate was successfully performed with the help of PTC–PDS in two-phase system. The kinetic features, the rate of polymerization (R_p) of n-butylacrylate, were increased with an increasing concentration of monomer, initiator and catalyst. The hydrogen ion concentration and ionic strength of the medium do not show any appreciable effect on the R_p. The reaction rate increases with an increasing temperature. The phase transfer-catalyzed polymerization of n-butylacrylate follows the order of unity with respect to monomer, initiator and catalyst. Based on the results obtained, a suitable mechanism has been proposed. The obtained results showed that the rate of polymerization was more efficient than the reported. The polymer obtained by free-radical polymerization of butylacrylate was confirmed by FT-IR spectral analysis. The versatile nature and use of PTC resulted in high reaction rate in heterogeneous (two-phase) system. This could be a practical interest in the preparation of polymers and organic compounds in which reagents incompatibility problem often occurs and in which PTC is frequently used as a way to bring the reactant together.

Acknowledgments VM is grateful to the Science and Engineering Research Board (DST-SERB), New Delhi, for the young scientist start-up research Grant (SB/FT/CS-008/2014) and also thanks the Management of B S Abdur Rahman University, Chennai-600 048 for the support.

Compliance with ethical standards

Conflict of interest The authors declare that there is no conflict of interest regarding the publication of this article.

References

1. Bannard RAS (1976) Phase transfer catalysis and some of its applications in organic chemistry. Verlag Chemie, West Berlin
2. Izatt RM, Christensen JJ (1978) Synthetic multidentate macrocyclic compounds. Academic Press, New York
3. Fiamegos YC, Stalikas CD (2005) Phase transfer catalysis in analytical chemistry. Anal Chimica Acta 550:1–12
4. Ellis SR, Pletcher D, Brooks WM, Healy KP (1983) Electrosynthesis in systems of two immiscible liquids and a phase transfer catalyst. V. The anodic chlorination of naphthalene. J Appl Electrochem 13:735–741
5. Laurent E, Rauniyar G, Thomalla M (1984) Anodic substitutions in emulsions under phase transfer catalysis conditions. I. Cyanation of dimethoxybenzenes. J Appl Electrochem 14:741–748
6. Laurent E, Rauniyar R, Jomalla M (1985) Anodic substitution in emulsions under phase transfer catalysis conditions. II. Acetoxylation of dimethoxybenzenes. J Appl Electrochem 15:121–127
7. Do JS, Do YL (1994) Indirect anodic oxidation of benzyl alcohol in the presence of phase-transfer catalyst in a CSTER: kinetics of oxidation of benzaldehyde and effect of concentration of benzyl alcohol. Elctrochim Acta 39:2299–2309
8. Tan SN, Dryfe RA, Girault HH (1994) Electrochemical study of phase-transfer catalysis reactions: the Williamson ether synthesis. Helv Chim Acta 77:231–242
9. Goren Z, Willmer I (1983) Photochemical and chemical reduction of vicinal dibromides via phase transfer of 4,4′-bipyridinium radical: the role of radical disproportionation. J Am Chem Soc 105:7764–7765
10. Kitamura T, Kobayshi S, Taniguchi H (1984) Photochemistry of vinyl halides. Heterocycles from reaction of photogenerated vinyl cations with azide anion. J Org Chem 49:4755–4760
11. Cook FL, Brooker RW (1982) Polymer syntheses employing phase transfer catalysis. Polym Prepr Am Chem Cos Div Polym Chem 23:149
12. Hurduc N, Surpateanu G, Bulacovshchi V (1992) Phase transfer catalysis of polycondensation processes—V. Internal rotation energies in polyethers based on oxetane and bisphenol-A. Eur Polym J 28:1589–1591
13. Hurduc N, Surpateanu G, Bulacovshchi V (1993) Phase transfer catalysis in the polycondensation processes. Polym Bull 30:69–73
14. Tarang JY, Shih JS (1994) Crown ether as the phase-transfer catalyst for free radical polymerization of hydrophobic vinyl monomers. J Chin Chem Soc 41:81–87
15. Jane YS, Shih JS (1994) Crown ether phase-transfer catalysts for polymerization of phenylacetylene. J Mol Catal 89:29–40
16. Dariusz B, Mateusz G, Grzegorz B, Jacek W, Stanislaw W (2010) Preparation of polymers under phase transfer catalytic conditions. Org Process Res Dev 14:669–683
17. Chiantore O, Trossarelli L, Lazzari M (2000) Photooxidative degradation of acrylic and methacrylic polymers. Polymer 41(5):1657–1668
18. Gohsh S, Krishnamurti N (2000) Use of glycidyl methacrylate monomers for developing cross-linkable pressure sensitive adhesives. Eur Polym J 36:2125–2131

19. Meyer T, Keurentjes J (2005) Handbook of polymer reaction engineering. Wiley-VCH, New York
20. Savitha S, Vajjiravel M, Umapathy MJ (2006) Polymerization of butyl acrylate using potassium peroxydisulfate as initiator in the presence of phase transfer catalyst—a kinetic study. Int J Polym Mater 55(8):537–548
21. Balakrishnan T, Damodarkumar S (2000) Phase transfer catalysis: free radical polymerization of acrylonitrile using peroxymonosulphate/tetrabutylphosphonium chloride catalyst system: a kinetic study. J Appl Polym Sci 76:1564–1571
22. Umapathy MJ, Mohan D (1999) Studies on phase transfer catalysed polymerization of acrylonitrile. Hung J Ind Chem 27(4):245–250
23. Dharmendirakumar M, Konguvelthehazhnan P, Umapathy MJ, Rajendran M (2004) Free radical polymerization of methyl methacrylate in the presence of phase transfer catalyst—a kinetic study. Int J Polym Mater 53:95–103
24. Umapathy MJ, Balakrishnan T (1998) Kinetics and mechanism of polymerization of methyl methacrylate initiated by phase transfer catalyst—ammonium perdisulfate system. J Polym Mater 15(3):275–278
25. Umapathy MJ, Mohan D (1999) Studies on phase transfer catalyzed polymerization of glycidyl methacrylate. J Polym Mater 16(2):167–171
26. Umapathy MJ, Mohan D (2001) Phase transfer polymerization of butyl methacrylate using potassium peroxydisulfate as initiator—a kinetic study. Ind J Chem Tech 8(6):510–514
27. Umapathy MJ, Malaisamy R, Mohan D (2000) Kinetics and mechanism of phase transfer catalyzed free radical polymerization of methyl acrylate. J Macromol Sci Pure Appl Chem A37(11):1437–1445
28. Vajjiravel M, Umapathy MJ, Bharathbabu M (2007) Polymerization of acrylonitrile using potassium peroxydisulfate as an initiator in the presence of a multisite phase-transfer catalyst: a kinetic study. J Appl Polym Sci 105:3634–3639
29. Vajjiravel M, Umapathy MJ (2008) Synthesis and characterization of multi-site phase transfer catalyst: application in radical polymerisation of acrylonitrile—a kinetic study. J Polym Res 15(1):27–36
30. Vajjiravel M, Umapathy MJ (2008) Free radical polymerisation of methyl methacrylate initiated by multi-site phase transfer catalyst—a kinetic study. Colloid Polym Sci 286:729–738
31. Vajjiravel M, Umapathy MJ (2009) Kinetics and mechanism of multi-site phase transfer catalyzed radical polymerization of ethyl methacrylate. Int J Polym Mater 58:61–76
32. Vajjiravel M, Umapathy MJ (2010) Multi-site phase transfer catalyzed radical polymerization of n-butyl methacrylate: a kinetic study. Chem Eng Commun 197:352–365
33. Vajjiravel M, Umapathy MJ (2010) Synthesis, characterization and application of a multi-site phase transfer catalyst in radical polymerization of n-butyl methacrylate—a kinetic study. Int J Polym Mater 59:647–662
34. Vajjiravel M, Umapathy MJ (2011) Kinetics of radical polymerization of glycidyl methacrylate initiated by multi-site phase transfer catalyst–potassium peroxydisulfate in two-phase system. J Appl Polym Sci 120:1794–1799
35. Yoganand KS, Kamali TV, Divya K, Navamani P, Umapathy MJ (2012) Role of phase transfer catalyst in radical polymerization of butyl acrylate using peroxomonosulphate and disulphate as water soluble initiator-a comparative kinetic study. Asian J Chem 24:3699–3703
36. Brandrup J, Immerugut EH (1975) Polymer handbook. Wiley-Interscience, New York
37. Zhu Mengping (2014) Integration of phase transfer catalysis into aqueous transfer hydrogenation. Appl Catal A General 479:45–48
38. Qiangqiang Z, Jie S, Baojiang L, Jinxin H (2015) Anion exchange cycle of catalyst in liquid–liquid phase-transfer catalysis reaction: novel autocatalysis. Chem Eng J 262:756–765
39. Sankarshana T, Yadagiri E, Murthy JSN (2014) Phase transfer catalysis: oxidation of 2-methyl-1-butanol. Chin J Chem Eng 22(9):1000–1004
40. Balakrishnan T, Damodarkumar S (2000) Phase transfer catalysed free radical polymerization: kinetics of polymerization of butyl methacrylate using peroxymonosulphate/tetrabutyl phosphonium chloride catalyst system. Ind J Chem 39A:751–755
41. Vivekanand PA, Wang M-L, Hsieh Y-M (2013) Sonolytic and silent polymerization of methacrylic acid butyl ester catalyzed by a new onium salt with bis-active sites in a biphasic system—a comparative investigation. Molecules 18:2419–2437

7

Electrochemical approach of Kalmegh leaf extract on the corrosion behavior of aluminium alloy in alkaline solution

Namrata Chaubey[1] · Vinod Kumar Singh[1] · M. A. Quraishi[2]

Abstract

Introduction The effect of Kalmegh leaf extract (KLE) on the corrosion behavior of aluminium in 1 M NaOH solution was studied using electrochemical impedance spectroscopy (EIS) and potentiodynamic polarization (PDP) studies.

Results Inhibition efficiency ($\eta\%$) increased with increasing the concentration of extract and maximum inhibition efficiency observed is 82.24 and 82.45% from EIS and PDP studies, respectively, at higher concentration.

Conclusion The adsorption of the inhibitor on aluminium surface was in accordance with the Langmuir adsorption isotherm. Potentiodynamic polarization study showed mixed type inhibition with predominantly cathodic effect. SEM and AFM study was carried out to support the experimental inhibition data.

Keywords Aluminium alloy · Corrosion · EIS · SEM

Introduction

Aluminium, with high energy density (8.1 kW h kg^{-1}) and an electrode potential of 2.35 V vs. standard hydrogen electrode (SHE) are known for wide range of applications in the various industries such as automotive, aerospace, construction and electrical power generation [1]. The behavior of aluminium was extensively studied in the context of corrosion where the research work was focused on the protection of the metal from anionic attacks. Aluminium dissolution was also studied in term of hydrogen production, but one of the advantageous applications of aluminium corrosion is metal/air batteries in which aluminium is used as anode material in alkaline medium. The metal/air batteries of aluminium have been widely used for electric vehicle propulsion [2]. When exposed to alkaline medium, aluminium, suffers substantial corrosion attack which induces fuel consumption during standby and columbic loss on discharge. It reduces the efficiency of battery and sometimes causes explosion as a result of hydrogen build up [3]. To improve and optimize the performance of these batteries, the corrosion rate of aluminium alloy must be reduced by the use of inhibitors which can raise the hydrogen evolution over potential without significant reduction in the rate of aluminium oxidation.

A survey of literature reveals that various types of organic and inorganic compounds have been used for the protection of aluminium corrosion in alkaline solution [4–7]. However, the use of chemical inhibitors has been limited because of being synthetic chemicals, highly expensive, and toxic to the environment. Therefore, it is worthwhile to give attention towards a very cheap and environmentally safe research for corrosion inhibition of aluminium in alkaline solution. In view of this, various plants extracts such as, *Damissa* [8], *Lupinus varius* [9], *Mesembryanthemum nodiflorum* [10], *Raphia hookeri* [11], *Phyllanthus amarus* [12], *Cantaloupe* [13], *Gossipium hirsutum* [14], livestock dung [15], *Gum Arabic* [16] and *Vigna unguiculata* [17] have been proved as efficient inhibitors by various researchers on aluminium in alkaline solution. Plant extracts are incredibly rich source of naturally synthesized organic compounds that can be extracted

✉ M. A. Quraishi
maquraishi.apc@itbhu.ac.in; maquraishi@rediffmail.com

[1] Department of Chemistry, Udai Pratap Autonomous College, Varanasi 221002, India

[2] Department of Chemistry, Indian Institute of Technology, Banaras Hindu University, Varanasi 221005, India

using simpler techniques with low cost [18]. As a contribution to the current interest on environmentally friendly inhibitors, this study investigates the inhibition effect of Kalmegh (*Andrographis paniculata*) leaves extract on AA in 1 M NaOH and observed maximum inhibition efficiency is 82.24% at higher concentration.

Kalmegh (*f*. Acanthaceae) is an annual herbaceous plant, native to India and Srilanka. The plant is known in northeastern India as "king of bitter". It is one of the bitter plants that are used in traditional medicine and is a great benefit to Unani, Ayurveda and Homeopathy. The plant leaves were chosen due to (a) presence of water soluble active constituents (b) ease of availability and (c) environmental friendliness.

Kalmegh (*Andrographis paniculata*) leaf extract is composed of numerous naturally occurring organic compounds. Andrographolide, Neoandrographolide, Paniculide-A, Paniculide-B, Paniculide-C, have been isolated from the whole plant and leaves which contain multiple bonds through which they get adsorbed on the AA surface. As reported in literature, Andrographolide is found to be major constituent extracted from the leaves of the plant which is a bicyclic diterpenoid lactone (given in Fig. 1) [19, 20].

The objective of this investigation is to evaluate the corrosion inhibition effect of KLE on aluminium alloy in 1 M NaOH solution. The inhibition performance is examined by potentiodynamic polarization (PDP) and electrochemical impedance spectroscopy (EIS). The experimental results were complemented well with SEM and AFM investigation.

Experimental details

Materials and test solution

The corrosion test was performed on the aluminium alloy coupons having the composition given in Table 1. The test solution, 1 M NaOH was prepared by dissolving 40 g of NaOH in 1000 ml of double distilled water.

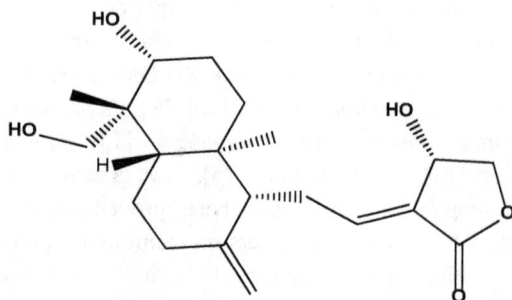

Fig. 1 Molecular structure of andrographolide

Preparation of inhibitor solution

Kalmegh plants were collected from the campus of Banaras Hindu University. Leaves were dried and grind to powdered form. The 5 g of powder was added to 500 ml 1 M NaOH solution in a round bottom flask and refluxed for 1 h. Thereafter, the mixture was cooled and filtered. The residue of all the leaves were dried and weighed. The volume of the filtrate was maintained up to 100 ml which was used as stock solution. The different concentrations of stock solution were taken for the corrosion test.

Electrochemical experiments

The AA coupons used in electrochemical tests were mechanically cut into $7.0 \times 1.0 \times 0.035$ cm dimension. Testing systems consisted of a three electrode cell configuration. AA with an area of 1 cm^2 was used as working electrode. A platinum wire and a saturated calomel electrode (SCE) i.e., (Cl$^-$|(4 M) Hg$_2$Cl$_2$ (s)| Hg (l) | Pt) were used as counter and reference electrodes, respectively. The three electrode cell connected to the Gamry Potentiostat/ Galvanostat (Model 300) instrument. All the tests were performed in the absence and presence of different concentration of KLE in 1 M NaOH solution at 303 K. The data obtained from electrochemical measurements was analyzed using Echem analyst 5.0 software. The tests were performed after 15 min immersion of AA in 1 M NaOH solution in the absence and presence of inhibitor. EIS measurements were carried out at OCP over a frequency range of 10^5–10^{-2} Hz using a 10 mV sine wave AC voltage. Finally, the Potentiodynamic polarization test was carried out by sweeping the electrode potential from -0.25 to $+0.25$ V vs. OCP at a scan rate of 1 mV/s.

Surface analysis

The surface morphologies of AA samples after exposure to 1 M NaOH for 3 h in the absence and presence of KLE were examined by SEM and AFM. The SEM of the AA surface was performed at an accelerating voltage of 5000 V and 5000 X magnification using FEI Quanta 200F microscope. The AFM was performed using NT-MDT multimode, Russia, controlled by solver scanning probe microscope controller.

Results and discussion

Electrochemical measurement

Potentiodynamic polarization study

The effect of KLE on the corrosion rate of AA in 1 M NaOH was studied using Tafel polarization technique.

Table 1 Chemical composition (wt%) of the AA used

	Si	Fe	Cu	Mn	Mg	Zn	Cr	Ti	V	Ga	Al
	0.77	0.93	0.02	0.11	0.01	0.01	0.05	0.02	0.01	0.01	Balanced

Figure 2 represents the potentiodynamic polarization curves for AA in 1 M NaOH at different concentrations of KLE at 303 K. The decrease in corrosion rate occurs by shifting the anodic curves to more positive potentials and cathodic curves to more negative potentials, and to the lower values of corrosion current densities (Fig. 2). Table 2 shows the values of electrochemical parameters i.e., corrosion potential (E_{corr}), corrosion current density (i_{corr}), cathodic Tafel constant (β_c) and anodic Tafel constant (β_a) along with percentage inhibition efficiency ($\eta\%$).

The inhibition efficiency (IE%) was calculated by following equation [21]:

$$\eta\% = \frac{i_0 - i}{i_0} \times 100 \qquad (1)$$

where, i_0 and i are the corrosion current densities in the absence and presence of inhibitor, respectively.

The data of Table 2 and Fig. 2 shows that the i_{corr} value is higher in NaOH but the presence of KLE causes a prominent decrease in the corrosion rate i.e., prominently shifts the cathodic curves to lower values of current densities. The maximum decrease in i_{corr} value (16.9 mA cm^{-2}) and maximum $\eta\%$ (82.4) is observed at 1.0 g L^{-1}.

It has been reported that anodic dissolution of aluminium in the alkaline medium takes place through a stepwise addition of surface hydroxyl species, culminating in the chemical dissolution of Al (OH)$_3$ in the presence of surface oxide film. The overall anodic reaction taking place in the corrosion of aluminium in the alkaline solution is represented as under [22–25]

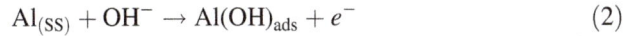

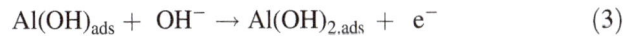

$$Al_{(SS)} + OH^- \rightarrow Al(OH)_{ads} + e^- \qquad (2)$$

$$Al(OH)_{ads} + OH^- \rightarrow Al(OH)_{2.ads} + e^- \qquad (3)$$

$$Al(OH)_{2.ads} + OH^- \rightarrow Al(OH)_{3.ads} + e^- \qquad (4)$$

$$Al(OH)_{3.ads} + OH^- \rightarrow Al(OH)_4^- + e^- \qquad (5)$$

The cathodic reaction on the film covered electrode surface is the reduction of water [26]:

$$2H_2O + 3e^- \rightarrow H_{2+}2OH^- \quad \text{(cathodic reaction)} \qquad (6)$$

However, the linear cathodic and anodic polarization curves indicate that the presence of KLE does not affect the mechanism of corrosion reaction at anodic and cathodic sites. No noticeable shift occurs in the values of anodic and cathodic Tafel constants in inhibited system as compared to blank. Thus, KLE behaves as mixed type inhibitor. Moreover, the addition of KLE shifts the corrosion potential (E_{corr}) slightly in the negative direction and reduces both the anodic and cathodic current densities. Thus, the inhibitor behaves as mixed type with predominantly cathodic.

Electrochemical impedance spectroscopy

The corrosion behavior of AA is investigated using electrochemical impedance technique at different concentration of KLE in 1 M NaOH solution at 303 K. With the help of Nyquist plot in Fig. 3a, it was found that the diameter is increased with increasing the concentration of KLE and it may be attributed to increasing the resistance but the shape remains same throughout the concentrations, indicating that there is no change in corrosion mechanism occurring through the KLE. The impedance spectra (Fig. 3a) is characterized by a capacitive time constant at higher frequency (HF), second capacitive time constant at lower frequency (LF), separated by an inductive time constant at medium frequency (MF) values.

The capacitive loop at HF is ascribed to the formation of protective (oxide) layer. According to Brett [27], the first capacitive time constant is associated with the reaction of aluminium oxidation at the metal/oxide/electrolyte interface. In this process, the formation of Al$^+$ ions at the metal/oxide interface and their migration through oxide layer to the oxide/solution interface occur due to high electric field strength, where they become oxidized to Al^{3+} [28]. This is attributed to the fact that these processes determined by capacitive time constant could either be suggested by overlapping of time constants or by the assumption that one process dominates and, therefore, excludes the other

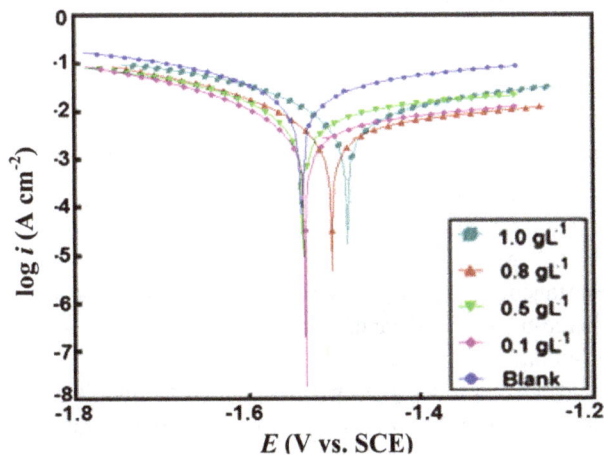

Fig. 2 Tafel curves for AA in 1M NaOH in absence and presence of different concentrations of KLE at 303 K

Table 2 Potentiodynamic polarization parameters for AA in 1 M NaOH in the absence and presence of different concentration of KLE at 303 K

KLE (g L^{-1})	Tafel polarization				
	i_{corr} (mA cm^{-2})	E_{corr} (V/SCE)	β_a (mV/dec)	β_c (mV/dec)	η (%)
0.0	96.3	−1.508	1001	504	–
0.2	30.2	−1.518	1230	358	70.0
0.5	28.8	−1.516	919	353	70.1
0.8	21.6	−1.518	1018	298	77.5
1.0	16.9	−1.520	1202	266	82.4

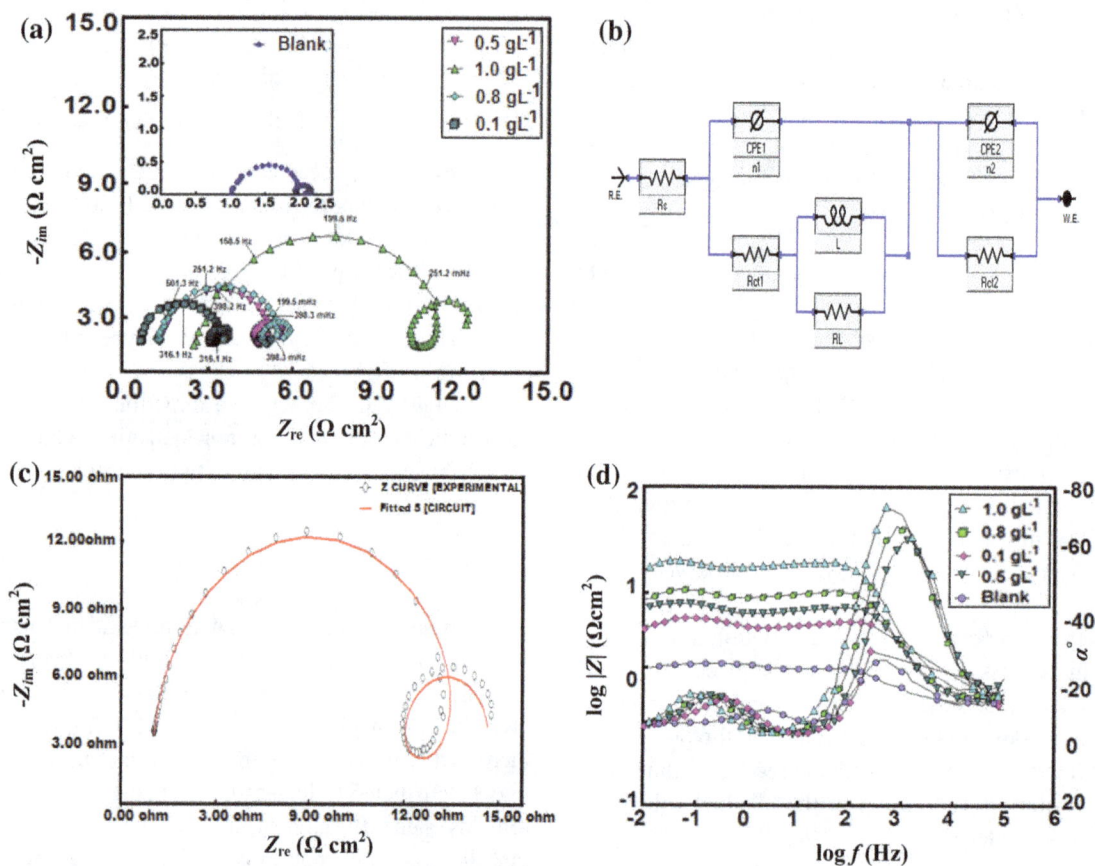

Fig. 3 a Nyquist plots for AA in 1 M NaOH without and with different concentrations of KLE at 303 K. **b** Electrical equivalent circuit used for the analysis of impedance spectra. **c** Simulated and experimentally generated EIS (Nyquist) plot. **d** Bode (log *f* vs. log |*Z*|) and phase angle (log *f* vs. α) plots of impendence spectra for AA in 1 M NaOH in absence and presence of different concentration of KLE at 303 K

processes [29]. The inductive loop at intermediate frequencies imputed to relaxation of the adsorbed intermediate species (OH$^-$) in the oxide layer, present on the metal surface [30]. The presence of inductive loop is reported in literature [31–35]. The second time constant of LF arises due to the adsorption and incorporation of hydroxide ions into the oxide film [36]

The impedance data is best described using an equivalent circuit mode displayed in Fig. 3b. Figure 3c clearly explains the fitting of an equivalent circuit model in Nyquist plot. The model consists of solution resistance

(R_s), inductance (L), charge transfer resistance (R_{ct}) parallel to constant phase element (*CPE/Q*).

This circuit includes another constant phase element (*CPE$_2$*) which is placed in parallel to charge transfer resistance element R$_{ct2}$. The R$_{ct2}$ value is the measure of charge-transfer resistance corresponds to the Al$^+$ → Al^{3+} reaction.

According to the reported mechanism Al dissolves into the solution in the form of Al^{3+} through the generation of Al$^+$ or Al^{2+} intermediate species [37]. Therefore, the polarization resistance, R_p, might be represented by the sum of R_{ct1} and R_{ct2} in the equivalent circuit. Hence, $\eta\%$ is represented using R_p in following equation:

$$\%IE = \frac{R_{p(inh)} - R_p}{R_{p(inh)}} \tag{7}$$

$R_{p\ (inh)}$ and R_p is polarization resistance with or without inhibitor.

Both R_{ct} and R_p value increases significantly with addition of KLE due to slower corrosion of electrode. The data in Table 3 reveals that increase in the values of charge transfer resistance is associated with a decrease in the double-layer capacitance at the whole concentration range. It may be stated that the constituents of KLE adsorbed on metal surface by replacing the water molecules at the metal surface which intern causes the decrease in the C_{dl} values. Thus, the rate of hydrogen evolution is reduced [38].

The double layer capacitance (C_{dl}) term is used to characterize the double layer at metal/solution interface by displaying the non ideal capacitive behavior. C_{dl} is calculated by the following relation [39]:

$$C_{dl} = Q \times (2\Pi f_{max})^{a-1} \tag{8}$$

A stepwise dissolution model has been proposed in Tafel measurements which require the stepwise addition of hydroxyl ions to metal/oxide interface. The inflow of hydroxyl ion is followed by the outflow of Al^{3+} ion across the interface. The adsorption of constituents of KLE is often a displacement reaction involving removal of adsorbed hydrated hydroxyl ions from the metal surface which is ascribed to the dielectric relaxation i.e., substitution of hydrated hydroxyl ions (high dielectric constant) with inhibitor molecules (low dielectric constant).

In the Bode spectra, three time constants are evident, namely, two time constants at high frequency (HF) low frequency (LF) regions and other time constant at middle frequency (MF) (Fig. 3d). Bode plot (S) and phase angle ($\alpha°$) are used to describe the nature of pure capacitive behavior. In other words, the values of S and $\alpha°$ should be $-1°$ and $-90°$ for an ideal capacitor. However, this study shows the deviation from the ideal capacitive behavior at intermediate frequencies. In this case, the maximum slope

value reaches up to -0.84 and the maximum phase angle is $-77°$.

Adsorption isotherm

Adsorption process occurs through the replacement of water molecules by the inhibitor molecules at the metal surface within the electrical double layer to produce less pronounced dielectric effect.

$$Inhibitor_{(sol)} + nH_2O_{(ads)} \leftrightarrow Inhibitor_{(ads)} + nH_2O_{(sol)}$$

The inhibitors may get adsorbed on the surface of aluminium and a protective film is formed. This restricts the diffusion of ions to or from the metal surface and hence retards the overall corrosion process. The interactions of the adsorbed inhibitor molecules with the metal surface may prevent the metal atoms from participating in the anodic reaction of the corrosion. This simple blocking effect decreases the number of metal atoms participating and hence decreases the corrosion rate.

Adsorption phenomenon is described to understand the nature of corrosion inhibition and it can be deduced in the term of adsorption isotherm. By fitting the various adsorption isotherms (including Freundlich, Temkin, Langmuir and Frumkin), Langmuir isotherm is best fitted and can be expressed by the following equation [40]:

$$\frac{C_{inh}}{\theta} = \frac{1}{K_{ads}} + C_{inh} \tag{9}$$

where $K_{(ads)}$ is adsorption equilibrium constant, C denotes the concentration of inhibitor and θ represents the surface coverage.

The plots of C/θ and C for the aluminium surface with different concentration of KLE give a straight line (Fig. 4) suggesting the adsorption of KLE constituents on the metal surface follows the Langmuir adsorption isotherm. It was found that R^2 and slope value obtained from Langmuir plots are close to 1, which suggests that KLE inhibitor occupies one active site on the metal surface. The adsorption equilibrium constant (K_{ads}) is associated with

Table 3 Electrochemical impedance parameters for AA in 1 M NaOH in the absence and presence of different concentration of KLE at 303 K

KLE (g L^{-1})	R_s (Ω)	Q_1 (S Ω^{-1} cm^{-2})	n	$(R_{ct})_1$ (Ω cm^2)	L (H cm^2)	R_L (Ω cm^2)	Q_2 (S Ω^{-1} cm^{-2})	$(R_{ct})_2$ (Ω cm^2)	R_p (Ω cm^2)	C_{dl} (μF cm^{-2})	η (%)
0.0	1.023	500×10^{-6}	0.975	0.849	0.221	0.121	39.8×10^{-6}	0.188	1.037	413.8	–
0.2	1.230	174×10^{-6}	0.979	1.934	0.189	2.268	51.2×10^{-6}	1.402	3.336	95.56	69.6
0.5	1.023	151×10^{-6}	0.981	2.002	0.186	3.434	62.1×10^{-6}	1.739	3.741	78.05	72.2
0.8	1.102	146×10^{-6}	0.989	2.987	0.199	2.340	68.8×10^{-6}	2.001	4.988	58.6	79.3
1.0	1.034	104×10^{-6}	0.991	3.723	0.198	2.022	77.2×10^{-6}	2.114	5.837	49.9	82.24

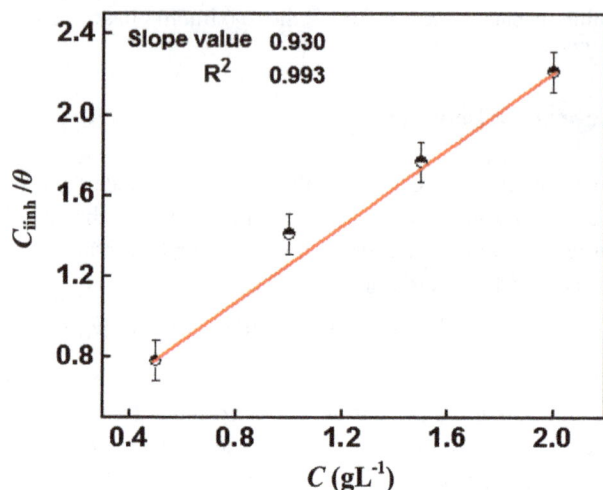

Fig. 4 Langmuir isotherm plot for adsorption of KLE molecule on AA in 1 M NaOH

Table 4 Thermodynamic parameters for the adsorption of KLE molecules on AA at different concentration in 1 M NaOH at 303 K

Inhibitors	Temperature (K)	K_{ads} 10^3 (g^{-1})	$G_{ads}°$ (KJ mol^{-1})
0.2	303	2.1	−15.21
0.5	303	3.2	−18.58
0.8	303	4.0	−20.33
1.0	303	4.3	−21.07

standard free energy of adsorption $\Delta G°_{(ads)}$ by the following equation [41]:

$$K_{ads} = \frac{1}{C_{(solvent)}} \exp\left(\frac{\Delta G°_{ads}}{RT}\right) \quad (10)$$

where R is universal gas constant, T is the absolute temperature and C is the concentration of water (1000 g L^{-1}). The values of K_{ads} is representing here in g^{-1} L. So, in equation, the concentration of water is taken in g L^{-1} in place of 55.5 mol L^{-1}.

The values of $K_{(ads)}$ and $\Delta G_{(ads)}°$ were calculated and given in Table 4. It is seen that the negative value of $\Delta G°$ is

found in all cases. In literature, it has been shown that the values of $\Delta G_{(ads)}°$ up to −20 kJ mol^{-1} are consistent with electrostatic interaction between charged molecules and a charged metal surface (physical adsorption), while those around −40 kJ mol^{-1} or higher corresponds to the charge sharing or charge transfer from the inhibitor molecules to the metal surface to form a co-ordinate type of bond (chemisorption) [42]. In this study, it is clear from the Table 4 that the values of $\Delta G_{(ads)}°$ is in the range of −15 to −21 kJ mol^{-1} i.e., the inhibitor adsorbed on the metal surface is in accordance with physical adsorption.

Surface morphology study

The SEM micrograph of corroded metal surface in NaOH is displayed in Fig. 5a. It can be seen that surface damage appears due to aggressive attack of alkaline media. In contrast, after treating the surface with inhibitor, the smoothness of the aluminium surface (Fig. 5b) results due to the formation of adsorption film on it.

Atomic force microscope (AFM) was used to investigate the corrosion inhibition ability of the extract to characterize the microstructure of AA surface. Figure 6 depicts three-dimensional AFM images of AA surface after 3 h exposure in 1 M NaOH at 303 K. In uninhibited system, the AA surface was fairly damaged due to dissolution in corrosive medium (Fig. 6a) with maximum height scale of 600 nm. The maximum height scale of inhibited AA surface (Fig. 6b) was 150 nm which indicates the smoothness of metal surface after treating with KLE.

Conclusion

1. Aqueous extract of KLE is an environmentally benign good corrosion inhibitor for AA in alkaline. Inhibition efficiency increases with increasing the concentration of extracts.

2. Tafel polarization indicates cathodic type inhibition through KLE.

Fig. 5 SEM micrographs of **a** uninhibited and **b** inhibited AA sample containing 1.0 g L^{-1} of KLE in 1 M NaOH

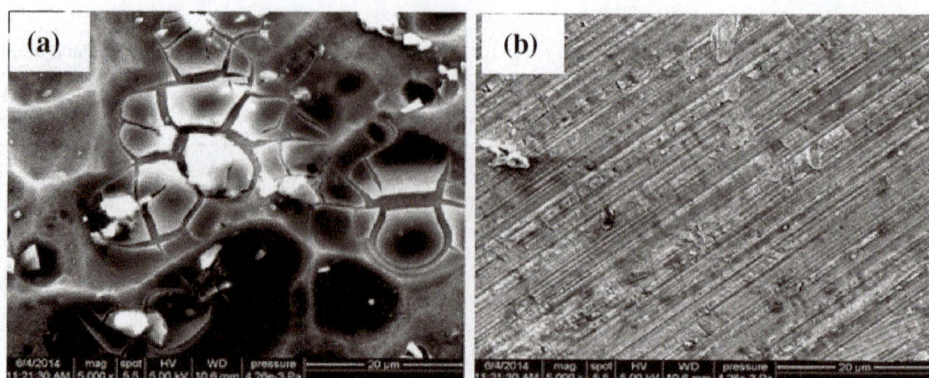

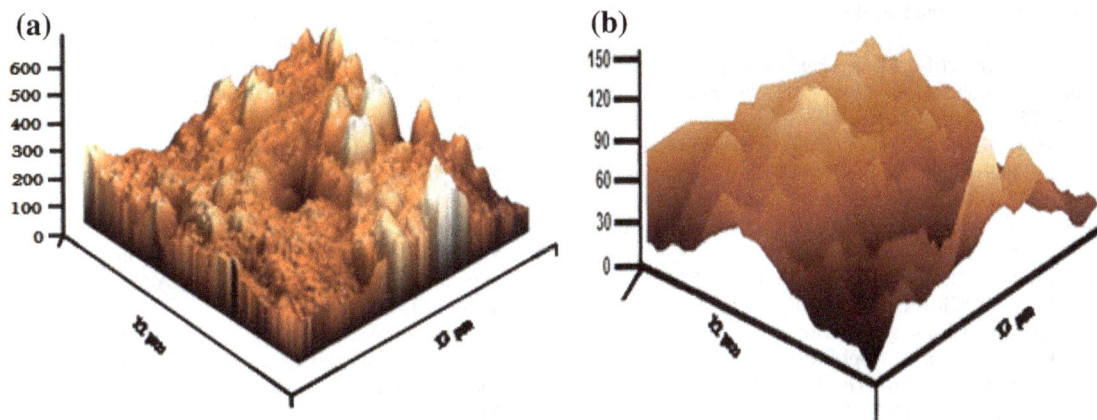

Fig. 6 AFM images of **a** uninhibited and **b** inhibited AA sample containing 1.0 g L^{-1} of KLE in 1 M NaOH

3. Adsorption of the KLE molecule on the AA surface in NaOH obeys the Langmuir's isotherm.
4. EIS indicates that increase in R_{ct} and decrease in C_{dl} is observed which is explained by decrease in local dielectric constant and or an increase in the electrical double layer thickness due to the adsorbed inhibitor molecules at the metal/solution interface.

Acknowledgements Authors are highly thankful to Prof. V. B. Singh, Head (Department of chemistry), B.H.U. for providing SEM and AFM facilities for successful completion of my research work.

References

1. Chaubey N, Yadav DK, Singh VK, Quraishi MA (2015) A comparative study of leaves extracts for corrosion inhibition effect on aluminium alloy in alkaline medium. Ain Shams Eng J. doi:10.1016/j.asej.2015.08.020
2. Amin MA, Abd EI-Rehim SS, El-Sherbini EEF, Hazzazi OA, Abbas MN (2009) Polyacrylic acid as a corrosion inhibitor for aluminium in weakly alkaline solutions. Part I: Weight loss, polarization, impedance EFM and EDX studies. Corros Sci 51:658–667
3. Oguzie EE (2007) Corrosion inhibition of aluminium in acidic and alkaline media by Sansevieria trifasciata extract. Corros Sci 49:1527–1539
4. Soliman HN (2011) Influence of 8-hydroxyquinoline addition on the corrosion behavior of commercial Al and Al-HO411 alloys in NaOH aqueous media. Corros Sci 53:2994–3006
5. Mercier D, Barthes-Labrousse MG (2009) The role of chelating agents on the corrosion mechanisms of aluminium in alkaline aqueous solutions. Corros Sci 51:339–348
6. Onuchukwu AI (1990) The inhibition of aluminium corrosion in an alkaline medium II: Influence of hard bases. Mater Chem Phys 24:337–341
7. Pyun SI, Moon SM, Ahn SH, Kim SS (1999) Effects of Cl^{-}, NO$_3^{-}$ and SO$_4^{2-}$ ions on anodic dissolution of pure aluminium in alkaline solution. Corros Sci 41:653–667
8. Abdel-Gaber AM, Khamis E, Abo-ElDahab H, Adeel S (2008) Inhibition of aluminium corrosion in alkaline solutions using natural compound. Mater Chem Phys 109:297–305
9. Irshedat MK, Nawafleh EM, Bataineh TT, Muhaidat R, Al-Qu-

daha MA, Alomary AA (2013) Investigations of the Inhibition of Aluminium Corrosion in 1 M NaOH Solution by Lupinus varius l. extract. Port Electrochim Acta 31:1–10
10. Al Shboula TMA, Jazzazi TMA, Bataineh TT, Al-Qudah MA, Alrawashdeh AI (2014) Inhibition of corrosion of aluminium in NaOH solution by leave extract of Mesembryanthemum nodiflorum. Jordan J Chem 9:149–158
11. Umoren SA, Obot IB, Ebenso EE, Obi-Egbedi NO (2009) The Inhibition of aluminium corrosion in hydrochloric acid solution by exudate gum from Raphia hookeri. Desalination 247:561–572
12. Abiola OK, Otaigbe JOE (2009) The effects of *Phyllanthus amarus* extract on corrosion and kinetics of corrosion process of aluminium in alkaline solution. Corros Sci 51:2790–2793
13. Emran KM, Ahmed NM, Torjoman BA, Al-Ahmadi AA, Sheekh SN (2014) Cantaloupe extracts as eco friendly corrosion inhibitors for aluminium in acidic and alkaline solutions. J Mate Environ Sci 5:1940–1950
14. Abiola OK, Otaigbe JOE, Kio OJ (2009) *Gossipium hirsutum* L. extracts as green corrosion inhibitor for aluminium in NaOH solution. Corros Sci 51:1879–1881
15. Umoren SA, Inam EI, Udoidong AA, Obot IB, Ubong UM, Kim KW (2015) Humic acid from livestock dung: ecofriendly corrosion inhibitor for 3SR aluminum alloy in alkaline medium. Chem Eng Commun 202:206–216
16. Umoren SA, Obot IB, Ebenso EE, Okafor PC, Ogbobe O, Oguzie EE (2006) Gum arabic as a potential corrosion inhibitor for aluminium in alkaline medium and its adsorption characteristics. Anti-corros Methods Mater 53:277–282
17. Umoren SA, Obot IB, Akpabio LE, Etuk SE (2008) Adsorption and corrosive inhibitive properties of Vigna unguiculata in alkaline and acidic media. Pigm Resin Technol 37:98–105
18. Chaubey N, Singh VK, Quraishi MA (2015) Effect of some peel extracts on the corrosion behavior of aluminium alloy in alkaline medium. Int J Ind Chem. doi:10.1007/s40090-015-0054-8
19. Singh A, Meena AK, Sudeep Meena, Pant P, Padhi MM (2012) Studies on standardisation of *Andrographis paniculata* nees and identification by HPTLC using andrographolide as marker compound. Int J Pharm Pharm Sci 4:197–200
20. Singh A, Singh VK, Quraishi MA (2010) Aqueous extract of Kalmegh (Andrographis paniculata) leaves as green inhibitor for mild steel in hydrochloric acid solution. Int J Corros. doi:10.1155/2010/275983
21. Ansari KR, Quraishi MA (2015) Experimental and quantum chemical evaluation of Schiff bases of isatin as a new and green

corrosion inhibitors for mild steel in 20% H_2SO_4. J Taiwan Inst Chem E 54:145–154

22. Kyung KL, Kim KB (2014) Electrochemical impedance characteristics of pure Al and Al-Sn alloys in NaOH solution. Corros Sci 43:561–575

23. Awad SA, Kamel KHM, Kassab A (1979) Corrosion behavior of aluminium in NaOH solutions. Electroanal Chem 105:291–294

24. Amin MA, Abd. EI-Rehim SS, F Essam, EI-Sherbini Mohsen NA (2009) Polyacrylic acid as a corrosion inhibitor for aluminium in weakly alkaline solutions. Part I: Weight loss, polarization, impedance EFM and EDX studies. Corros Sci 51:658–667

25. Wang JB, Wang JM, Shao HB, Zhang JQ, Cao CN (2007) The corrosion and electrochemical behavior of pure aluminium in alkaline methanol solutions. J Appl Electrochem 37:753–758

26. Aksut KCE, Abbas A (2000) The behavior of aluminum in alkaline media. Corros Sci 42:2051–2067

27. Brett CMA (1989) Studies on aluminium corrosion in hydrochloric acid solution. Portug Electrochim Acta 7:123–126

28. Khaled KF (2010) Electrochemical investigation and modeling of corrosion inhibition of aluminum in molar nitric acid using some sulphur-containing amines. Corros Sci 52:2905–2916

29. Lenderink HJW, Linden MVD, Wit JHWD (1993) Corrosion of aluminium in acidic and neutral solutions. Electrochim Acta 38:1989–1992

30. Ahamed I, Khan S, Ansari KR, Quraishi MA (2011) Primaquine: a pharmaceutically active compound as corrosion inhibitor for mild steel in hydrochloric acid solution. J Chem Pharm Res 3:703–717

31. Bessone JC, Mayer C, Juttner K, Lorenz WJ (1983) AC impedance measurements of aluminum barrier type oxide films. Electrochim Acta 28:171–175

32. Frers SE, Stefnel MM, Mayer C, Chierchie T (1990) AC-Impedance measurements on aluminium in chloride containing solutions and below the pitting potential. J Appl Electrochem 20:996–999

33. Bessone JB, Salinas DR, Mayer CE, Ebert E, Lorenz WJ (1992) An EIS study of aluminium barrier-type oxide films formed in different media. Electrochim Acta 37:2283–2290

34. Brett CMA (1990) The application of electrochemical impedance techniques to aluminium corrosion in acidic chloride solution. J Appl Electrochem 20:1000–1003

35. Macdonald DD (1990) Review of mechanistic analysis by electrochemical impedance spectroscopy. Electrochim Acta 35:1509–1525

36. Prabhu D, Rao P (2014) Corrosion behaviour of 6063 aluminium alloy in acidic and in alkaline media. Arab J Chem. doi:10.1016/j.arabjc.2013.07.059

37. Al-Kharafi FM, Badawy WA (1998) Inhibition of corrosion of Al 6061, aluminum, and an aluminum-copper alloy in chloride-free aqueous media: Part 2-Behavior in basic solutions. Corrosion 54:377–385

38. Ansari KR, Quraishi MA (2014) Bis-Schiff bases of isatin as new and environmentally benign corrosion inhibitor for mild steel. J Ind Eng Chem 20:2819–2829

39. Chaubey N, Savita Singh VK, Quraishi MA (2015) Corrosion inhibition performance of different bark extracts on aluminium in alkaline solution. JAAUBAS. doi:10.1016/j.jaubas.2015.12.003

40. Ansari KR, Quraishi MA, Singh A (2015) Pyridine derivatives as corrosion inhibitors for N80 steel in 15% HCl: Electrochemical, surface and quantum chemical studies. '. Measurement 76:136–147

41. Chaubey N, Singh VK, Quraishi MA (2016) Alstonia Scholaris bark as an environmentally benign corrosion inhibitor for aluminium alloy in sodium hydroxide solution. J Mater Environ Sci 7:2453–2467

42. Madkour LH, Elroby SK (2015) Inhibitive properties, thermodynamic, kinetics and quantum chemical calculations of polydentate Schiff base compounds as corrosion inhibitors for iron in acidic and alkaline media. Int J Ind Chem. doi:10.1007/s40090-015-0039-7

8

Carbonate formation on ophiolitic rocks at different pH, salinity and particle size conditions in CO_2-sparged suspensions

Riza A. Magbitang[1,2] · Rheo B. Lamorena[1,2]

Abstract Mineral carbonation is a promising CO_2 sequestration strategy that offers a long-lasting and environmentally safe solution. In this study, the effect of pH, salinity and particle size in the mineral carbonation process was investigated. Ultramafic–mafic rock samples were collected from different ophiolite rock sampling sites in Luzon Island, Philippines, and these were used in mineral carbonation reaction. Dissolution experiments were conducted by exposing powdered rock samples in suspensions sparged with CO_2 for 60 days at ambient conditions (25 °C and 1 bar). Carbonation reactions were observed at various pH conditions (4, 6, and 10) and particle sizes (62–125 and 250–420 µm). In separate experiments, the effects of pH and salinity were studied in experimental set-ups containing 5 % $MgCl_2$ maintained at low and high pH. Inductively coupled plasma-mass spectrometry (ICP-MS) was used to monitor concentrations of metals that could participate in the mineralization reaction (Mg, Al, Ca, and Fe) during exposure to CO_2. X-ray diffraction (XRD) analysis was used to confirm the formation of carbonate minerals. Results indicate an enhancement in the carbonation process upon varying pH and salinity of the system, while there is a negligible difference in the mineral carbonation reaction at the range of particle sizes used in this study.

Keywords Ophiolitic rocks · Carbon mineralization · Carbon dioxide · Carbon capture · Storage

Introduction

Mineral carbonation is a very promising CO_2 sequestration method and is already considered viable in different types of geological frameworks. Mineral carbonation has many advantages due to the fact that (1) there are abundant sources in the environment of Ca and Mg, which are needed to store excess CO_2 as carbonate minerals [17]; (2) it reduces the risks associated with pressurized geologic injection of CO_2, such as possible leakage, contamination of potable ground water, and other unpredicted seismogeological incidents [3]; and (3) the end-products of this process do not pose environmental threats [16].

The reactions involved in the mineral carbonation process in rock samples are: first (1) conversion of $CO_{2(g)}$ into $CO_3{}^{2-}{}_{(aq)}$; then (2) dissolution of minerals to release the metal cations (Ca^{2+} and Mg^{2+}); finally (3) the combination of carbonate anion and metal cation to form the carbonate minerals [24, 25]. In these reactions, the mineral dissolution reaction is the rate-determining step [35], thus enhancing mineral dissolution reaction will also yield faster mineralization.

Flow injection pilot tests and other feasibility studies have recognized the full potential of basaltic deposits for CO_2 storage [4, 5, 22, 33, 36]. Other rock formations such as the Samail Ophiolite of the Sultanate of Oman have been considered in several studies for natural CO_2 sequestration sites [14, 26]. More rock formations found in an archipelagic settings has also been considered [2]. These geological frameworks have a high CO_2 sequestration capacity through mineral fixation and solubility trapping [7].

✉ Rheo B. Lamorena
  rheolamorena9@yahoo.com; rheolamorena@gmail.com

[1] Natural Sciences Research Institute, College of Science, University of the Philippines Diliman, 1101 Quezon City, Philippines

[2] Institute of Chemistry, College of Science, University of the Philippines Diliman, 1101 Quezon City, Philippines

However, though these geological frameworks have significant potential, physical and chemical activation treatments are still necessary to facilitate in stepping up the reaction rates of metal ion dissolution for ensuing carbonation [1]. These interventions vary from heat or steam treatments, to leaching agents and complexing agent usages [1, 13, 21, 29]. One prospect to include in considering an injection host is the different types of geological frameworks with nearby alkaline water reservoirs.

The method of CO_2 injection into serpentinitic aquifers poses a feasible method in improving the interaction of CO_2 with dissolved metal cations. Geochemical data on spring waters involving serpentinitic rocks and ultramafic rocks have shown this potential [7]. Alkaline springs are associated with ophiolitic rocks and assumed to equilibrate with atmospheric CO_2 and eventually form calcite, brucite and aragonite [6, 30, 34]. Salinity and pH are important features of these alkaline springs and would prove beneficial in the dissolution step. Salinity and pH are important factors in carbon mineralization reactions; the latter parameter is important evidently because H^+ concentration influence formation of HCO_3^- (dissolution of CO_2 in water), which will then react to the available metals in the system [8, 28]. Changes in pH would be observed once CO_2 is injected into reaction mixtures such as brine solutions, which will in turn influence the uptake of divalent cations, and eventually promote carbonate formation [18, 19]. Salinity is also a parameter to study since an increase in salinity might result to low dissolution of CO_2 [8, 32]. The presence of nearby alkaline springs would eliminate the required chemical interventions in the dissolution of metal ions from the host minerals. Experimental studies on the assessment of carbonation reactions using these types of rocks in the presence of alkaline solutions are quite limited [20].

Particle size also plays an important role in mineral dissolution reaction because, theoretically, a smaller particle size will result in faster dissolution and will increase the availability of divalent metal cations. On the contrary, reducing the particle size will require additional energy in the over-all process, and the enhancement may or may not be sufficient to compensate for the additional energy input [10, 12].

Small-scale or laboratory scale investigations must constantly be conducted to continuously develop energy-conscious and economically feasible CO_2 sequestration processes on carbonation reactions. In this study, suspensions of different ophiolitic rock samples were prepared to evaluate different experimental conditions operating at minimal energy input (ambient pressure and temperature conditions) on the formation of carbonates. Gaseous CO_2 was directly bubbled into the suspensions to produce CO_2-sparged aqueous solutions. Salinity and pH were considered to resemble the alkaline spring waters in natural settings. A parametric study was done on these two factors as part of artificial enhancement pre-treatment steps on carbonation rates. Observed alterations on the mineral surfaces (in all suspensions) are attributed to carbonate formation at different pH and salinity of rock suspensions as well as particle sizes of the rock samples in CO_2-sparged aqueous solutions.

Materials and methods

Chemicals

All chemicals and reagents used in this study were of analytical reagent grade unless otherwise specified. High purity CO_2 (99.9 %, Linde Phils.) was used in the carbonation reaction, Hydrochloric acid (RCI Labscan, Thailand) and sodium hydroxide (HiMedia Laboratories, India) were used to adjust the pH of the system. Magnesium chloride (HiMedia Laboratories, India) was used to prepare the required salinity for aqueous carbonation experiments. All dilutions and solution preparations for the ICP-MS analysis were done using ultrapure water prepared using a Millipore Direct-Q5, Ultrapure Water Purification System (18.2 M cm resistivity, Merck Millipore). The rock powder was decanted then air dried prior to XRD analysis.

Sample collection initial characterization

Ophiolitic rock samples were collected from different locations in Luzon as shown in Fig. 1; namely Angat Ophiolites (AO), Zambales Ophiolite Complex (ZOC) and Camarines Norte Ophiolite Complex (CNOC). The obtained rock samples were submitted to the Mines and Geosciences Bureau, Petrolab, for thin-section petrographic analysis.

The rock samples were crushed and powdered prior to carbon mineralization experiments. Laboratory mesh sieves were used for the particle size-controlled experiments. Fractions of fine (62–125 μm) and (250–420 μm) coarse particle sizes were chosen in this study.

Mineral carbonation experiments

The experimental design for the carbonation experiment was based on a simple direct aqueous carbonation method, wherein water and grounded rock suspension was supplied with CO_2 to facilitate both the mineral dissolution reactions as well as the formation of HCO_3^- species. Baseline carbonation studies were previously conducted [20] on the mineral rock samples by sparging the powdered rock-water mixture with CO_2, under natural pH and ambient

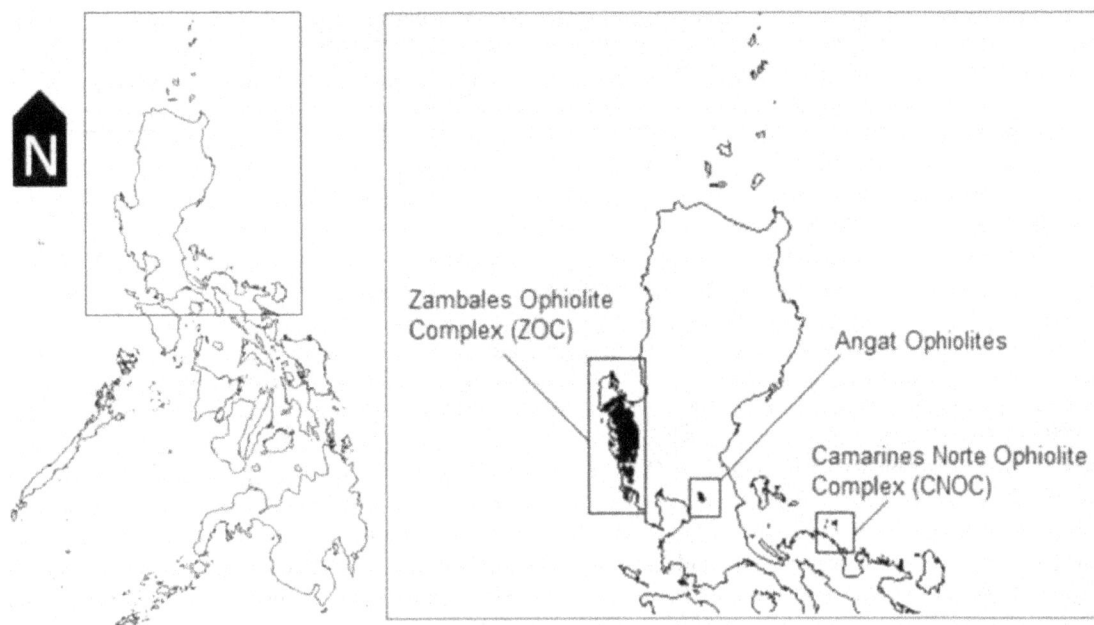

Fig. 1 Locations of rock formations around Luzon regions where the samples were collected

conditions (1 bar and 298 K). In addition, mineral carbonation experiments were also conducted under high and low pH (pH 13 and 1, respectively). To further study the effect of pH in mineral carbonation reactions, carbonation experiments were conducted under intermediate pH (pH 4, 6 and 10). Increase or decrease in the concentrations of Mg, Al, Ca and Fe were monitored to determine the dissolution of the rock samples at different pH values.

Mineral carbonation experiments were done by adding 10 mL water to 10 g powdered rock samples in 20 mL vials and sparging (5 mL/min) the suspensions with CO_2 for 60 days.

pH-only and pH-salinity experiments

The effect of pH was studied by maintaining the carbonation reaction vials at pH 4, 6, and 10. The effects of both pH and salinity were observed using; (1) a high pH condition (pH $\sim$ 13) with 5 % $MgCl_2$, and (2) a low pH condition (pH $\sim$ 1) also with 5 % $MgCl_2$, these were compared against suspensions wherein the pH and salinity were not modified.

Bubbling of CO_2 into suspensions was also conducted for 60 days. The suspensions were prepared as follows: (1) unaltered condition (no modification in pH and salinity of the suspensions); (2) low pH with 5 % $MgCl_2$; and (3) high pH with 5 % $MgCl_2$, these were all compared with the X-ray diffraction pattern of the untreated rock sample (i.e., not purged with CO_2). The water used for the "unaltered" set-up is taken from the streams near the sampling site of the rock samples. The pH and salinity were not modified.

Grain size range used for pH, and pH-salinity experiments was 62–420 µm.

Grain size only experiments

For the particle size-controlled experimental batch, carbonation experiments were done on fine (particle size: 62–125 µm) and coarse (particle size: 250–420 µm) rock samples. No quantification measurements were conducted to determine the amount of CO_2 uptake in all experiments.

Instrumentation

The CO_2–water–rock interaction that occurred upon bubbling the powdered rock samples with CO_2 was studied by monitoring the metal concentration in the aqueous part of the system as well as the change in the mineral composition of the rock samples. An Agilent 7500cx inductively coupled plasma-mass spectrometry (ICP-MS) was used to quantify the metal content of the samples. Shimadzu MAXima_X XRD-7000 X-ray diffractometer (XRD) was used to monitor the change in mineral composition of the rock samples.

Results and discussion

Ophiolitic rock compositions

Mineral composition of each of the ophiolitic rock samples used in the investigation is summarized in Table 1.

Table 1 Mineral composition of rock samples based on thin-section petrographic analysis

Sample source	Major mineral components[a]		Rock type
	Mineral name	Chemical formula[b]	
Angat ophiolites	Plagioclase (60 %)	$(Na, Ca)(Si, Al)_3O_8$	Diabase
	Pyroxene (20 %)	$(Ca, Mg, Fe, Mn, Na, Li)(Al, Mg, Fe, Mn, Cr, Sc, Ti)(Si, Al)_2O_6$	
Zambales ophiolite complex	Serpentine (73 %)	$(Mg, Al, Fe, Mn, Ni, Zn)_{2-3}(Si, Al, Fe)_2O_5(OH)_4$	Serpentinite
	Relict Augite (15 %)	$(Ca, Mg, Fe)_2(Si, Al)_2O_6$	
Camarines Norte ophiolite complex	Relict Enstatite (38 %)	$(Mg, Fe)SiO_3$	Metagabbro
	Relict Augite (30 %)	$(Ca, Mg, Fe)_2(Si, Al)_2O_6$	
	Tremolite (20 %)	$Ca_2Mg_5Si_8O_{22}(OH)_2$	

[a] Major mineral components: more than 10 % of the total mineral components; minor mineral components (10 % and below) include chlorites, clinozoisite, opaques and clays

[b] [23] Materials Data, Inc., Livermore, CA USA (http://www.materialsdata.com)

Samples utilized are mostly of magnesium and calcium bearing minerals. Angat ophiolites are composed mainly of plagioclase (60 %) and considered as an Altered Diabase rock. Zambales ophiolites are composed mostly of serpentine (73 %) and identified as Serpentinite rock. Camarines Norte ophiolites are a mixture of Enstatite (38 %), Augite (30 %) and Tremolite (20 %). It was classified as an Altered Gabbro rock.

Carbonate formation under different pH only conditions

Availability of Ca^{2+} and Mg^{2+} ions is one of the controlling factors for the development of in situ CO_2 mineralization [30]. Dissolution of these metals into CO_2-sparged solutions would be easily available for CO_2 consumption [26]. Ophiolitic rock samples are rich with these divalent cations, thus, the decrease of metal concentrations in the purged suspensions indicates elemental scavenging by CO_2. Results from ICP-MS analyses, as seen in Fig. 2, showed a distinctive trend wherein an increase in the metal concentration in the aqueous layer was observable in the 15th and 30th day of CO_2 exposure, which could be accounted for the dissolution of the minerals from the rock samples. As expected, Mg^{2+} ions were detected prominently indicating the possibility of extracted Mg^{2+} ions dissolved into the aqueous layer of ZOC suspensions. As observed from petrographic analyses, Mg^{2+} ions are prominent in ZOC suspensions mainly because of the intrinsic mineral composition. At all pH conditions, Mg^{2+} ions showed increased concentrations on the 15th day of bubbling CO_2 into the suspensions. As can be seen in Fig. 3, the intensity of the carbonate mineral peak is not as significant compared to the original 'serpentine peaks' (for the carbonation experiment carried under ambient conditions). And for the "enhanced" carbonation set-up, the

composition is still not entirely changed into newly formed carbonate minerals as the presence of the original peaks attributed to the original mineral components of the rock samples are still evident. The "enhanced" carbonation set-up is a suspension that is considered as more favorable condition for carbonation reaction. Thus, formation of $MgCO_3$ is not easily formed, even though the Mg^{2+} ions are dissolved in the suspensions and available for formation reactions. This indicates that $MgCO_3$ mineral, in an extended carbonation period, may eventually form in these rocks. The formation of $MgCO_3$ might have been suppressed in early reaction times by the presence of the $MgCl_2$ in the suspensions. A need to investigate the influence of Mg^{2+}/Ca^{2+} molar ratios in the $MgCO_3$ and $CaCO_3$ formation should be considered in future studies.

The identification of the source of Mg^{2+} ions among the ophiolite rocks which contribute to the new $MgCO_3$ formed is not easily discerned from the carbonation experiments, since the rock samples are already composed of a mixture of minerals. Based on the dissolution rates of enstatite, augite and serpentine, with respective log K: -9.02, -6.82, and -5.70 values [27], it can be inferred that serpentine undergoes faster dissolution than enstatite and augite, thus there are more available Mg^{2+} ions from serpentine than the other minerals.

Aluminum, iron and calcium ions showed a similar trend in suspensions of Angat ophiolites indicating that these ions may be easily available for new mineral formation during carbonation process. A decrease in the metal concentrations (Mg, Al, Ca, and Fe) was observed during the 45th and 60th day of bubbling which could indicate the consumption of CO_2 and mineralization could have taken place. For the pH-controlled carbonation vessels, the process of mineral dissolution (of the original mineral components of the rock samples) and re-precipitation (formation of carbonate minerals) were observed by

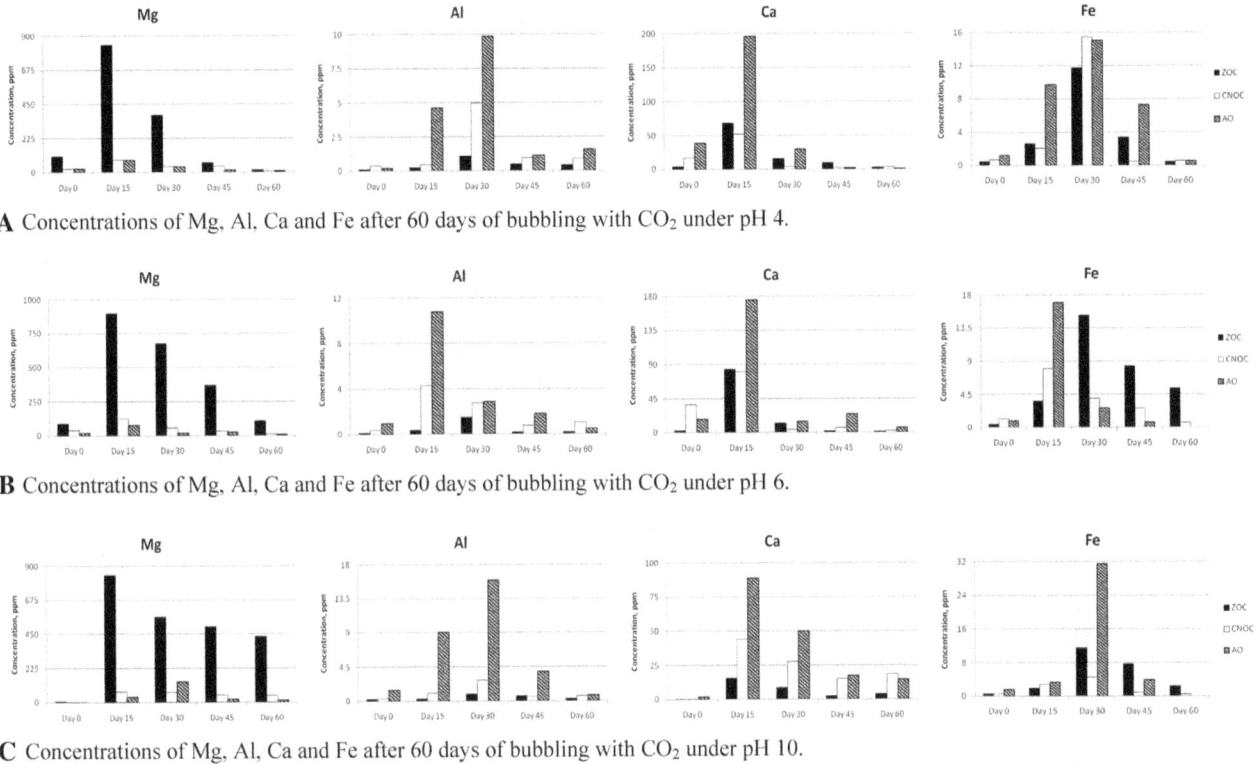

A Concentrations of Mg, Al, Ca and Fe after 60 days of bubbling with CO_2 under pH 4.

B Concentrations of Mg, Al, Ca and Fe after 60 days of bubbling with CO_2 under pH 6.

C Concentrations of Mg, Al, Ca and Fe after 60 days of bubbling with CO_2 under pH 10.

Fig. 2 Concentrations of Mg, Al, Ca and Fe after 60 days of bubbling with CO_2 at **a** pH 4, **b** pH 6, and **c** pH 10 in different suspensions. (*ZOC* zambales ophiolites complex, *CNOC* Camarines Norte ophiolites complex, *AO* Angat ophiolites)

monitoring the changes in the concentration of metals (that participate in the reaction) in the aqueous component of the rock-water-CO_2 mixture. With these observations, carbonate formation or new mineral formation on Zambales ophiolites are still considered favorable at working pH (pH = 4, 6 and 10) compared with Angat and Camarines Norte ophiolites.

Carbonate formation under different pH and salinity conditions

Another experimental batch was prepared to investigate the effects of both pH and salinity on the mineral carbonation reaction. Salinity is considered to reduce solubility trapping due to reduction in solubility of CO_2 and change in the partial molar volume of water [11]. The metal concentrations were monitored in this set-up (Fig. 2) and the carbonate phases resulting from the carbonation procedure were visible from the XRD spectra. Significant changes on the rock surfaces were observed as seen from Fig. 3. An observable change in the mineral content in the powdered rock sample is the appearance of peak at $2\theta = 29.42$. This corresponds to a d-spacing = 3.035, which is a characteristic of the carbonate mineral, Calcite [$CaCO_3$]. This peak was evident in all the samples subjected to mineral carbonation. The difference in the intensity is attributed to the

relative amount of the formed mineral in the powdered rock sample. Using a similar rationale for the new $MgCO_3$ mineral formation, it would be difficult to distinctly identify the source of Ca^{2+} ions for the formation of $CaCO_3$. Between the two minerals, tremolite and plagioclase, with respective log K: -8.40, and -7.5 values [27], dissolution of plagioclase would be faster, and thus could be the easier source of Ca^{2+} ions.

Aside from the appearance of characteristic peaks of carbonate based minerals, the decrease in the intensity of peak at $2\theta \sim 12.1$ (which corresponds to d-spacing ~ 7.31) could be an evidence of the dissolution of serpentine group of minerals (i.e., lizardite [$Mg_3Si_2O_5(OH)_4$]) which is one of the original mineral component of the rock samples. An additional peak at $2\theta = 30.84$ (d-spacing = 2.899) appeared in the sample subjected to high pH with 5 % $MgCl_2$, this could be assigned as the characteristic peak of hydromagnesite [$Mg_5(CO_3)_4(OH)_2 \cdot 4H_2O$], which is very probable due to the high alkalinity at the given experimental conditions. Given these results, an enhancement in the carbonation reaction was apparent in suspensions 2 and 3 (low pH + 5 % $MgCl_2$ and high pH + 5 % $MgCl_2$, respectively) as compared with suspension 1 (unaltered pH and salinity). Though the calcite peak was visible in all three suspensions, the intensity is highest in suspension 3, which could mean that the formation of carbonate mineral

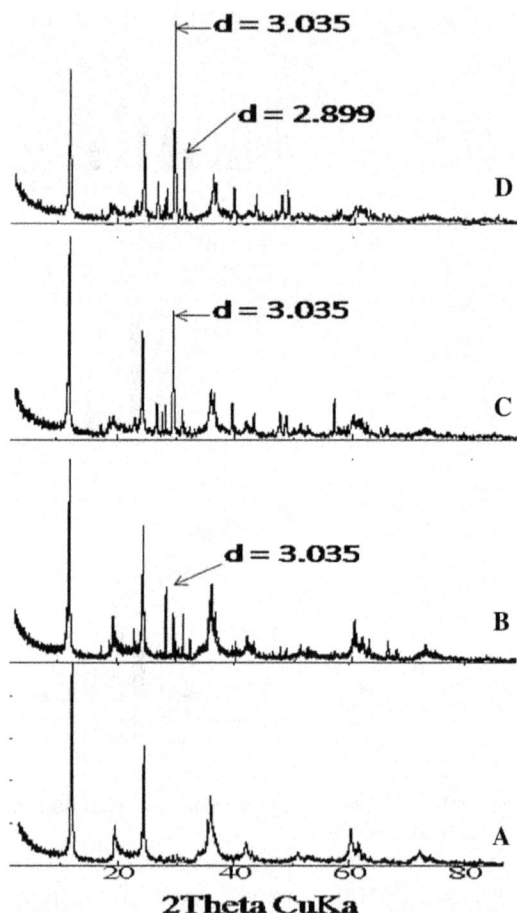

Fig. 3 X-ray diffraction patterns of rock samples after 60 days of subjecting to mineral carbonation experiments. **a** Untreated: not subjected to carbonation experiments; **b** baseline: ambient conditions, no modification in pH and salinity; **c** low pH, 5 % $MgCl_2$; **d** high pH, 5 % $MgCl_2$. Peak at 2.899 is assigned to hydromagnesite while peak at 3.035 is assigned to calcite. *Y*-axes are all on the same scale

is more favorable in this condition. In addition, more diverse types of carbonate minerals could be formed in highly alkali suspensions due to the participation of OH^- (as seen in the formation of hydromagnesite $[Mg_5(CO_3)_4(OH)_2 \cdot 4H_2O]$).

It is likely that other non-carbonate precipitation or competing reactions have occurred. The presence of non-carbonate phases strongly indicates an incomplete carbonation process. The suspensions and other experimental conditions must be further refined to achieve a maximum sequestration-efficient procedure.

Carbonate formation at different rock particle size

Particle size is also considered as an important factor affecting carbonation as it is generally accepted that the smaller the average particle size, the larger the mineral surface area per unit mass and the higher the reactivity of the particles become [15]. Results of the XRD analyses, as

shown in Fig. 4, for rock samples with fine (62–125 μm) and (250–420 μm) coarse particle sizes showed that the change in the mineral composition is similar in both experimental group. In addition to the previously mentioned formation of calcite and hydromagnesite, Dolomite $[CaMg(CO_3)_2]$ and Pirssonite $[Na_2Ca(CO_3)_2 \cdot 2H_2O]$ could also be formed in the carbonation reaction as indicated by the peaks that appeared at $2\theta = 30.98$ (d-spacing = 2.886) and $2\theta = 34.98$ (d-spacing = 2.565), which correspond to their respective d-spacings. The intensity of the peaks in the fine particle size set-up is slightly higher than the peaks in coarse particle size set-up. This could be an evidence that smaller particles indeed undergo faster mineral dissolution. Thus, the formation of new carbonate minerals is also slightly favorable in this condition. The coarse or fine fractions can influence additional energy input to enhance surface interactions [31]. Nonetheless, their difference is not significant and it can consequently be concluded that, within the particle sizes used in this study (62–420 μm), the effect of particle size on mineral carbonation reactions is negligible and is indicative only of the viable working range. Understanding of particle size would introduce an additional variable to be considered in identifying prospective injection points in Luzon ophiolite rocks sites.

Environmental implications of carbon mineralization

The experimental findings and methods used could be further developed to CO_2 mineralization strategies at potential sites where carbon footprint could be significant, such as Bicol and Leyte, Philippines. Large geothermal power plants are in use [9] in these areas and possibly, tons of CO_2 could be emitted from geothermal operations. However, mineral exploration within the vicinity of the geothermal plants is yet to be done to determine if available ophiolitic rocks are close to the site for CO_2 injection. In addition, instead of emitting CO_2 into the atmosphere, pipes could be connected to nearby alkaline springs for direct channeling of CO_2 and consequently produce the CO_2-sparged solutions. Mineral carbonation process is a relatively young technology, where many areas of this technology has yet to be explored, primarily the quantification of CO_2 sequestered, as well as other parameters affecting carbonation reaction (i.e., pressure and temperature).

A future research investigation will be undertaken to assess the physical and chemical characteristics (e.g., stability) of the carbonate phases and non-carbonate phases formed after the carbonation procedure. This will give a better direction on identifying the potential commercial

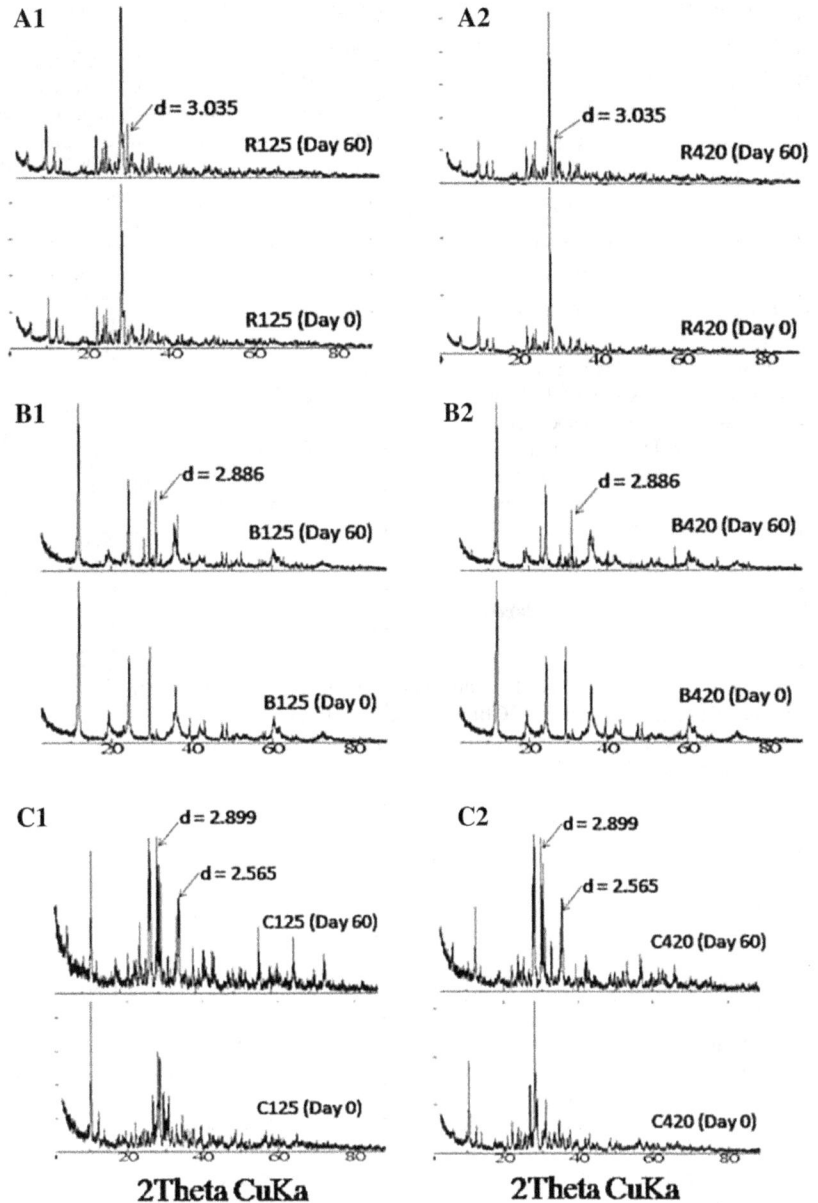

Fig. 4 X-ray diffraction patterns of rock samples from **a** Angat Ophiolites, **b** Zambales Ophiolite complex, and **c** Camarines Norte Ophiolite Complex, before (day 0) and after (60 days) subjecting to mineral carbonation experiments (**1**) Fine (particle size: 62–125 μm); (**2**) coarse (particle size: 250–420 μm). *Y*-axes are all on the same scale

markets (e.g., construction materials) of these high value products.

Several research and development gaps still need to be addressed before mineral carbonation can be integrated to established power plant systems. Deployment of in situ carbonation should be given a serious chance since rock formations (ophiolites) containing high concentrations of Ca and Mg are abundant in many locations in the world.

Conclusions

A CO_2 mineral sequestration procedure that utilizes alkaline spring samples was proposed. Investigations on bubbling of CO_2 into alkaline springs would be beneficial to decrease the energy input in carbon sequestration studies. Bubbling of CO_2 is applied to varying experimental conditions including pH, salinity and particle sizes using different sources of ophiolitic rocks. Salinity and pH were shown to significantly enhance the mineral carbonation reaction from an unaltered system (natural pH and salinity). The formation of carbonate minerals such as calcite [$CaCO_3$], hydromagnesite [$Mg_5(CO_3)_4(OH)_2 \cdot 4H_2O$], Dolomite [$CaMg(CO_3)_2$] and Pirssonite [$Na_2Ca(CO_3)_2 \cdot 2H_2O$] was confirmed in the XRD analysis of samples purged with CO_2 at various conditions for 60 days. The observation of a newly formed carbonate-based mineral in the CO_2-sparged suspensions is a good indication that the method employed is suitable for improving the interaction of CO_2 and metal cations in aqueous suspensions.

The experiment with controlled particle sizes gave similar results in the x-ray diffraction analysis, hence it can be inferred that within the particle size range used in this study, 62–420 μm, the effect of particle sizes in mineral carbonation process is negligible. The results are yet to be applied at a scale sufficient to show their industrial viability and additional parameters, such as presence of organic constituents, should be considered in future studies. An interference study with other gases (e.g., of other green house gases) is also an interesting work in the future.

Acknowledgments The authors would like to thank Natural Sciences Research Institute (CHE-12-2-01) and Office of the Vice-Chancellor for Research and Development (Project HJR-12-483), University of the Philippines Diliman for their financial support. We are also grateful for the technical assistance of Dr. Alyssa Peleo-Alampay of the Nanoworks Laboratory, National Institute of Geological Sciences, University of the Philippines.

Compliance with ethical standards

Funding This study was funded by the Natural Sciences Research Institute (CHE-12-2-01) and the Office of the Vice-Chancellor for Research and Development (Project HJR-12-483) in the University of the Philippines at Diliman.

Conflict of interest The authors declare that they have no conflict of interest.

References

1. Alexander G, Maroto-Valer M, Gafarova-Aksoy P (2007) Evaluation of reaction variables in the dissolution of serpentine for mineral carbonation. Fuel 86:273–281
2. Arcilla CA, Pascua CS, Alexander WR (2011) Hyperalkaline groundwaters and tectonism in the Philippines: significance of natural carbon capture and sequestration. Energy Procedia 4:5093–5101
3. Barros N, Oliveira G, Lemos de Sousa MJ (2012) Environmental impact assessment of carbon capture and sequestration: General overview. "IAIA12 Conference Proceedings" Energy Future: The Role of Impact Assessment. In: 32nd Annual Meeting of the International Association for Impact Assessment, 27 May–1 June 2012, Centro de Congresso da Alfândega, Porto, Portugal
4. Big Sky Carbon, Sequestration Partnership. http://www.bigskyco2.org/research/geologic/basaltproject. Accessed 26 Feb 2015
5. CarbFix. http://www.or.is/en/projects/carbfix. Accessed 26 Feb 2015
6. Chavagnac V, Ceuleneer G, Monnin C, Lansac B, Hoareau G, Boulart C (2013) Mineralogical assemblages forming at hyperalkaline warm springs hosted on ultramafic rocks: a case study of Oman and Ligurian ophiolites. Geochem Geophys Geosyst 14:2474–2495
7. Cipolli F, Gambardella B, Marini L, Ottonello G, Zuccolini MV (2004) Geochemistry of high-pH waters from serpentinites of the Gruppo di Voltri (Genova, Italy) and reaction path modeling of CO_2 sequestration in serpentinite aquifers. Appl Geochem 19(5):787–802
8. Dilmore RM, Allen DE, McCarthy Jones JR, Hedges SW, Soong Y (2008) Sequestration of dissolved CO_2 in the oriskany formation. Environ Sci Technol 42:2760–2766
9. First Gen Corporation. http://www.firstgen.com.ph/OurAssets.php. Accessed 20 Feb 2015
10. Gerdemann SJ, O'Connor WK, Dahlin DC, Penner LR, Rush H (2007) Ex situ aqueous mineral carbonation. Environ Sci Technol 41:2587–2593
11. Ghanbari S, Al-Zaabi Y, Pickup GE, Mackay E, Gozalpour F, Todd AC (2006) Simulation of CO_2 storage in saline aquifers. Chem Eng Res Des 84(9):764–775
12. Huijgen WJJ, Ruijg GJ, Comans RNJ, Witkamp GJ (2006) Energy consumption and net CO_2 sequestration of aqueous mineral carbonation. Ind Eng Chem Res 45:9184–9194
13. Jonckbloedt RLC (1998) Olivine dissolution in sulphuric acid at elevated temperatures—implications for the olivine process, an alternative waste acid neutralizing process. J Geochem Exp 62:337–346
14. Keleman PB, Matter J (2008) In situ carbonation of peridotite for CO_2 storage. Proc Natl Acad Sci 105(45):17295–17300
15. Kodama S, Nishimoto T, Yamamoto N, Yogo K, Yamada K (2008) Development of a new pH-swing CO_2 mineralization process with a recyclable reaction solution. Energy 33:776–784
16. Lackner KS, Wendt CH, Butt DP, Joyce EL, Sharp DH (1995) Carbon dioxide disposal in carbonate minerals. Energy 20:1153–1170
17. Lal R (2008) Carbon sequestration. Philos Trans R Soc B 363:815–830. doi:10.1098/rstb.2007.2185
18. Liu Q, Maroto-Valer MM (2012) Studies of pH buffer systems to promote carbonate formation for CO_2 sequestration in brines. Fuel Process Technol 98:6–13
19. Liu Q, Maroto-Valer MM (2010) Investigation of the pH effect of a typical host rock and buffer solution on CO_2 sequestration in synthetic brines. Fuel Process Technol 91:1321–1329
20. Magbitang RA, Lamorena, RB (2013) Mineral carbonation process in ophiolite complexes during the CO_2-purged aqueous solutions exposure. In: Proceedings of the Annual International Conference on Chemistry, Chemical Engineering, and Chemical Process (CCECP 2013). doi:10.5176/2301-3761_CCECP.38
21. Maroto-Valer MM, Fauth DJ, Kuchta ME, Zhang Y, Andresen JM (2005) Activation of magnesium rich minerals as carbonation feedstock minerals for CO_2 sequestration. Fuel Process Technol 86:1627–1645
22. Matter JM et al (2011) The CarbFix pilot project-storing carbon dioxide in basalt. Energy Proc 4:5579–5585
23. MDI-MINERAL (2011) Materials data Inc. LiveMArtormore, CA USA (XRD mineralogy database)
24. O'Connor WK, Dahlin DC, Nilsen DN, Rush GE, Walters RP, Turner PC (2001) Carbon dioxide sequestration by direct mineral carbonation: Results from recent studies and current status. In: 1st Annual DOE Carbon Sequestration Conference, Washington, D.C., May 2001
25. Olajire AA (2013) A review of mineral carbonation technology in sequestration of CO_2. J Petrol Sci Tech 109:364–392
26. Olsson J, Stripp SLS, Gislason SR (2014) Element scavenging by recently formed travertine deposits in the alkaline springs from the Oman Samail Ophiolite. Miner Mag 78(6):1479–1490
27. Palandri JL, Kharaka YK (2004) A compilation of rate parameters of water-mineral interaction kinetics for application to geochemical modeling. US Geological Survey, Open File Report 2004-1068
28. Park AA, Fan LS (2004) CO_2 mineral sequestration: physically activated dissolution of serpentine and pH swing process. Chem Eng Sci 59:5241–5247
29. Park AA, Jadhav R, Fan LS (2003) CO_2 mineral sequestration: chemically enhanced aqueous carbonation of serpentine. Can J

Chem Eng 81:885–890

30. Sanna A, Uibu M, Caramanna G, Kuusik R, Maroto-Valer MM (2014) A review of mineral carbonation technologies to sequester CO_2. Chem Soc Rev 43:8049–8080

31. Santos RM, Verbeeck W, Knops P, Rijnsburger K, Pontikes Y, Gerven TV (2013) Integrated mineral carbonation reactor technology for sustainable for carbon dioxide sequestration: CO_2 energy reactor. Energy Proc 37:5884–5891

32. Shafeen A, Croiset E, Douglas PL, Chatzis I (2004) CO_2 sequestration in Ontario, Canada. Part I: storage evaluation of potential reservoirs. Energy Convers Manage 45:2645–2659

33. Snaebjornsdottir SO, Wiese F, Fridriksson T, Armansson H, Einarsson GM, Gislason SR (2014) CO_2 storage potential of basaltic rocks in Iceland and the oceanic ridges. Energy Proc 63:4585–4600

34. Streit E, Kelemen P, Eiler J (2012) Coexisting serpentine and quartz from carbonate-bearing serpentinized peridotite in the Samail Ophiolite, Oman. Contrib Miner Petrol 164(5):821–837

35. Tier S, Revitzer H, Eloneva S, Fogelholm C, Zevenhoven R (2007) Dissolution of natural serpentinite in mineral and organic acids. Int J Miner Process 83:36–46

36. Van Pham T, Aagaard P, Hellevang H (2012) On the potential for CO_2 mineral storage in continental flood basalts-PHREEQC bach and 1D-diffusion reaction simulation. Geochem Trans 13(5):1–12

9

Production of environmentally adapted lubricant basestock from jatropha curcas specie seed oil

Matthew C. Menkiti[1,2] · Ocholi Ocheje[2] · Chinedu M. Agu[2]

Abstract Jatropha curcas seed oil was studied for the synthesis of trimethylolpropane based biolube basestock via chemical transesterification of Jatropha methyl ester with trimethylolpropane (TMP) using calcium hydroxide catalyst. Reactions temperatures ranged between 80 and 160 °C and methyl esters to TMP mole ratios ranged between 3:1 and 7:1. Product analysis, ester groups and physio-chemical properties were obtained by gas chromatography, Fourier transform infrared spectroscopy and American Society for testing and material standard methods, respectively. Gibbs free energy indicated that the reaction was spontaneous with a second order rate constant of $1.00E-01$ (%wt/ wt min C)$^{-1}$ and kinetic energy of 13.57 kJ/mol. Jatropha biolubricant (JBL) had the following properties: viscosity of 39.45 and 8.51 cSt at 40 and 100 °C, respectively; viscosity index of 204, pour point of -12 °C and flash point of 178 °C. Temperature and mole ratio were the main factors that influenced the reaction. JBL properties complied with ISO VG 32 standard and could be applied as lube basestock with minor modifications.

Keywords Jatropha curcas · Biolubricant · Transesterification · Trimethylolpropane

Abbreviations

ASTM	American Society for testing and material
CJO	Crude jatropha oil
DE	Diesters
FAME	Fatty acid methyl ester
FTIR	Fourier transform infrared
GC	Gas chromatography
ISO	International Standard Organisation
JBL	*Jatropha* biolubricant
JME	*Jatropha* methyl ester
JTE	*Jatropha* triester
JTMPE	*Jatropha* trimethylolpropane esters
ME	Monoester
PE	Polyol ester
PUFA	Polyunsaturated fatty acid
TE	Triesters
TER	Transesterification
TMP	Trimethylolpropane
VG	Viscosity grade
VI	Viscosity index

Introduction

Strong environmental concerns and growing regulations on contamination and pollution of the environment by petroleum based lubricants have increased the need for renewable and biodegradable lubricants [1]. There have been lots of active research and development in this area due to increasing pressure from public demand, industrial concern and government agencies.

✉ Matthew C. Menkiti
cmenkiti@yahoo.com; matthew.menkiti@ttu.edu

1   Civil, Environmental and Construction Engineering Department, Texas Tech University, Lubbock, TX, USA

2   Chemical Engineering Department, Nnamdi Azikiwe University, Awka, Nigeria

The oleochemical esters are a growing interest with respect to the base lubricants industry. Their advantages compared to mineral base oil include low toxicity, higher biodegradability, renewability, high flash point, low volatility, high additive solvency power, high added value, good lubricity (due to molecule polarity), high viscosity index due to the double bonds and molecular linearity. However, the main disadvantages of these organic compounds are oxidative instability, hydrolytic instability, low temperature properties. These disadvantages can be minimized by additives, but the biodegradability, toxicity and the price can be endangered. Thus, the chemical synthesis of these compounds seems to be a veritable choice towards eco-friendly basestocks. The additives that could be used include anti-oxidant, anti-wear, anti-corrosion, etc., which are associated with low biodegradability. However, the additives industry is working hard to develop biodegradable additives [2].

Development works reported by many studies on novel high performance biodegradable lubricants focus on reducing the market price, ecological compatibility, processes as well as technical performances [3]. Several studies focus on improving the performance of vegetable oils through modification of structures to improve their properties [4–9]. Due to its structure, unmodified vegetable oil suffers from inadequate oxidative stability, poor corrosion protection, poor hydrolytic stability and poor low temperature performance. One of the techniques that could improve the properties of the vegetable oil is to change the structure of the oil by converting it to a new type of ester called polyol ester (PE). This process eliminates the hydrogen atom on the β-carbon of the vegetable oil structure, thus providing the esters with high degree of thermal stability, seldom found in vegetable oil [10].

The transesterification (TER) process for biolubricant synthesis can be catalysed chemically or enzymatically. Equations (1) and (2) illustrate a two-stage base catalysed transesterification for biolubricant synthesis.

Stage one

$$H_2C - OCOR_1$$
$$|$$
$$H C - OCOR_2 \qquad + \qquad 3CH_3OH$$
$$|$$
$$H_2C - OCOR_3$$

Triglyceride   Methanol

$$CH_2 - OH \qquad\qquad R_1COOCH_3$$
$$|$$
Catalyst
$$CH - OH \qquad + \qquad R_2COOCH_3$$
$$|$$
$$CH_2 - OH \qquad\qquad R_3COOCH_3$$

Glycerol   Methyl esters

(1)

Stage two

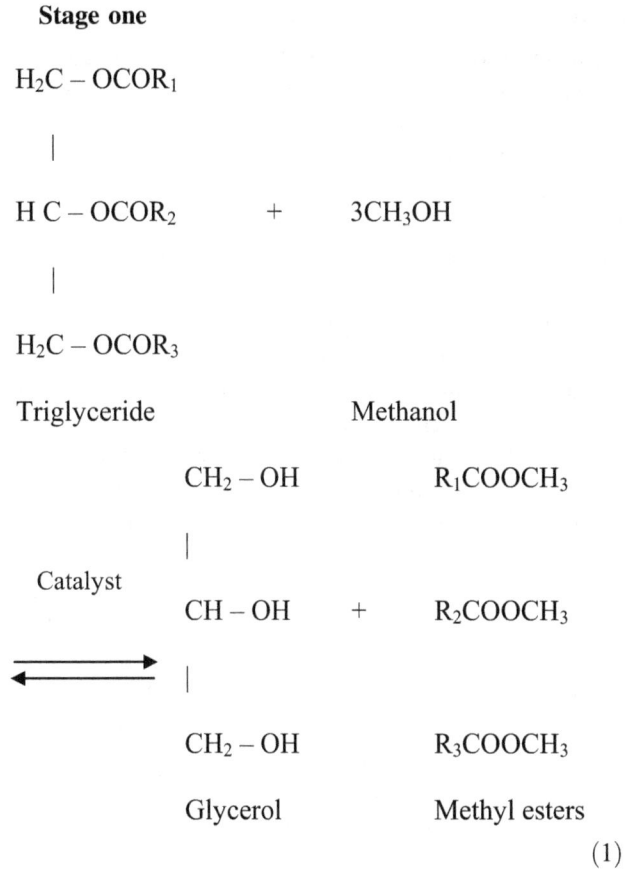

Trimethylolpropane   Methyl esters

R_1COOCH_3
$+$ R_2COOCH_3
R_3COOCH_3

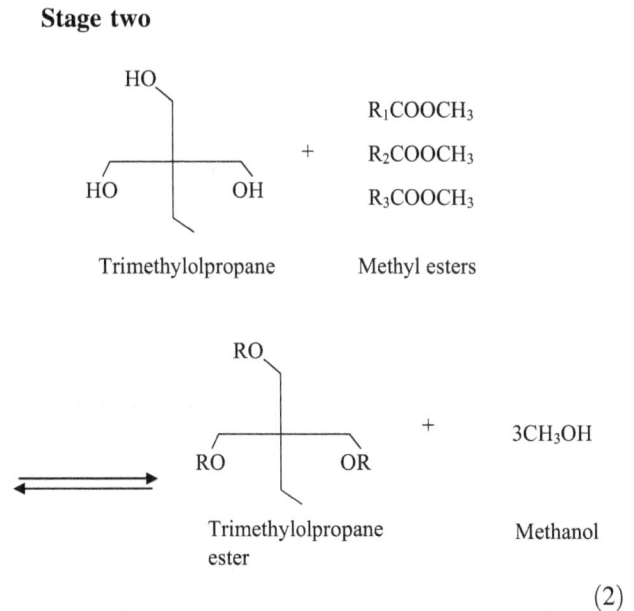

Trimethylolpropane   Methanol
ester

$+$ $3CH_3OH$

(2)

The objective of this work is to investigate the use of Nigerian Jatropha oil as a feedstock for the production of biolubricants in a two-step base catalysed reaction of Jatropha oil methyl ester (JME) with trimethylolpropane (TMP). JME was first synthesized from crude Jatropha oil (CJO) by transesterification, after which JME was then washed and purified before used in the synthesis of Jatropha biolubricant (JBL) [11]. Process parameters that were systematically investigated include the reaction temperature, reaction time and the molar ratio of methyl ester to trimethylolpropane. Furthermore, the kinetics of the reaction was also discussed. Jatropha curcas is a shrub belonging to the family of *Euphorbiaceae*. It is a plant with many attributes, multiple uses and great potential. It is a native of tropical America, but now thrives in many parts of the tropics and sub-tropics in Africa and Asia. Jatropha curcas has the potential to become one of the world's key energy crops. At present, it is globally taking the centre stage as the oil seed of choice in biolubricant production [12–18].

The composition of Jatropha curcas oil from two different locations in Nigeria consists of main fatty acids, such as palmitic acid (14.69–14.68%), stearic acid (0–5.23%), oleic acid (0–6.06%) and linoleic acid (79.08–80.07%) [19], in addition to high percentage of unsaturated fatty acid.

This makes the oil suitable for biolubricant production because the presence of double bond will lower the melting point, which would enhance the low temperature performance of the biolubricants. Also after chemical modification, drawbacks such as instability at high temperature would be overcome thereby improving its lubricating properties.

However, the chemical compositions of the oil vary according to the climate and locality. It has been reported that there is a large variability in different accessions of Jatropha curcas from diverse agro climatic regions [20]. Accordingly, Kaushik et al [20] reported values of kinematic viscosity at 30 °C for Jatropha oil from different countries (Variety Capeverde-39cSt, Variety Nicaragua-37cSt and Variety Nigeria-17cSt). Similarly, a number of authors have carried out research works on Nigerian jatropha curcas oil from different locations across the country and obtained, to some extent, varying physicochemical properties and fatty acid composition depicted in references [19, 21–23].

Although Jatropha seed has been used as a starting material for biolubricant synthesis [20, 24–27], the process has been usually driven using conventional homogeneous catalysis. In the current study, heterogeneous catalyst was employed to drive the process. The use of heterogeneous catalyst offered the advantages of no soap production (as by-product) and easy separation of catalyst from the reaction products. In addition, in this light, inherent complementary process thermodynamics would be investigated for the consideration of the feasibility of the reaction.

Jatropha oil is non-edible due to the presence of anti-nutritional substances such as phorbol esters [24] and thus makes it suitable as biodiesel and biolubricant feedstock [25]. In Nigeria, Jatropha can grow very well and already abundant in the country, but mainly for border demarcation of farm lands [28]. It is widely cultivated in the tropics as a live fence (hedge) around farm lands, since the toxins in the plant deter animals [29]. The tree has a life span of up to 30–40 years and could grow on a wide range of land types, including non-arable, marginal and waste lands, and need not compete with vital food crops for agricultural land. Crude Jatropha oil is not edible and its price may not be distorted by competing food uses [28]. The oil content of the seed ranges between 50 and 60% [30–33].

Materials and methods

Materials

Jatropha curcas samples were collected from Idah, Kogi State, Nigeria. The seeds were cleaned, shelled and air-dried in the shade for few days. All the chemicals and reagents used for this work were of analytical grade.

Extraction of oil from jatropha seed sample

Shelled and air-dried Jatropha seeds were crushed and tied in a white piece of cloth. This was later soaked in hexane in a tightly sealed bucket for 3 days, before collecting the extract through filtration. The cloth containing the crushed Jatropha seeds was further rinsed with fresh hexane to extract more oil. The hexane contained in the extracted Jatropha oil was removed by distillation while the crude oil extract was collected in a beaker [34].

Synthesis of jatropha methyl ester

Jatropha oil extracted from Jatropha seed was transesterified to form Jatropha Methyl Ester (JME). In this method, a mixture of 300 g of Jatropha oil, 100 g methanol and 1% wt/wt orthophosphoric acid catalyst were poured into continuously stirred reactor equipped with a water-cooled reflux condenser and heated up to 65 °C for 90 min. The mixture was dosed with 0.2 molar solution of sodium trioxocarbonate IV, which on neutralizing the acid catalyst, stopped the reaction. The neutralized mixture was later transferred to a separating funnel and subsequently allowed to stand overnight to ensure complete separation of methyl esters and glycerol phases. Glycerol phase (bottom phase) was emptied into a clean container and then allowed to stand. The obtained JME was heated at 65 °C to remove methanol. Entrained catalyst in the JME was removed by successive rinses with hot distilled water. Finally, water present in the JME was eliminated by oven-heating at 70 °C [34].

Synthesis of jatropha biolubricant

This was as described by Surapoj et al. [35] with modifications. A process flow for the synthesis is shown in Fig. 1.

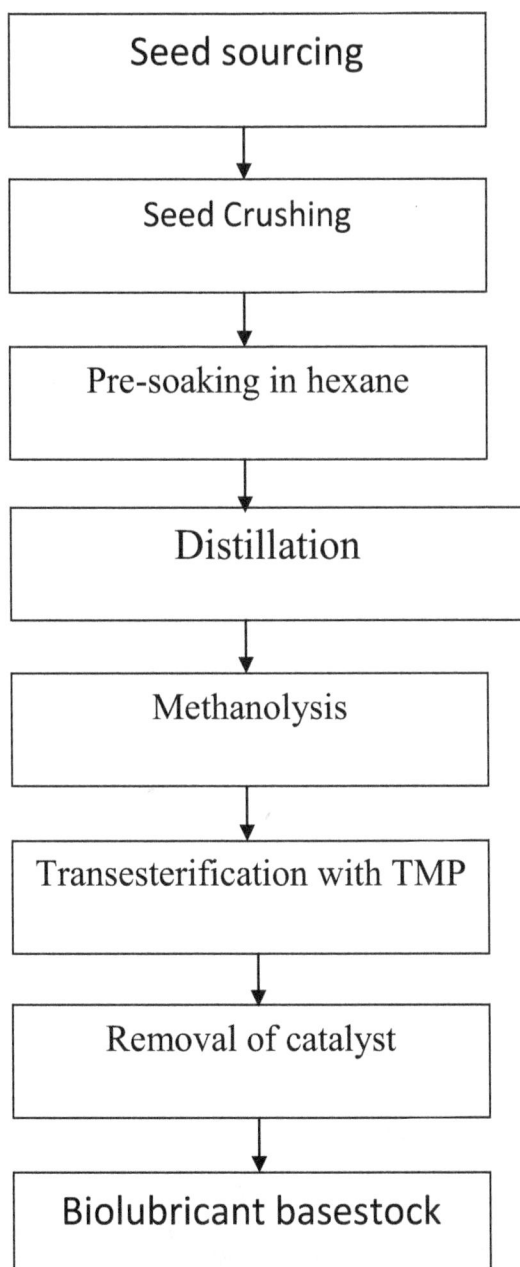

```
┌──────────────────────────────┐
│      Seed sourcing           │
└──────────────────────────────┘
              │
              ▼
┌──────────────────────────────┐
│      Seed Crushing           │
└──────────────────────────────┘
              │
              ▼
┌──────────────────────────────┐
│   Pre-soaking in hexane      │
└──────────────────────────────┘
              │
              ▼
┌──────────────────────────────┐
│        Distillation          │
└──────────────────────────────┘
              │
              ▼
┌──────────────────────────────┐
│        Methanolysis          │
└──────────────────────────────┘
              │
              ▼
┌──────────────────────────────┐
│  Transesterification with TMP│
└──────────────────────────────┘
              │
              ▼
┌──────────────────────────────┐
│     Removal of catalyst      │
└──────────────────────────────┘
              │
              ▼
┌──────────────────────────────┐
│    Biolubricant basestock    │
└──────────────────────────────┘
```

Fig. 1 Flow sheet process for biolubricant synthesis

In this method, TMP was initially heated using a transesterification experimental set-up comprising 50-mL three-necked round-bottom flask equipped with a water-cooled reflux condenser, a thermometer, Kipp's apparatus and a thermofisher scientific 50094711(THERMO SCIENTIFIC CIMAREC I MONO DIRECT) stirrer operated at 1000 rpm. The TMP contained in the flask was heated to and kept at 110 °C for 15 min, while being stirred at 1000 rpm under CO_2 flow. 110 °C was maintained to evolve moisture from the TMP. Using the same experimental set-up, a $Ca(OH)_2$ catalysed batch transesterification reactions between JME (FAME) and already cooled TMP were conducted at the following JME–TMP ratios: 3:1, 4:1, 5:1, 6:1 and 7:1. Each of the stated JME–TMP ratios was subjected to transesterification at individual temperatures of 80, 100, 120, 140 and 160 °C. Each of the individual experimental runs at a given particular ratio and temperature was monitored and samples were collected at reaction times intervals of 1, 2, 3, 4 and 5 h for analyses. At the end of each reaction, the product mixture was brought to room temperature and filtered to separate the solid catalyst from the liquid mixture (JBL). The filtered Jatropha bio-based stock was analysed using the GC to determine the product composition. Pour point, viscosities, flash point and viscosity index were also determined by appropriate analysis [34]. The unreacted methyl ester was not removed before measuring the properties of the biobased TMP ester to eliminate potential conjugation reaction at destructive high temperature (180–200 °C) molecular distillation (needed to remove unreacted methyl ester) involving poly unsaturated fatty acid (PUFA) and also improve wear resistance of the biobased TMP ester [6, 36]. Meanwhile, specified measurements of product (JBL) properties were selectively done for only JME–TMP mole ratio of 4:1 at 140 °C as indicated in "Lubricating properties of jatropha biolubricant".

Analysis of transesterification product

Functional groups present in JBL were identified through Fourier transform infrared resonance (FTIR). Samples were collected at hourly reaction times intervals (for a total period of 5 h) and analysed for JME, monoester (ME), diesters (DE), triesters (TE) and TMP by gas chromatography. The yield of each product was determined from the GC chromatogram calibrated against the known samples according to the procedure described by Yunus et al. [37].

Lubricating characteristics

The following named lubricating characteristics were determined based on the corresponding referred American Society for testing and material (ASTM) procedures: Pour

Point [38]; Kinematics Viscosities [39]; Viscosity Index [40] and Flash Point [41].

Results and discussion

Fourier transform infrared (FTIR) analysis of the JBL

The FTIR analysis (Fig. 2) was performed to determine the functional groups present in the biolubricant. Sample's discernable peaks ranged between 4000 and 700 cm^{-1}. The peaks located at 1228.3599 and 1171.4302 cm^{-1} corresponding to C–O stretching, indicated the presence of esters. The band at 1737.808 cm^{-1} was for C=O stretching, usually present in the esters. The peak at 2890.222 cm^{-1} (C–H stretching and CH$_2$ stretching) indicated the presence of carbohydrate and nucleic acids. The peak at 1737.808 cm^{-1} (C=O stretching) showed the presence of fats. For O–H stretching and phenol O–H stretching characterized by 3463.213 cm^{-1}, the presence of water and phenols were indicated [33, 42].

Time lined transesterification of TMP with JME

Figures 3 and 4 and S1–S3 indicate the progress of transesterification reactions at different times. It was observed that transesterification proceeded stepwise, in which ME formation first reached a maximum value. This was followed by a steady formation of DE. At the point of

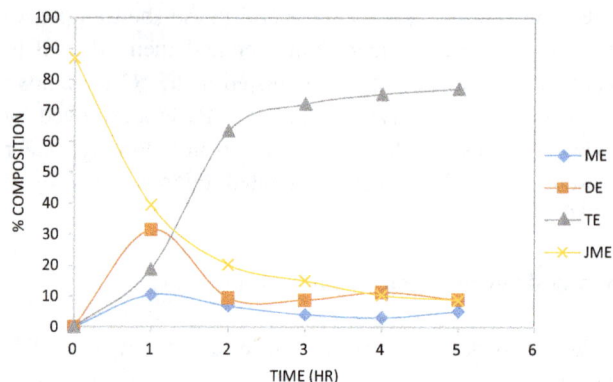

Fig. 3 TER between TMP and JME of 140 °C and 3:1 molar ratio using 1 wt% Ca(OH)$_2$ catalyst

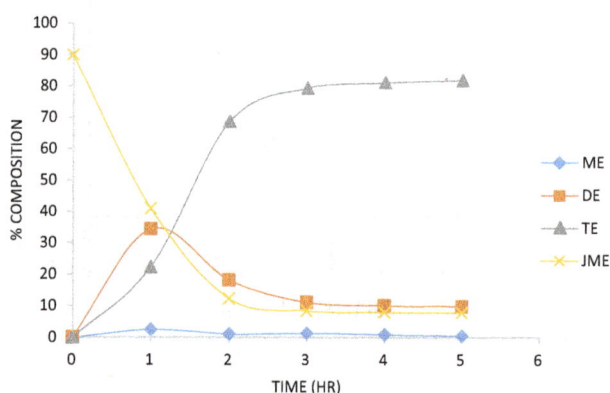

Fig. 4 TER between TMP and JME of 140 °C and 4:1 molar ratio using 1 wt% Ca(OH)$_2$ catalyst

Fig. 2 FTIR spectrum of Jatropha TMP ester

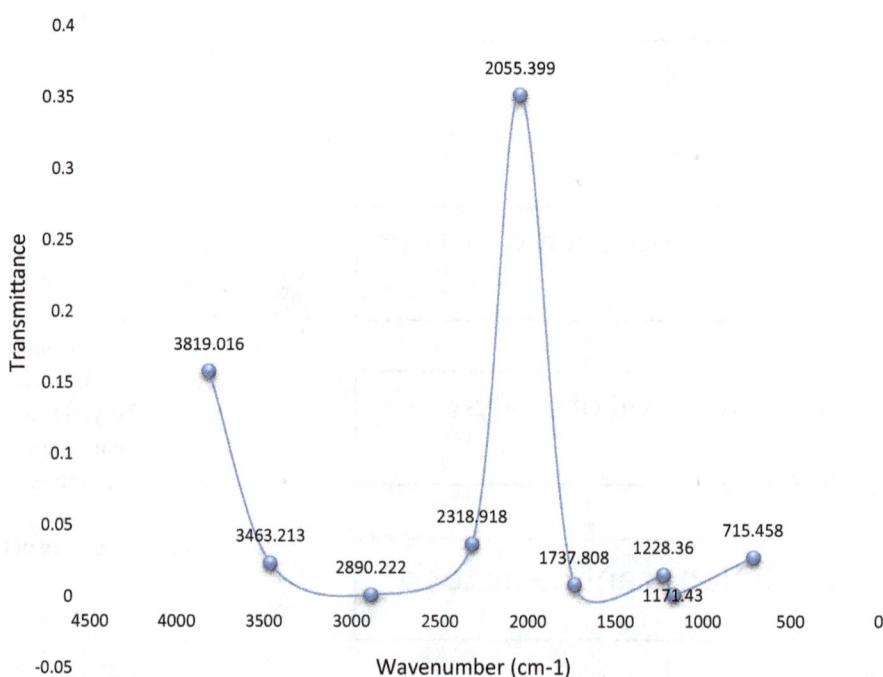

maximum formation of DE, TE increased rapidly. This was because the transesterification stepwise process preceded with the formation of intermediates products, before the commencement of the formation of final desired product, which was the TE [37]. Initially, ME, which was a single branch polyol ester was formed during the reaction. The increasing amount of ME, however, would immediately undergo conversion to form DE, which would react with JME to produce TE. Concentration of TE would rise with the decrease of DE and ME concentrations. Similar reaction mechanism has been reported earlier by other researchers [34, 43, 44].

Effects of temperature

To determine the effect of temperature, a series of experiments was conducted at JME–TMP molar ratios of 3:1, 4:1, 5:1, 6:1 and 7:1. Catalyst (Ca(OH)$_2$) amount was fixed at 1.0% wt/wt of reaction mixture. The reactions monitored for 1–5 h were carried out at 80, 100, 120, 140 and 160 °C for each of the mole ratio to observe the effect of temperature on the transesterification products. Representative results of the syntheses are shown in (Figs. 5, 6 and S4–S6) and (Figs. 6, 7 and S7–S9). Figures 4 and 5 and S4–S6 show the influence of temperature on the production profile of Jatropha trimethylolpropane esters (JTMPE or JBL) consisting of mono ester (ME), diester (DE), trimester(TE) and unreacted JME. Figures 5 and 6 and S4–S6 indicated that as the temperature increased, the TE composition increased, until at about 140 °C, after which the increase in TE composition became marginal. This was because at higher temperature, the amount of FAME in the reactor was low as a result of vaporization, enhancing the occurrence of the reverse reaction. By re-condensing the FAME (i.e. JME in this study) vapour back into the reactor, the reverse reaction would be contained and the esterification of DE to TE would prevail. Hence, water used in the condenser should be cold enough to ensure condensation of

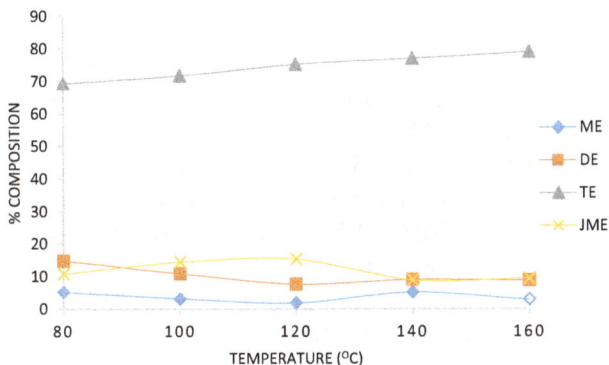

Fig. 6 Effects of temperature on JTMPE composition at 4:1 JME–TMP mole ratio and 1% catalyst loading for 5 h

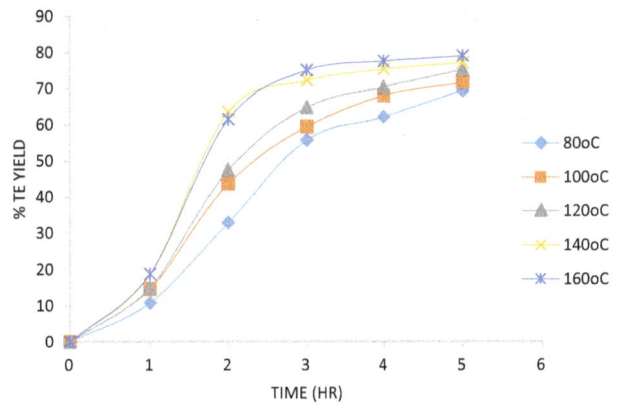

Fig. 7 Temporal yield of JTE at various temperatures, 3:1 JME–TMP mole ratio and 1% catalyst loading

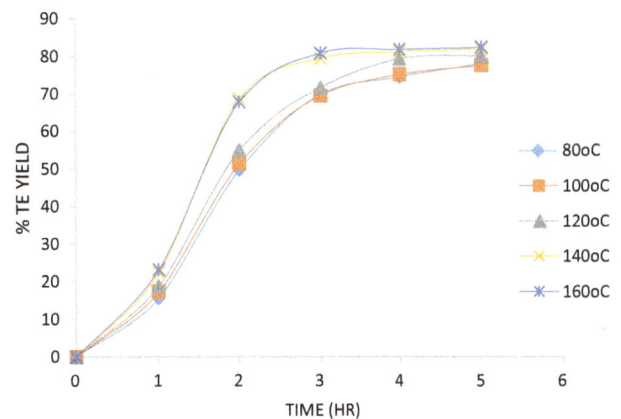

Fig. 8 Temporal yield of JTE at various temperatures, 4:1 JME–TMP mole ratio and 1% catalyst loading

the vaporized FAME back to the reactor [34, 45, 46]. The amount of ME, DE and JME, would be considered insignificantly unchanged with temperature variation. Figures 7 and 8 and S7–S9 depict the influence of temperature specifically on temporal yield profile of Jatropha triester (JTE), which was the major active ingredient in JBL.

Fig. 5 Effects of temperature on JTMPE composition at 3:1 JME–TMP mole ratio and 1% catalyst loading for 5 h

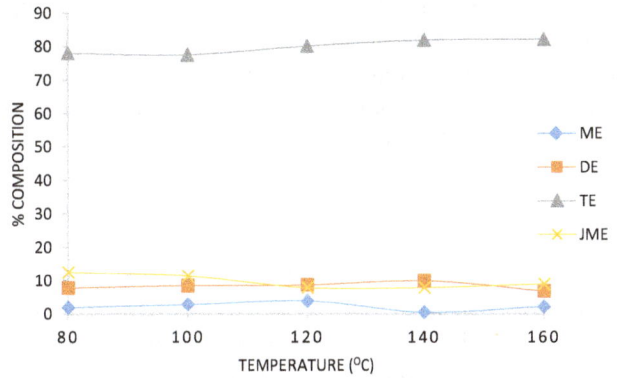

Figures 7 and 8 and S7–S9 indicated that generically, the yield increased with increase in time for all the temperatures considered. Expectedly, least yield was obtained at 100, since least successful molecular collision for product formation would be the case.

Effects of mole ratio

Transesterification, being a reversible reaction, could be driven to enhance the yield of triesters using excess amount of FAME or TMP. Excess FAME was chosen over TMP due to its relative lower cost. The same pattern of experiments as described in "Effects of temperature" applied here for the study of influence of mole ratio on the % composition of transesterification products. Figures 9 and 10 and S10–S12 represent the results of the effect of mole ratio on % composition of JTMPE at 120 and 140 °C, respectively. On the other hand, Figs. 11 and 12 and S13–S15 represent the results of the temporal yield of JTE at various mole ratios for temperatures of 120 and 140 °C,

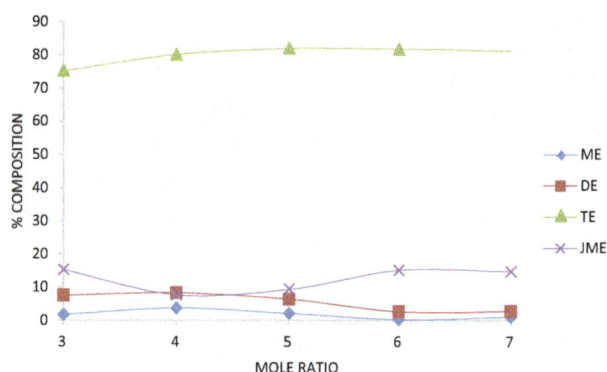

Fig. 9 Effects of JME:TMP mole ratio on % composition of JTMPE at 120 °C and 1% catalyst loading

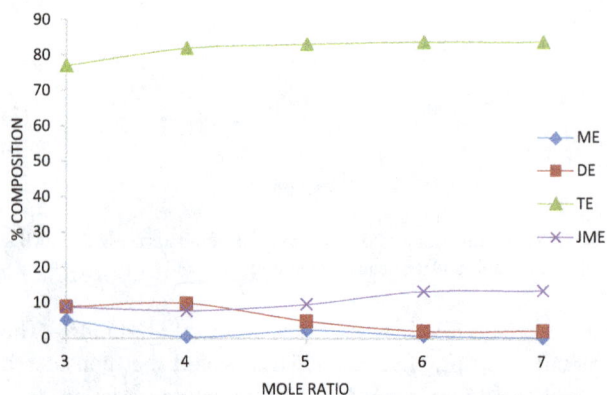

Fig. 10 Effects of JME:TMP mole ratio on % composition of JTMPE 140 °C and 1% catalyst loading

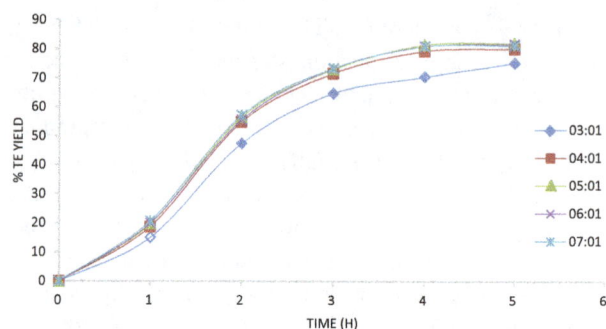

Fig. 11 Temporal yield of JTE at various mole ratios, 120 °C and 1% catalyst loading

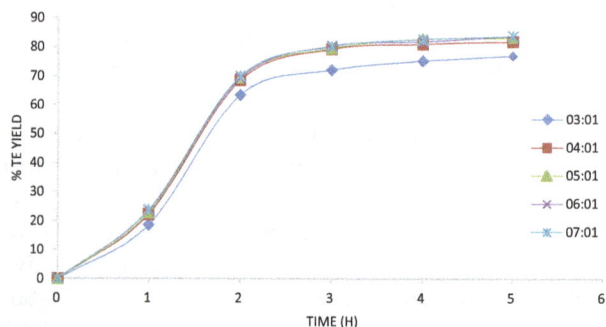

Fig. 12 Temporal yield of JTE at various mole ratios, 140 °C and 1% catalyst loading

respectively. Figures 9 and 10 and S10–S12 indicate that when molar ratio of FAME: TMP was increased, TE yield increased. Generally, a better product yield was obtained by keeping the molar ratio of reactants higher than the stoichiometric values since the reaction was driven more towards completion. This is in consonance with Le Chatelier's principle. According to Le Chatelier's principle, an excess of TMP would increase the TE yield by shifting the equilibrium to the right. However, excess JME was used in this study due to its relatively lower cost compared to TMP. However, from Figs. 1, 2, 3, 4, 5, 6, 7, 8, 9, 10, 11 and 12 and S13–S15, it was obvious that increasing the molar ratio above 4:1 provided marginal gain in TE yield. This could be due to onset of reverse reaction that retarded the conversion of DE to TE [38]. In the chemical and enzymatic transesterification of rapeseed methyl ester with TMP, maximum conversion was obtained at 3.3:1 and 3.5:1, respectively [44]. Yunus et al. [37] reported optimal transesterification process at palm oil methyl ester (POME)—TMP ratio of 3.8:1, 120 °C, 0.9% sodium methoxide catalyst and 20 Mbar [44]. In the report [37], conversion to TE increased from 83 to 86% as the ratio was increased from 3.5:1 to 3.7:1 [34]. Table 1 shows TE yield at various combinations of temperature and JME–TMP mole ratios.

Table 1 Percentage TE yield at various combinations of temperature and JME–TMP mole ratio

Temp (°C)	Mole ratio				
	03:01	04:01	05:01	06:01	07:01
80	69.71	77.93	77.87	79.32	80.25
100	71.71	77.48	77.59	77.97	76.79
120	75.25	80.12	81.97	81.82	81.27
140	77.06	81.92	83.11	83.77	83.83
160	78.98	82.26	82.94	83.64	84.12

Kinetics of transesterification of FAME and TMP

Transesterification reaction is a consecutive and reversible reaction being driven by excess FAME and a catalyst. The reaction could be represented as follows [26, 34]:

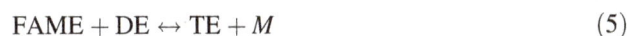

$$FAME + TMP \leftrightarrow ME + M \tag{3}$$

$$FAME + ME \leftrightarrow DE + M \tag{4}$$

$$FAME + DE \leftrightarrow TE + M \tag{5}$$

Overall reaction:

$$3FAME + TMP \leftrightarrow TE + 3M \tag{6}$$

Kinetics study was conducted to determine the reaction order and rate constants for the reaction time ranging from 1 to 50 min. The best JME:TMP ratio of 4:1 obtained from the batch transesterification study was used for the kinetic study while considering TMP as the limiting reactant. Figures 13 and 14 show the kinetics plots for the transesterification process at various temperatures for the first 50 min.

The best kinetic model for the data appears to be a second order kinetics. A model was developed based on the kinetics of decreased concentration of trimethylolpropane

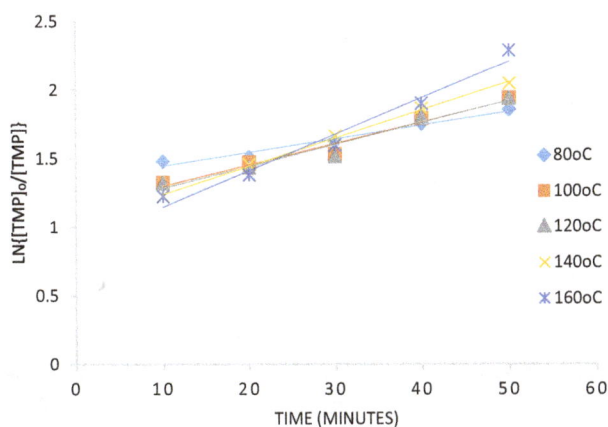

Fig. 13 First order kinetics plot for the synthesis of JBL at various temperatures (4:1 JME–TMP mole ratio)

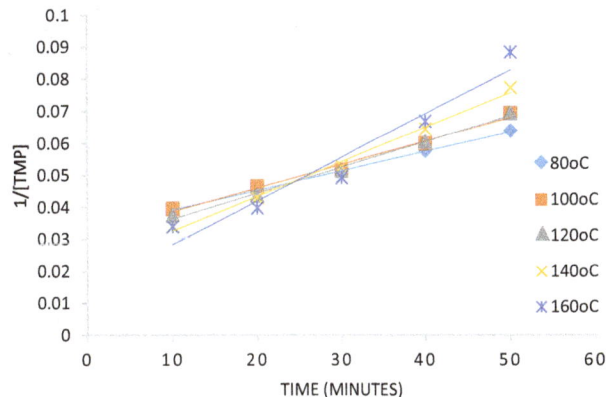

Fig. 14 Second order kinetics plot for the synthesis of JBL at various temperatures (4:1 JME–TMP mole ratio)

Table 2 Rate constant, K_{TMP} and regression values at various temperatures (JBL)

Temperature (°C)	First order		Second order	
	$K_{TMP,1st}$ (%wt/wt min)	R^2	$k_{TMP,2nd}$ (%wt/wt min)$^{-1}$	R^2
80	9.90E−03	0.9697	6.00E−04	0.9977
100	1.55E−02	0.9674	7.00E−04	0.988
120	1.60E−02	0.9596	8.00E−04	0.9941
140	2.04E−02	0.9983	1.10E−03	0.9938
160	2.64E−02	0.9703	1.40E−03	0.9419

[26, 34]. The second order reaction rate law for this study would be as follows:

$$\frac{1}{TMP} = K_{TMP}t + \frac{1}{TMP_0} \tag{7}$$

where K_{TMP} is the overall second order rate constant, t is the reaction time, TMP_0 is the initial trimethylolpropane concentration, TMP is final concentration of trimethylolpropane.

For transesterification reaction, a plot of reaction time versus 1/[TMP] will be straight line if the model would be valid. The rate constants for different temperatures were calculated from the slope and presented in Table 2, together with the corresponding R^2 values for both first and second order kinetics.

The first-order kinetic equation can be expressed as:

$$\ln([TMP_0/TMP]) = Kt \tag{8}$$

There was an increase in k at higher temperature. This indicated an increase in the reaction rate as the temperature increased for the temperature range studied. Table 3 summarizes the overall rate for both the first and second order kinetics at different temperatures.

Table 3 Comparison between the overall first and second order kinetics with temperature dependence

Rate of reaction	Overall 1st order		Overall 2nd order	
	(%wt/wt min °C)	R^2	(%wt/wt min °C)$^{-1}$	R^2
k_{TMP}	1.895E−02	0.9479	1.00E−01	0.9469

Activation energy

The activation energy of the transesterification reaction was determined based on the well-known Arrhenius Eq. (9):

$$\log_{10}k = (-E^{\pm}/2.303RT) + \log_{10}A \tag{9}$$

where R is the gas constant (kJ/molK), T is temperature in Kelvin. The logarithm of the reaction rate data was plotted as a function of reciprocal of temperature and the activation energy was calculated from the slope of the regression line and presented in Table 4. Activation energy may be defined as the minimum energy required for starting a chemical reaction. The activation energy for JBL synthesis was 13.57 kJ/mol. Mohammad et al. have reported activation energy of 1.65 kJ/mol for Jatropha biolubricant synthesis [26]. The differences in the activation energies could be due to disparities in the reaction temperatures.

A high temperature is needed for high activation energy reaction while low temperature is needed for low activation energy reaction. In kinetic analysis, often times when low activation energy is accompanied with high temperature, it results in high rate constant and hence speeding up the reaction [47]. However, both are within the same range of sensitivity toward temperature [47].

Transesterification thermodynamics

Thermodynamic parameters were systematically determined. The Gibbs free energy (ΔG) of a reaction is a measure of the thermodynamic driving force that makes a reaction to occur. A negative value for ΔG indicates that a reaction can proceed spontaneously without external inputs, while a positive value indicates that it will not. The enthalpy (ΔH) is a measure of the actual energy that is liberated or taken when the reaction occurs. If it is negative, then the reaction gives off energy (exothermic), while if it is positive the reaction requires energy (endothermic). The entropy (ΔS) is a measure of the change in the

Table 4 Activation energy for the synthesis of jatropha biolubricants (JBL)

JBL (this work)		JBL [26]	
E (kJ/mol)	R^2	E (kJ/mol)	R^2
13.57	0.986	1.65	0.861

Table 5 Thermodynamic parameters for JBL synthesis

T (K)	ΔG (KJ/mol)	ΔS (J/mol K)	ΔH (KJ/mol)
353	−30.63		
373	−31.89		
393	−33.16	−23.94	13.52
413	−33.75		
433	−34.52		

possibilities for disorder in the products compared to the reactants. For example, if a solid (an ordered state) reacts with a liquid (a somewhat less ordered state) to form a gas (a highly disordered state), there is normally a large positive change in the entropy for the reaction.

The thermodynamic properties such as free energy (ΔG), enthalpy change (ΔH) and entropy change were determined by applying the second order reaction constant K {(%wt/wt min)$^{-1}$} in the Van 't Hoff Eq. (11) [34]:

$$\Delta G^{\circ} = -RT \ln k \tag{10}$$

$$\ln k = \frac{\Delta S^{\circ}}{R} - \frac{\Delta H^{\circ}}{RT} \tag{11}$$

$\ln k$ was plotted against $1/T$ to determine the enthalpy change, ΔH and entropy change, ΔS as presented in Table 5. The positive value of ΔH in Table 5 indicates an endothermic reaction. The negative value for the Gibbs free energy confirmed the feasibility of the process and that the degree of spontaneity increased with increased temperatures as ΔG° became more negative. The increasing negative value of ΔG° with an increase in temperature indicated that the transesterification reaction became more favourable at higher temperatures. The negative ΔS value means the disorder of the system decreases [34].

Lubricating properties of jatropha biolubricant

The basic functions of a lubricant are friction and wear reduction, heat removal and contaminant suspension. Apart from important application in internal combustion engines, vehicles and industrial gear boxes, compressors, turbines or hydraulic systems, there are vast numbers of other applications, which mostly require specifically tailored lubricants. The physicochemical and performance requirements define a lubricant identity and its ability to perform these functions. The basic properties of JBL are shown in Table 6.

Table 6 Properties of JBL (at 140 °C, 4:1 JME–TMP mole ratio)

Property	Units	JBL	CJO	Method
KV @ 40 °C	cSt	39.45	17.15	ASTM D 445
KV @ 100 °C	cSt	8.51	4.83	ASTM D 445
Viscosity Index (VI)	–	204	233	ASTM D 2270
Pour Point (PP)	°C	−12	−7	ASTM D 97
Flash Point	°C	178	92	ASTM D 93

The most important property of a lubricant is the viscosity. Loosely defined, the viscosity is the fluid's ability to resist motion. Kinematic viscosities (KV) for lubricant are usually determined at 40 and 100 °C. JBL was found to have kinematic viscosities of 39.45 and 8.51 cSt at 40 and 100 °C, respectively. In an earlier work KV of 35.55 and 7.66 cSt were reported at 40 and 100 °C, respectively, for sesame oil based trimethylolpropane ester [34]. Similarly, 35.43 and 7.93 cSt KV were reported for Sesame oil TMP ester [48]. Other researchers have reported the following kinematic viscosities at 40 °C: 39.7–54.1 cSt for TMP esters of palm and palm kernel oils [6], 43.9 cSt for TMP esters of Jatropha curcas oil [26] and 11.2–36.1 cSt for TMP esters of 10-undecenoic acid [49]. They also reported the following kinematic viscosities at 100 °C: 7.7–9.8 cSt [6], 8.7 cSt [11] and 3.2–7.3 cSt [49]. These reports show a good comparison between JBL and other seed oil based lubricant.

An ideal lubricant for most purposes is one that maintains a constant viscosity throughout temperature changes. Variation in the viscosity with change in temperature of a lubricant is determined by viscosity index (VI). The importance of the VI can be shown easily by considering automotive lubricants. Oil having a high VI resists excessive thickening when the engine is cold, and consequently, promotes rapid starting and prompt circulation. It resists excessive thinning when the motor is hot and thus provides full lubrication and prevents excessive oil consumption. A very high VI is a desirable property of lubricant due to its ability to resist oxidation and thermal exposure [50]. The VI of an oil may be determined if its viscosity at any two temperatures is known. The viscosity index of JBL shown in Table 6 was found to be 204 using the kinematic viscosity values at 40 and 100 °C. Sripada [44] reported VI of

193 and 204 for the synthesis of biolubricant from methyl oleate and canola biodiesel, respectively. Similarly, Gryglewicz et al. [51] recorded high VI range of 209–235 for the synthesis of neopentyl glycol and trimethylolpropane esters of olive oil, rapeseed oil and lard fatty acids. Yunus et al. [6] found that TMP esters of palm and palm kernel oils exhibited very high VI of 167–187. According to Ghazi et al. [11] and Rao et al. [49], VI of 180 for Jatropha curcas-derived TMP esters and 162–172 for polyol esters of 10-undecenoic acid were obtained. Furthermore, Åkerman et al. [52] reported that TMP esters of C_5–C_{18} fatty acids had VI of 80–208. When compared with previous results, JBL had shown a relatively favourable VI.

Climatic conditions are important consideration when selecting lubricants. Therefore, viscosities and pour point values are important parameters needed to assess the performance of lubricants [50]. Oil thickens as the temperature falls. At a certain temperature, it no longer flows by its own weight. This temperature is called the pour point. The pour point depends on, e.g. the viscosity and chemical structure of the oil. Oil extracted from Jatropha seed was tested for pour point and viscosity and further compared with JBL. While that of crude Jatropha oil (CJO) was −7 °C, the pour point of JBL was −12 °C, indicating an improvement in pour point as a result of the transesterification reaction. This was due to the presence of polyol group in the TMP and the absence of beta-hydrogen in the final product [8, 53].

Flash point measures the readiness of the oil to ignite momentarily in air and is a consideration for the fire hazard of the oil. Flash point of JBL showed great improvement over the CJO and thus justified the chemical modification of the oil.

The ISO viscosity classification is recommended for industrial applications. The reference temperature of 40 °C represents the operating temperature in machinery. Each subsequent Viscosity grade (VG) within the classification has approximately a 50% higher viscosity, whereas the minimum and maximum values of each grade ranges ±10% from the mid-point. Lubricants are usually identified by their grades. Table 7 presents specification of ISO viscosity grades and the properties of JBL. These specifications were earlier used by Mohammed et al. [39].

Table 7 ISO viscosity grade requirement and properties of JBL (at 140 °C, 4:1 JME–TMP mole ratio)

Property	ISO VG32 [39]	ISO VG46 [39]	ISO VG68 [39]	ISO VG100 [39]	JBL
Viscosity (Cst)					
40 °C	>28.8	>41.1	>61.4	>90	39.45
100 °C	>4.1	>4.1	>4.1	>4.1	8.51
Viscosity Index (VI)	>90	>90	>198	>216	204
Flash Point (°C)					178
Pour point (°C)	<−10	<−10	<−10	<−10	−12

Conclusion

Production of environmentally adapted lubricant basestock through transesterification of JME and TMP using calcium hydroxide catalyst had been achieved. The ester group was confirmed by FTIR. The effects of process parameters such as time, temperature and mole ratio on the synthesis of JBL indicated that temperature and mole ratio were the main factors that affected the transesterification process. After about 3 h of reaction, there was no remarkable increase in yield. It can be concluded that the optimal condition for the esterification of TMP ester was found at 3 h, 140 °C and JME:TMP of 4:1 at 1% wt/wt catalyst loading. Gibbs free energy indicated that the reaction was spontaneous with a second order rate constant of $1.00E{-}01$ $(\%wt/wt\ min\ °C)^{-1}$ and kinetic energy of 13.57 kJ/mol. The pour point for JBL was -12 °C, viscosity of 39.45 and 8.51 cSt at 40 and 100 °C, respectively, with the Viscosity Index of 204. JBL properties complied with ISO VG 32 standard. The resulting properties indicated at the conditions of the experiment, that JBL has a high potential for production of lubricants with slight modifications.

References

1. Fox N, Shachowiak G (2007) Vegetable oil based lubricant. Rev Oxid Tribol Int 40:1035–1044
2. Lal K, Carrick V (1993) Performance testing of lubricants based on high oleic vegetable oils. J Synth Lubr 11(3):189–206
3. Ruzaimah NMK, Suzana Y (2010) Modeling of reaction kinetics for transesterification of palm-based methyl esters with trimethylolpropane. Bioresour Technol 10:5877–5884
4. Hwang HS, Erhan SZ (2001) Modification of epoxidized soybean oil for lubricant formulations with improved oxidative stability and low pour point. JAOCS 78(12):1179–1184
5. Boyde S (2002) Green lubricants. Environmental benefits and impacts of lubrication. Green Chem 4:293–307
6. Yunus R, Fakhrulrazi A, Ooi TL, Iyuke SE, Idris A (2003) Preparation and characterization of trimethylolpropane esters from palm kernel oil methyl esters. J Oil Palm Res 15(2):42–49
7. Lathi PS, Mattiasson B (2007) Green approach for the preparation of biodegradable lubricant base stock from epoxidized vegetable oil. Appl Catal B 69:207–212
8. Sharma BK, Doll KM, Erhan SZ (2008) Ester hydroxy derivatives of methyl oleate: tribological, oxidation and low temperature properties. Bioresour Technol 99:7333–7340
9. Campanella A, Rustoy E, Baldessari A, Baltanas MA (2010) Lubricants from chemically modified vegetable oils. Bioresour Technol 101:245–254
10. Wagner H, Luther R, Mang T (2001) Lubricant base fluids based on renewable raw materials. Their catalytic manufacture and modification. Appl Catal A 221:429–442
11. Mohd. Ghazi TI, Gunam Resul MFM, Idris A (2009) Bioenergy. II. Production of biodegradable lubricant from jatropha curcas and trimethylolpropane. Int J Chem React Eng 7:A68
12. Amit KJ, Amit S (2012) Research approach & prospects of non edible vegetable oil as a potential resource for biolubricant—a review. Adv Eng Appl Sci Int J 1(1):23–32
13. Shahabuddin M, Masjuki HH, Kalam MA (2013) Development of eco-friendly biodegradable biolubricant based on jatropha oil. Centre for Energy Sciences, Faculty of Engineering, University of Malaya, Kuala Lumpur
14. Nurdin S, Misebah FA, Yunus RM, Mahmud MS, Sulaiman AZ (2014) Conversion of Jatropha curcas oil to ester biolubricant using solid catalyst derived from saltwater clam shell waste (SCSW). World Academy of Science, Engineering and Technology. Int J Chem Mol Nucl Mater Metall Eng 8(9):1033–1039
15. Shahabuddin M, Masjuki HH, Kalam MA (2013) Experimental investigation into tribological characteristics of biolubricant formulated from Jatropha oil. In: 5th BSME international conference on thermal engineering. Procedia engineering, vol 56, pp 597–606
16. Aji MM, Kyari SA, Zoaka G (2015) Comparative studies between bio lubricants from jatropha oil, neem oil and mineral lubricant (Engen Super 20W/50). Appl Res J 1(4):252–257
17. Arbain N, Salimon J (2009) Synthesis and characterization of ester trimethylolpropane based Jatropha curcas oil as biolubricant base stock. J Sci Technol 47–58
18. Amit S, Rehman A, Khaira HK (2012) Potential of non edible vegetable oils as an alternative lubricants in automotive applications. Int J Eng Res Appl 2(5):1330–1335
19. Inekwe UV, Odey MO, Gauje B, Dakare AM, Ugwumma CD, Adegbe ES (2012) Fatty acid composition and physicochemical properties of Jatropha Curcas oils from Edo and Kaduna states of Nigeria and India. Ann Biol Res 3(10):4860–4864
20. Kaushik N, Kumar K, Kumar S, Kaushik N, Roy S (2007) Genetic variability and divergence studies in seed traits and oil content of Jatropha (Jatropha curcas L.) accessions. Biomass Bioenerg 31:497–502
21. Wilson P (2010) Biodiesel production from Jatropha curcas: a review. Sci Res Essays 5(14):1796–1808. Available online at http://www.academicjournals.org/SRE. ISSN:1992-2248 (©2010 Academic Journals)
22. Belewu MA, Adekola FA, Adebayo GB, Ameen OM, Muhammed NO, Olaniyan AM, Adekola OF, Musa AK (2010) Physicochemical characteristics of oil and biodiesel from Nigerian and Indian Jatropha curcas seeds. Int J Biol Chem Sci 4(2):524–529
23. Zaku SG, Emmanual SA, Isa AH, Kabir A (2012) Comparative studies on the functional properties of neem, jatropha, castor, and moringa seeds oil as potential feed stocks for biodiesel production in Nigeria. Glob J Sci Front Res Chem 12(7):2249–4626
24. Gubiz GM, Mittelbac M, Trabi M (1997) Biofuels and industrial products from Jatropha curcas. In: Symposium "Jatropha 97", Managua, Nicaragua
25. Muhammad FM, Gunam R, Tinia I, Mohd G, Azni I (2011) Temperature dependence on the synthesis of jatropha biolubricant. IOP Conf Ser Mater Sci Eng 17:012032. doi:10.1088/1757-899X/17/1/012032
26. Mohamad FMGR, Tinia IMG, Azni I (2012) Kinetic study of jatropha biolubricant from transesterification of jatropha curcas oil with trimethylolpropane: effects of temperature. Ind Crop Prod 38:87–92
27. Ghazi TIM, Resul MFG, Idris A (2009) Bioenergy II: production of biodegradable lubricant from jatropha curcas and trimethylolpropane. Int J Chem React Eng 7:1542–6580
28. Bugaje IM, Mohammed IA (2008) Biofuel production technology. Science and Technology Forum (STF), Zaria, Nigeria, 1st edn, 25-200
29. Ramesh D, Samapathrajan A, Venkatachalam P (2004) Production of biodiesel from Jatropha curcas oil using pilot plant. Retrieved from http://www.bioenergy.org.nz/documents/liquid
30. Tewari DN (2007) Jatropha and biodiesel. Ocean Book Ltd, New Delhi
31. Makkar HPS, Becker K (2009) Jatropha curcas, a promising crop

for the generation of biodiesel and value-added coproducts. Eur J Lipid Sci Technol 111:773–787

32. Wirawan SS (2009) Potential of *Jatropha curcas L*. In: Joint task 40/ERIA workshop, 28 Oct 2009, Tsukuba, Japan

33. Tigere TA, Gatsi TS, Mudita II, Chikuvire TJ, Thamangani S, Mavunganidze Z (2006) Potential of *Jatropha curcas* in improving smallholder farmers' livelihoods in Zimbabwe: an exploratory study of Makosa Ward, Mutoko district. J Sustain Dev Afr 8(3):1–9

34. Menkiti MC, Ocheje O, Oyoh KB, Onukwuli OD (2015) Synthesis and tribological evaluation of sesame oil-based trimethylolpropane ester. J Chin Adv Mater Soc 3(2):71–88

35. Surapoj K, Suchada B, Chawalit N (2013) Effects of transesterification conditions on synthesis of trimethylolpropane esters. In: Pure and applied chemistry international conference (PACCON 2013)

36. Yunus R, Fakhru'l-Razi A, Ooi T, Iyuke S, Perez J (2004) Lubrication properties of trimethylolpropane esters based on palm oil and palm kernel oils. Eur J Lipid Sci Technol 60:52–60

37. Yunus R, Ooi TL, Fakhru'l-Razi A, Basri S (2002) A simple capillary column gas chromatography method for analysis of palm oil based polyol esters. J Am Oil Chem Soc 79:1075–1080

38. ASTM Standards (1991) Standard test method for pour point of products D97, vol 05.02. American Society for Testing Materials, Philadelphia, pp 57–64

39. ASTM Standards (1991) Standard test method for kinematic viscosity of transparent and opaque liquids D445, vol 05.02. American Society for Testing Materials, Philadelphia, pp 169–176

40. ASTM Standards (1991) Standard practice for calculating viscosity index from kinematic viscosity at 40 and 100 °C, vol 05.01. American Society for Testing Materials, Philadelphia. doi:10.1520/D2270-93R98

41. ASTM Standards D93-13e1 (2013) Standard test methods for flash point by Pensky–Martens closed cup tester. ASTM International, West Conshohocken

42. Barbara S (2004) Analytical techniques in science. Infrared spectroscopy; fundamentals and application, 1st edn. Wiley, New York

43. Yunus RA, Fakhru'l-Razi OITL, Iyuke SE, Idris A (2003) Development of optimum synthesis method for transesterification of palm oil methyl esters and trimethylolpropane to environmentally acceptable palm oil-based lubricant. J Oil Palm Res 15(2):35–41

44. Sripada PK, Sharma RV, Dalai AK (2013) Comparative study of tribological properties of trimethylolpropane-based biolubricants derived from methyl oleate and canola biodiesel. Ind Crop Prod 50:95–103. doi:10.1016/j.indcrop.2013.07.018

45. Nagendramma P, Kaul S (2012) Development of ecofriendly/ biodegradable lubricants: an overview. Renew Sustain Energy Rev 16:764–774

46. Noureddini H, Zhu D (1997) Kinetic of transesterification of soybean oil. J Am Oil Chem Soc 74(11):1457–1463

47. Levenspiel O (1999) Chemical reaction engineering, 3rd edn. Wiley, New York, pp 13–37

48. Dodos GS, Zannikos F, Lois E (2011) Utilization of sesame oil for the production of bio-based fuels and lubricants. School of Chemical Engineering, Laboratory of Fuel Technology and Lubricants, National Technical University of Athens, 15780, Athens, Greece

49. Padmaja KV, Rao BV, Reddy RK, Bhaskar PS, Singh AK, Prasad RBN (2012) 10-Undecenoic acid-based polyol esters as potential lubricant base stocks. Ind Crop Prod 35:237–240

50. Rudnick L (2006) Synthetic, mineral oils, and bio-based lubricants: chemistry and technology. Taylor & Francis Group, New York

51. Gryglewicz S, Piechocki W, Gryglewicz G (2003) Preparation of polyol esters based on vegetable and animal fats. Bioresour Technol 87:35–39

52. Åkerman CO, Gaber Y, Ghani NA, Lämsä M, Kaul RH (2011) Clean synthesis of biolubricants for low temperature applications using heterogeneous catalysts. J Mol Catal B Enzym 72:263–269

53. Cermak SC, Isbell TA (2003) Synthesis and physical properties of estolide-based functional fluids. Ind Crop Prod 18:183–196

10

Corrosion inhibition of carbon steel in 1 M H$_2$SO$_4$ solution by *Thapsia villosa* extracts

A. Kalla[1] · M. Benahmed[1] · N. Djeddi[1] · S. Akkal[2] · H. Laouer[3]

Abstract Ethyl acetate extract (EAE) and butanolic extract (BE) of *Thapsia villosa* were investigated as corrosion inhibitors of Carbon Steel (CS) in 1 M H$_2$SO$_4$ using electrochemical impedance spectroscopy (EIS) techniques, potentiodynamic polarization and weight loss measurements. The effect of temperature on the corrosion behavior of CS was studied in the range of 20–40 °C. The experimental results show that EAE and BE are good corrosion inhibitors and the protection efficiency increased with increasing concentration of the extracts, but decrease with rise in temperature. The EAE and BE act as a mixed types inhibitors. The adsorption of extracts on CS surface follow Langmuir isotherm. The apparent energies, enthalpies and entropies of the dissolution process were discussed.

Keywords Carbon steel · Corrosion · EIS · Potentiodynamic polarization · Plant extract · Weight loss

✉ M. Benahmed
riad43200@yahoo.fr

[1] Laboratoire des Molécules Bioactives et Applications, Université Larbi Tébessi, Route de Constantine, 12000 Tébessa, Algeria

[2] Laboratoire de Phytochimie et Analyses physicochimiques et Biologiques, Département de Chimie, Faculté de Sciences exactes, Université Mentouri Constantine, Route d'Ain el Bey, 25000 Constantine, Algeria

[3] Laboratoire de Valorisation des Ressources Naturelles Biologiques, Département de Biologie et d'écologie végétale, Université Ferhat Abbas de Sétif 1, Sétif, Algérie

Introduction

Metals and alloys react electrochemically with the environment to form stable compounds, in which the loss of metals occurs. Metallic structures are exposed to conditions that facilitate corrosion processes. Furthermore, hydrochloric and sulfuric acids are widely used for pickling and de-scaling of carbon steel which promote the acceleration of metallic corrosion, causing ecological risks and economic consequences in term of repair, replacement and product losses [1]. Therefore, the prevention of the corrosion is vital not only for the protection of metals but also in decreasing the dispersion of the toxic compounds into the environment [2]. One of the best-known methods for corrosion protection is the use of inhibitors [3]. Organic compounds having functional groups such as -OR, -COOH, -SR and/or NR$_2$ have been reported to inhibit corrosion of metals in acid solutions [4]. The presence of oxygen, sulfur, nitrogen atoms and multiple bonds in organic compounds enhances their adsorption ability and corrosion inhibition efficiency [1]. However, most of these compounds are expensive, toxic and not biodegradable [5]. Therefore, alternative sources of products are preferred. Investigation of plant extracts as corrosion inhibitors is interesting because they are ecologically acceptable and not expensive. Extracts of some plants such as bupleurum lancifolium [6], *Limonium thouinii* [7], and *Punica granatum* [8] have been reported to inhibit the corrosion of metals in acid solutions.

Thapsia villosa, which belongs to the family Apiaceae, grow over a wide area in the West Mediterranean region, including Portugal, Spain, the south of France and the North West of Africa. *Thapsia villosa* has been used in folk medicine as a purgative [9]. *Thapsia villosa* is found to contain phenylpropanoid; 2,3-dlhydroxy-2-methylbutyrlc acids [9], sesquiterpenes [10], terpenes [11] and essential oils [12]. However, it has never been studied for the purpose of corrosion inhibition.

The aim of this work is to investigate the inhibitory effects of ethyl acetate and n-butanol extracts as corrosion inhibitors for carbon steel in acidic solution using the weight loss, potentiodynamic polarization curves and electrochemical impedance measurements.

Experimental

Material

ASTM A179 low carbon steel composed of (wt%): C 0.11 %, Mn 0.52 %, P 0.024 %, S 0.030 % and Fe balance was used in the present study. The steel specimens were taken from the Seamless cold-drawn tube of heat exchanger for petroleum refining. Each sheet was mechanically press cut into specimens in sizes of $3 \times 3 \times 0.2$ cm were used for gravimetric measurements, whereas specimens used for polarization and EIS measurements were imbedded in epoxy resin leaving a working area of 1.0 cm^2.

Preparation of plant extracts

Air-dried aerial parts of *Thapsia villosa* were macerated in methyl alcohol (70 %) at room temperature. The hydro alcoholic solutions were concentrated under reduced pressure to dryness and the residue was dissolved in hot water and kept in cold overnight. After filtration, the residue was successively treated with ethyl acetate and n-butanol. Then, the solvents were removed to afford ethyl acetate and n-butanol extracts [13, 14]. The ethyl acetate extract (EAE) and n-butanol extracts (BE) were then used directly in the experiments.

Solution

The aggressive solution of 1 M H$_2$SO$_4$ was prepared by H$_2$SO$_4$ 98 % (Merck) with distilled water. The concentration range of the EAE and BE employed varied from 100 to 800 ppm.

Electrochemical measurements

The Electrochemical experiments were carried out in the conventional three-electrode cell consisting of a CS as working electrode, a platinum rod as counter electrode and a saturated calomel electrode (SCE) as a reference electrode. Before measurement the working electrode was immersed in test solution at open circuit potential (OCP) for 30 min to ensure OCP to reach steady state.

Electrochemical impedance spectroscopy (EIS) was carried out at the OCP of each sample was immersed for 30 min over a frequency range of 100 kHz–10 mHz with a signal amplitude perturbation of 10 mV. Inhibition efficiency ($\eta_R\%$) was estimated using the following relation:

$$\eta_R \% = \frac{R'_p - R_p}{R'_p} \times 100 \qquad (1)$$

where R_p and R'_p are polarizations resistors in the absence and presence of the inhibitor, respectively.

The potential of potentiodynamic polarization curves was started from cathodic potential of -250 mV to anodic potential of $+250$ mV vs. OCP at a sweep rate of 1.0 mV s^{-1}. Inhibition efficiency ($\eta_p\%$) was defined as [6]:

$$\eta_p \% = \frac{i_{corr} - i'_{corr}}{i_{corr}} \times 100 \qquad (2)$$

where i_{corr} and i'_{corr} represent corrosion current density values in absence and presence of inhibitor, respectively.

All electrochemical measurements were performed using a computer-controlled instrument, Voltalab-PGZ 301 with Voltamaster (ver 7.0.8) software. The above electrochemical tests were conducted for each concentration of *Thapsia villosa* extracts at different temperatures. Each experiment was repeated at least three times to check the reproducibility.

Weight loss measurements

CS specimens were abraded with a series of SiC paper, washed with distilled water, degreased with acetone and dried with a cold air stream. Experiments were realized under total immersion in stagnant aerated condition at 20–40 °C. The specimens were weighed and suspended in beakers. After 7 h, these coupons were taken out, washed, dried and weighed accurately. From the weight loss data, the corrosion rate (CR) was calculated from the following equation [15]:

$$CR = \frac{W}{At} \qquad (3)$$

where W is the average weight loss, A is the total area of the specimen and t is immersion time (7 h). The inhibition efficiency (η_w) was calculated as follows:

$$\eta_w \% = \frac{CR - CR'}{CR} \times 100 \qquad (4)$$

where CR and CR$'$ represent the values of corrosion rate in absence and presence of inhibitor respectively.

Results and discussion

EIS measurements

Figure 1 shows the EIS response of CS in 1 M H_2SO_4 solution without and with various concentrations of EAE and BE at 20 °C, represented via Nyquist plots. Only one capacitive loop at the higher frequency range is observed which means that the corrosion of CS is controlled by the charge transfer process [16, 17]. The increasing diameter of loop obtained in 1 M H_2SO_4 in the presence of EAE and BE indicated the corrosion inhibition and the strengthening of inhibitor film [18]. These loops are not perfect semi circles which can be attributed to the frequency dispersion effect as a result of the roughness and inhomogeneous of metal surface [19, 20]. Due to non-ideal frequency response the capacitance is usually replaced by a constant phase element (CPE) [19], whose impedance is given by [21]:

$$Z_{CPE} = \frac{1}{Q(j\omega)^n} \tag{5}$$

where Q is the magnitude of the CPE, ω is the angular frequency ($\omega = 2\pi f$, where f is the AC frequency), j is the imaginary unit, and n is the deviation parameter of the CPE: $0 \leq n \leq 1$, for $n = 1$, Eq. (5) agrees to the impedance of an ideal capacitor, where Q is identified with the capacity.

A simple electrical equivalent circuit (EEC) has been proposed to model the experimental data. The EEC depicted in Fig. 2 is employed to analyze the impedance spectra, where R_1 represents the solution resistance, R_2 denotes the charge-transfer resistance, and a CPE instead of a pure capacitor represents the interfacial capacitance. The values of the interfacial capacitance C_{dl} can be calculated from CPE parameter and polarization resistor according to the following equation [22, 23]:

$$C_{dl} = R_p^{\frac{1-n}{n}} Q^{\frac{1}{n}} \tag{6}$$

where R_p is the polarization resistor. The values of parameters such as R_p, Q, n and χ^2, obtained from fitting the recorded EIS as well as the derived parameters C_{dl} are listed in Table 1. The Chi-squared (χ^2) is used to evaluate the precision of the fitted data. Inspection of Table 1 reveals that the χ^2 values are low, which indicates that the fitted data have good agreement with the experimental data. It is observed that R_p values increased and the C_{dl} values decreased with increasing inhibitors concentration. The increase in R_p values can be attributed to the adsorption of the inhibitors on the metal surface leading to the formation of protective film on the metal surface and thus decreases the extent of the dissolution reaction [24]. The decrease in the C_{dl} values may be due to the increase in the thickness of the electric double layer [1]. The inhibition efficiency ($\eta_p \%$) was achieved at (60 %) and (80 %) for EAE and BE, respectively.

Potentiodynamic polarization curves

Polarization curves were obtained for CS in 1 M H_2SO_4 solution without and with the inhibitor. Tafel plots obtained in different concentrations of EAE and BE solutions at

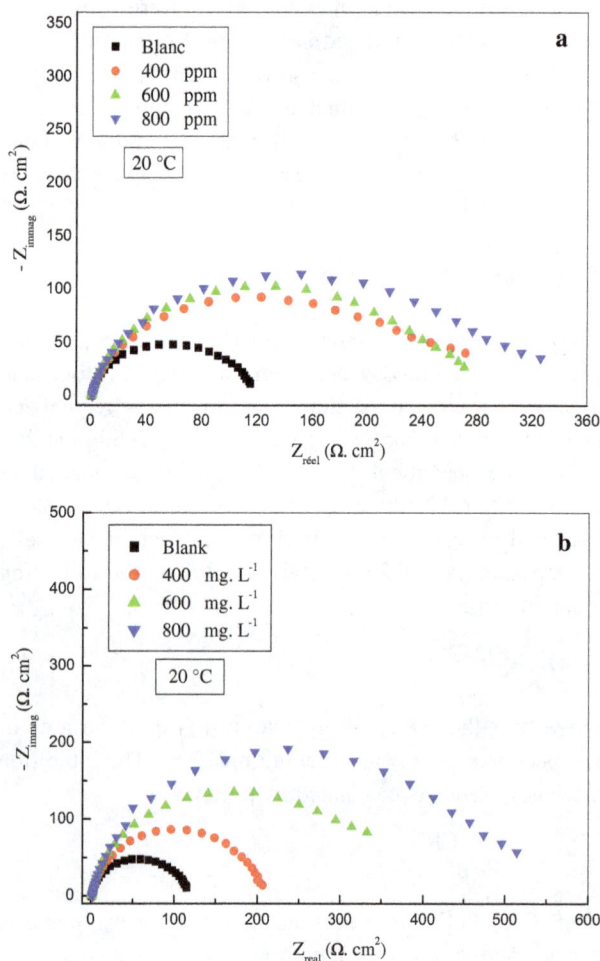

Fig. 1 Experimental Nyquist plots for in 1 M H_2SO_4 **a** with and without EAE, **b** with and without BE

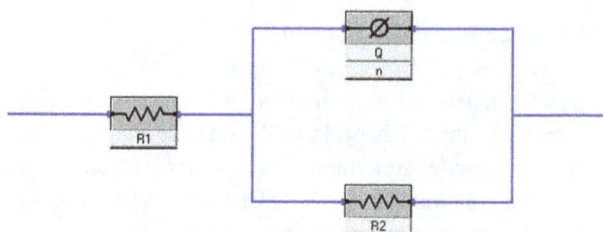

Fig. 2 Equivalent circuit used to fit the capacitive loop

Table 1 Impedance parameters and inhibition efficiency values for CS in 1 M H$_2$SO$_4$ containing different concentrations of EAE and BE at 20 °C

Electrochemical impedance parameters

C (ppm)	EAE							BE						
	R_p (Ω cm^2)	10^4 Q (Ω^{-1} S^n cm^{-2})	n	C_{dl} (μF cm^{-2})	χ^2 (10^{-2})	θ	η_R (%)	R_p (Ω cm^2)	10^4 Q (Ω^{-1} S^n cm^{-2})	n	C_{dl} (μF cm^{-2})	χ^2 (10^{-2})	θ	η_R (%)
0	129.5	10.34	0.715	287	1.38	–	–	129.5	10.34	0.715	287	1.38	–	–
400	270.0	7.46	0.765	139	0.58	0.5203	52.03	301.1	4.20	0.865	125	0.52	0.5699	56.99
600	293.4	4.42	0.864	128	1.05	0.5586	55.86	387.3	4.22	0.781	95	1.50	0.6656	66.56
800	331.2	1.94	0.882	114	0.55	0.6090	60.90	488.5	1.65	0.758	75	0.69	0.7349	73.49

20 °C were shown in Fig. 3. The electrochemical parameters including corrosion potential (E_{corr}), corrosion current density (i_{corr}), anodic and cathodic Tafel slopes (β_a and β_c), surface coverage values (θ) and inhibition efficiency ($\eta_p = \theta \times 100$) are presented in Table 2.

It is clear from Fig. 3 and Table 2 that, the addition of both EAE and BE to the acid solution causes a remarkable decrease in the corrosion rate predominantly shifts the cathodic curves to lower values of current densities; it may be due to the adsorption of organic compounds present in the extracts at the active sites of CS surface, retarding both metallic dissolution and hydrogen evolution reactions and consequently slowed down the corrosion process [25]. The structure and functional groups of the inhibitors play prominent roles during the adsorption process [1]. Inspection of Table 2 showed that both anodic and cathodic Tafel slopes do not change remarkably upon addition of EAE and BE, which indicates that the extracts act as a mixed type inhibitor for the corrosion of C steel. The values of inhibition efficiency ($\eta_p\%$) determined using potentiodynamic

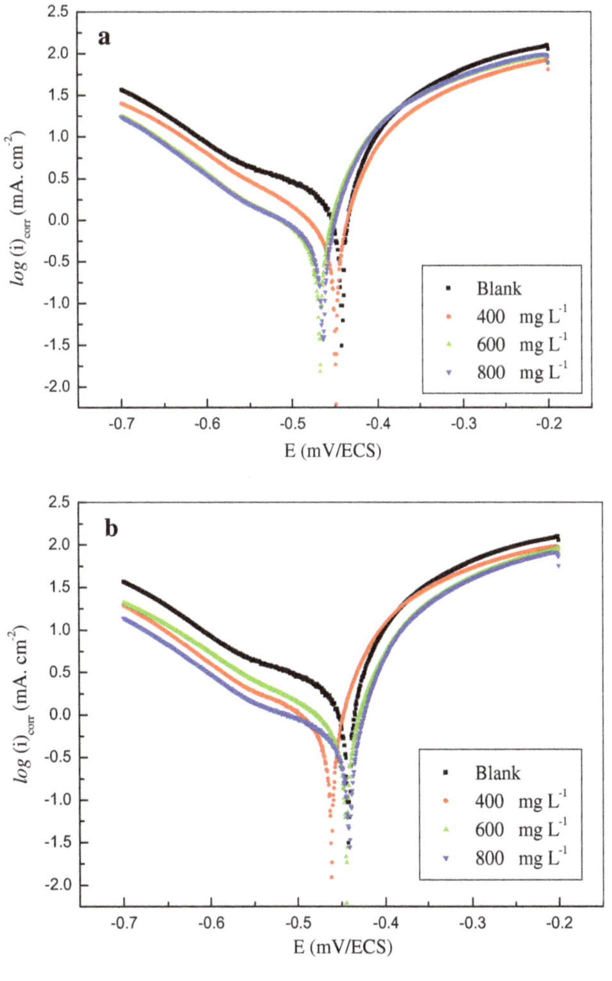

Fig. 3 Potentiodynamic polarization curves for CS in 1 M H$_2$SO$_4$ **a** with and without EAE, **b** with and without BE

Table 2 Polarization parameters and corresponding inhibition efficiency for the corrosion of CS in 1 M H₂SO₄ containing different concentrations of EAE and BE at 20 °C

C (ppm)	Polarization parameters											
	EAE						BE					
	$-E_{corr}$ (mV)	i_{corr} (mA cm⁻²)	β_a (mV dec⁻¹)	$-\beta_c$ (mV dec⁻¹)	θ	η_p (%)	$-E_{corr}$ (mV)	i_{corr} (mA cm⁻²)	β_a (mV dec⁻¹)	$-\beta_c$ (mV dec⁻¹)	θ	η_p (%)
0	441.2	1.0308	42.4	109.1	–	–	441.2	1.0308	42.4	109.1	–	–
400	501.3	0.5052	46.7	75.6	0.5099	50.99	460.6	0.4405	37.6	99.0	0.5727	57.27
600	458.7	0.4565	37.9	72.6	0.5571	55.71	443.7	0.3780	41.3	94.5	0.6333	63.33
800	463.1	0.4082	37.4	70.3	0.6040	60.40	440.3	0.2671	37.6	92.3	0.7409	74.09

polarization are in good agreement with those obtained from EIS measurements.

Weight loss measurements

Effect of concentration and temperature on corrosion rate and inhibition efficiency

The weight loss expressed as the corrosion rate (CR) for the CS specimens in 1 M H₂SO₄ solution containing different concentrations of *Thapsia villosa* extracts (EAE and BE) as a function of inhibitor concentration in the temperature range of 20–40 °C is showed in Fig. 4. Inspection of the plots revealed that CR decreases noticeably with increase in both of EAE and BE concentrations, indicating that the addition of plant extracts retard the dissolution process of CS.

In similar experimental conditions, the influence of temperature on CR was studied. The results presented in the Table 3 and Fig. 4 show that the CR increases with temperature both in uninhibited and inhibited solutions,

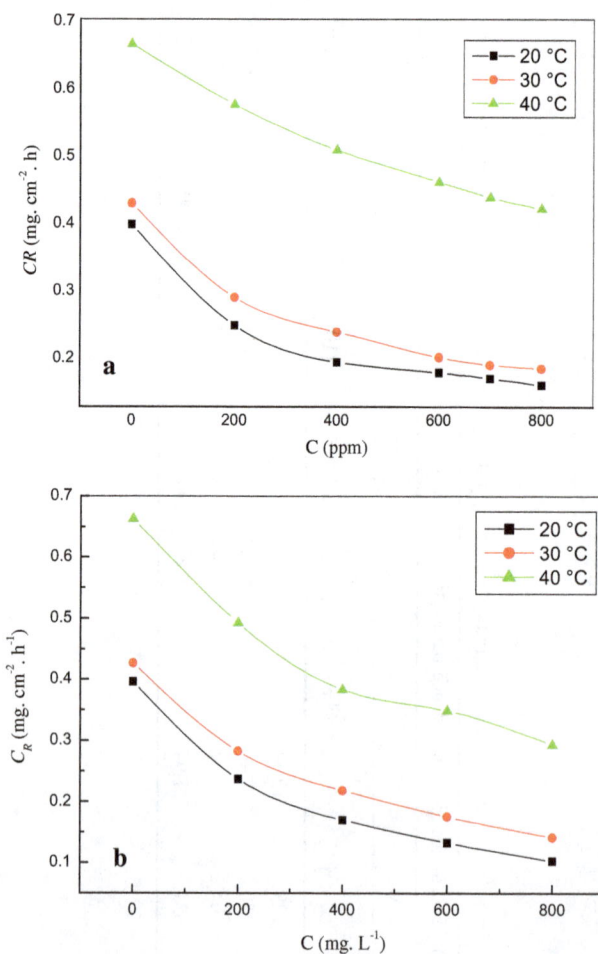

Fig. 4 Relationship between corrosion rate and concentration of: **a** EAE in 1 M H₂SO₄, **b** BE in 1 M H₂SO₄

Table 3 Corrosion parameters obtained from weight loss of CS in 1 M H_2SO_4 containing various concentrations of EAE and BE at different temperatures

| | C (ppm) | Temperature (°C) | | | | | | | | |
| | | 20 | | | 30 | | | 40 | | |
		CR (mg cm^{-2} h^{-1})	θ	η_w (%)	CR (mg cm^{-2} h^{-1})	θ	η_w (%)	CR (mg cm^{-2} h^{-1})	θ	η_w (%)
EAE	Blank	0.3957	–	–	0.4271	–	–	0.663	–	–
	200	0.2457	0.3791	37.91	0.2871	0.3278	32.78	0.5734	0.1351	13.51
	400	0.1914	0.5163	51.63	0.2357	0.4481	44.81	0.5057	0.2372	23.72
	600	0.1757	0.5560	55.60	0.1986	0.5350	53.50	0.4586	0.3083	30.83
	700	0.1671	0.5777	57.77	0.1871	0.5619	56.19	0.4357	0.3428	34.28
	800	0.1571	0.6030	60.30	0.1814	0.5753	57.53	0.4186	0.3686	36.86
BE	Blank	0.3957	–	–	0.4271	–	–	0.663	–	–
	200	0.2371	0.4008	40.08	0.2828	0.3378	33.78	0.4928	0.2567	25.67
	400	0.1700	0.5704	57.04	0.2186	0.4882	48.82	0.3828	0.3567	35.67
	600	0.1328	0.6644	66.44	0.1757	0.5886	58.86	0.3486	0.4742	47.42
	800	0.1028	0.7402	74.02	0.1414	0.6689	66.89	0.2928	0.5583	55.83
	800	0.1028	0.7402	74.02	0.1414	0.6689	66.89	0.2928	0.5583	55.83

and goes up more rapidly at the higher temperature; the rise in temperature usually accelerates the corrosion reactions which results in higher dissolution rates of the metal.

The variation of inhibition efficiency (η_w%) with temperature and plant extracts concentrations is shown in Table 3 and Fig. 5. It is clear from Fig. 5 that η_w% increases with the increase in EAE and BE concentration, while it decreased with increase in temperature. This can be attributed to increased rate of desorption of phytochemical compounds from the surface of CS with increasing temperature because these two opposite processes are in equilibrium [26, 27]. Several authors have reported similar observation and the plant extracts were believed to be physically adsorbed on the CS surface [26, 28, 29].

At the EAE concentration of 800 ppm, the maximum EI % in 1 M H_2SO_4 is 60 % at 20 °C; 57 % at 30 °C; and 37 % at 40 °C. While at the same concentration of BE, the maximum EI % in 1 M H_2SO_4 is 74 % at 20 °C; 67 % at 30 °C; and 56 % at 40 °C. The results indicate that both extracts are good inhibitors for CS in 1 M H_2SO_4 solution and the maximum inhibition efficiency was achieved using BE.

Adsorption isotherm

The decrease in CR by addition of EAE and BE is attributed to either adsorption of the plant component on the CS surface [30]. To evaluate the adsorption process of phytochemical components on the CS surface, Langmuir,

Temkin and Freundlich isotherms were obtained according to following equations:

$$\text{Langmuir: } \frac{C}{\theta} = \frac{1}{K_{ads}} + C \tag{7}$$

$$\text{Temkin: } \theta = \frac{1}{\alpha} \log K_{ads} C \tag{8}$$

$$\text{Freundlich: } \log \theta = \log K_{ads} + \alpha \log C \tag{9}$$

where: C is the concentration of inhibitor, K_{ads} is the adsorption equilibrium constant, θ is the surface coverage, α is the adsorbate parameter

The correlation coefficient (r^2), presented in the Table 4, was used to choose the isotherm that best fit experimental data. Best results from the plots were obtained for Langmuir adsorption isotherm, that suggests monolayer adsorption of both EAE and BE on the CS surface at all temperatures.

Figure 6a, b show the straight lines of C/θ versus C, deviate from unity for EAE at 20–40 °C, indicates that the interaction force between phytochemical compounds on the CS surface cannot be neglected [28, 31], and each molecule occupies more than one adsorption site on the metal surface [32]. A modified Langmuir adsorption isotherm could be applied to this phenomenon, which is given by following equation [33]:

$$\frac{C}{\theta} = \frac{n}{K_{ads}} + nC \tag{10}$$

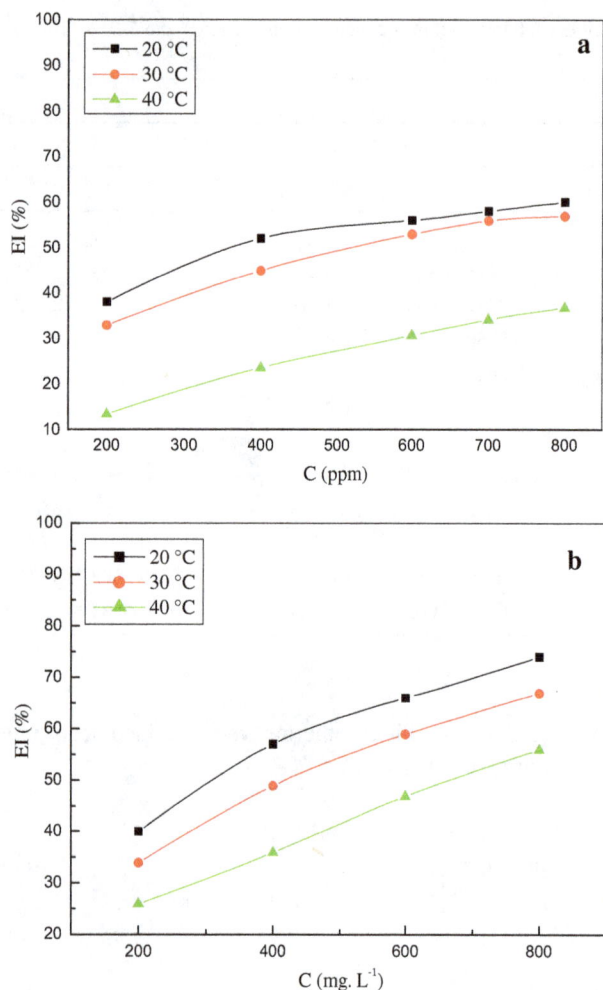

Fig. 5 Effect of temperature and concentration on the inhibition efficiency of: **a** EAE in 1 M H_2SO_4, **b** BE in 1 M H_2SO_4

Table 4 Correlation coefficient (r^2)

Isotherm	EAE			BE		
	20 °C	30 °C	40 °C	20 °C	30 °C	40 °C
Langmuir	0.999	0.999	0.999	0.999	0.999	0.998
Temkin	0.991	0.998	0.997	0.936	0.990	0.978
Freundlich	0.984	0.976	0.998	0.997	0.997	0.966

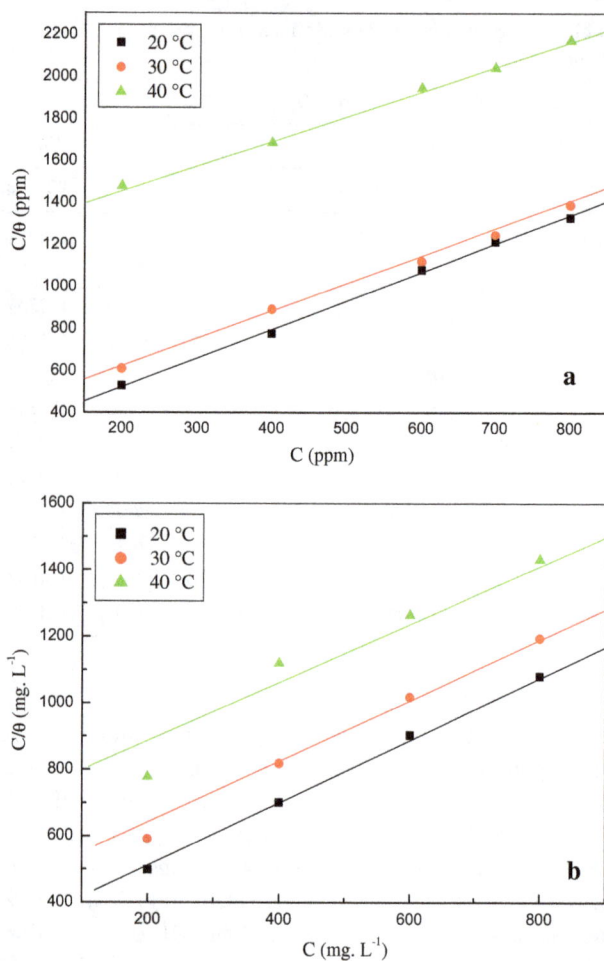

Fig. 6 Langmuir isotherm adsorption for CS in 1 M H_2SO_4 at different temperatures of **a** the EAE **b** of the BE

On the contrary, for the BE, the slope almost equals to unity, which suggests that the interaction of adsorbed species is negligible [34].

The adsorptive equilibrium constant (K_{ads}) listed in Table 5 was estimated from intercept of the Langmuir isotherm plot. The values of equilibrium constant decrease with rise in temperature, which may be attributed to desorption of inhibitor components at higher temperature [35, 36].

Table 5 Parameters of the linear regression between C/θ and C in 1 M H_2SO_4

T (°C)	Parameters of the linear regression					
	EAE			BE		
	Slope	r^2	K_{ads} (L mg^{-1})	Slope	r^2	K_{ads} (L mg^{-1})
20	1.36	0.999	5.47×10^{-3}	0.93	0.999	3.09×10^{-3}
30	1.27	0.999	3.49×10^{-3}	1.02	0.999	2.51×10^{-3}
40	1.15	0.999	0.93×10^{-3}	1.08	0.998	1.64×10^{-3}

Table 6 Polarization parameters and corresponding inhibition efficiency for the corrosion of CS in 1 M H_2SO_4 without and with EAE and BE at (20–40) °C

Polarization parameters at (20–40) °C

T (°C)	C (ppm)	EAE							BE					
		$-E_{corr}$ (mV)	i_{corr} (mA cm^{-2})	β_a (mV dec^{-1})	$-\beta_c$ (mV dec^{-1})	θ	η_p (%)	C ppm	$-E_{corr}$ (mV)	i_{corr} (mA cm^{-2})	β_a (mV dec^{-1})	$-\beta_c$ (mV dec^{-1})	θ	η_p (%)
20	Blank	441.2	1.0308	42.4	109.1	–	–	0	441.2	1.0308	42.4	109.1	–	–
	800	463.1	0.4082	37.4	70.3	0.6040	60.40	800	440.9	0.2700	37.6	92.3	0.7403	74
30	Blank	430.8	1.0998	43.8	136.0	–	–	0	430.8	1.0998	43.8	136.0	–	–
	800	425.3	0.4710	66.5	54.2	0.5717	57.17	800	471.1	0.3483	34.6	94.2	0.6689	66.89
40	Blank	430.0	1.1019	48.0	137.7	–	–	0	430.0	1.1019	48.0	137.7	–	–
	800	422.2	0.7017	45.6	108.0	0.3632	36.32	800	451.0	0.4630	39.8	96.4	0.5583	55.83

Table 7 Impedance parameters and inhibition efficiency values for CS without and with EAE and BE at (20–40) °C

Polarization parameters at (20–40) °C

T (°C)	C (ppm)	EAE							BE						
		R_p (Ω cm^2)	$10\,Q$ (Ω^{-1} S^n cm^{-2})	n	C_{dl} (μF cm^{-2})	χ^2 $(10^{-2})^2$	θ	η_R (%)	R_p (Ω cm^2)	$10^4\,Q$ Ω^{-1} S^n cm^{-2}	n	C_{dl} (μF cm^{-2})	χ^2 (10^{-2})	θ	η_R (%)
20	Blank	129.5	10.34	0.715	287	1.38	–	–	129.5	10.34	0.715	287	1.38	–	–
	800	331.2	1.94	0.882	114	0.55	0.6090	60.90	651.4	1.10	0.755	60	0.70	0.8012	75.12
30	Blank	110.5	10.60	0.717	302	2.14	–	–	110.5	10.60	0.717	302	2.14	–	–
	800	264.0	2.32	0.827	129	1.04	0.5814	58.14	357.9	1.98	0.700	91	1.73	0.6912	69.12
40	Blank	92.42	13.0	0.684	314	1.85	–	–	80.42	13.0	0.684	314	1.85	–	–
	800	144.6	6.01	0.620	201	0.87	0.3608	36.08	188.3	7.02	0.712	133	1.30	0.5729	57.29

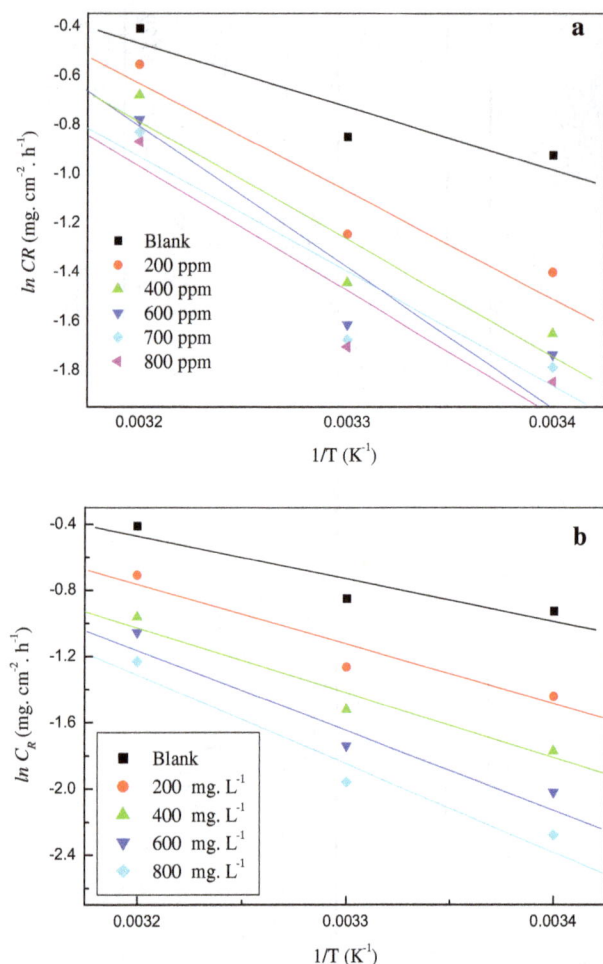

found that the corrosion current density (i_{corr}) increased but the polarization resistance (R_p) and the inhibition efficiency decreased with increasing temperature. The decrease in inhibition efficiency reveals that the film formed on the metal surface is less protective at higher temperatures, since desorption rate of the inhibitor is greater at higher temperatures [37].

The activation parameters were calculated from Arrhenius equation:

$$CR = A exp\left(-\frac{E_a^\circ}{RT}\right) \quad (11)$$

where CR is corrosion rate, E_a° is the apparent activation energy of the CS dissolution and A is the Arrhenius pre-exponential factor. The apparent activation energy was calculated from the plots of logarithm of CR versus $1/T$ (Fig. 7) and shown in Table 8. It can be seen in the Table 8 that E_a° is higher in the presence of the inhibitors than in their absence and increased with the increase in concentration of EAE and BE, which indicate a strong adsorption of the inhibitor molecules at the CS surface [1].

An alternative form of Arrhenius equation is the transition-state equation [38, 39]:

$$CR = \frac{RT}{N_A h} exp\frac{\eta S_a^\circ}{R} exp(-\frac{\eta H_a^\circ}{RT}) \quad (12)$$

where h is the plank's constant, N_A is Avogadro's number, ηS_a° the entropy of activation and ηH_a° is the enthalpy of activation. Figure 8 shows a plot of $ln CR/T$ vs. $1/T$. Straight lines were obtained with a slope of $-\frac{\eta H_a^\circ}{RT}$ and an intercepts of ($ln\frac{R}{N_A h} + \frac{\eta S_a^\circ}{R}$), from which the values of ηS_a° and ηH_a° were calculated and listed in Table 8. For both ethyl acetate and n-butanol extracts, the positive signs of enthalpies reflect the endothermic nature of the dissolution process [39, 40].

Fig. 7 Arrhenius plots related to the corrosion rate for CS in 1 M H$_2$SO$_4$ **a** EAE, **b** BE

Effect of the temperature

The effect of temperature on the rate of the CS corrosion process using electrochemical measurements was studied in 1 M H$_2$SO$_4$ alone and in the presence of EAE and BE. Corresponding data are given in Tables 6 and 7. It was

Table 8 The values of activation parameters E_a°, ΔH_a° and ΔS_a° for CS in 1 M H$_2$SO$_4$ at different concentrations of EAE and BE at different temperatures

Activation parameters E_a°, ΔH_a° ΔS_a° and at (20–40) °C						
C (ppm)	EAE			BE		
	E_a° (kJ mol^{-1})	ΔH_a° (kJ mol^{-1})	ΔS_a° (J mol^{-1} K^{-1})	E_a° (kJ mol^{-1})	ΔH_a° (kJ mol^{-1})	ΔS_a° (J mol^{-1} K^{-1})
Blank	21.45	18.71	−15.58	21.45	18.71	−15.58
200	35.23	32.49	27.06	30.42	27.69	10.59
400	40.39	37.64	42.49	33.74	31.00	19.25
600	39.88	37.13	39.82	40.12	37.37	38.74
800	39.84	37.09	39.24	43.51	40.76	48.13
1000	–	–	–	51.73	48.98	74.01

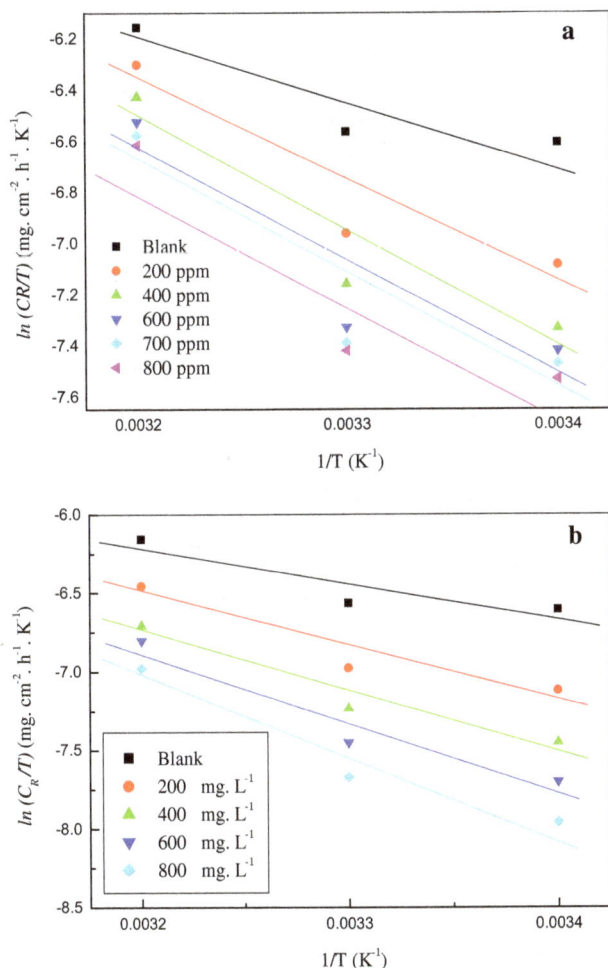

Fig. 8 Arrhenius plots of corrosion $\ln(CR/T)$ vs. $1/T$ for CS in 1 M H_2SO_4 **a** with and without EAE, **b** with and without BE

It is evident from data listed in Table 8 that, the values of E_a^0 are larger than corresponding values of ηH_a^0 indicating the corrosion process involved a gaseous reaction, simply the hydrogen evolution reaction, associated with a decrease in total reaction volume [39, 40]. Moreover, the difference value $E_a^0 - \eta H_a^0$ is 2.74 kJ/mol, which is approximately equal to the average value of RT. Therefore, this shows that the corrosion process is a unimolecular reaction as it is characterized by the equation:

$$E_a^0 - \eta H_a^0 = RT \tag{13}$$

Investigation of Table 8 reveals that the sign of ηS_a° is negative in free acid solution, whereas it becomes positive with the addition of both extracts, this suggest that the adsorption of organic inhibitor molecules is accompanied by desorption of water molecules from the steel surface [41]. Hence, the gain in entropy is attributed to the increase in solvent entropy and to more positive water desorption enthalpy [26]. The positive values of the entropy, related to

substitutional adsorption, can be attributed more to the increase of adsorbed inhibitor molecules rather than the decrease of water molecule desorption [41].

Conclusion

It can be concluded as follows:

- EAE and BE of *Thapsia villosa* act as good inhibitors for the corrosion of CS in 1 M H_2SO_4 solution.
- The inhibition efficiency of all electrochemical tests and weight loss measurements were in good agreement.
- The inhibition efficiency of CS in acid solution increased with increases in concentrations of the EAE and BE and decreases with rising temperature.
- The adsorption of organic molecules on the CS surface obeys the Langmuir adsorption isotherm.
- Potentiodynamic polarization measurements demonstrate that EAE and BE act as a mixed-type inhibitions.
- The apparent activation energy (E_a) of CS dissolution increases in presence of *Thapsia villosa* extracts.

References

1. Kumar CBP, Mohana KN (2014) Corrosion inhibition efficiency and adsorption characteristics of some Schiff bases at mild steel/hydrochloric acid interface. J Taiwan Inst Chem Eng 45:1031–1042
2. Solmaz R (2010) Investigation of the inhibition effect of 5-((E)-4-phenylbuta-1,3-dienylideneamino)-   1,3,4-thiadiazole-2-thiol Schiff base on mild steel corrosion in hydrochloric acid. Corros Sci 52:3321–3330
3. Doner A, Sahin EA, Kardas G et al (2013) Investigation of corrosion inhibition effect of 3-[(2-hydroxybenzylidene)-amino]-2-thioxio-thiazolidin-4-one on corrosion of mild steel in the acidic medium. Corros Sci 66:278–284
4. Abiola OK, Tobun Y (2010) *Cocos nucifera* L. water as green corrosion inhibitor for acid corrosion of aluminium in HCl solution. Chin Chem Lett 21:1449–1452
5. Chevalier M, Robert F, Amusant N et al (2014) Enhanced corrosion resistance of mild steel in 1 M hydrochloric acid solution by alkaloids extract from *Aniba rosaeodora* plant: electrochemical, phytochemical and XPS studies. Electrochim Acta 131:96–105
6. Benahmed M, Selatnia I, Achouri A et al (2015) Steel corrosion inhibition by *Bupleurum lancifolium* (Apiaceae) extract in acid solution. Trans Indian Inst Met 68:393–401
7. Benahmed M, Lafhal M, Djeddi N et al (2012) Inhibition of the corrosion of carbon steel in acid solution by the extract of *Limonium thouinii* (Plumbaginaceae). Adv Environ Biol 6:4052–4056
8. Behpour M, Ghoreishi SM, Khyatkashani M et al (2012) Green approach to corrosion inhibition of mild steel in two acidic solutions by the extract of *Punica granatum* peel and main constituents. Mater Chem Phys 131:621–633
9. Teresa JDP, Pascual MD, Arias A et al (1985) Helmanticine, a phenylpropanoid from *Thapsia villosa*. Phytochemistry 24(8):1773–1778
10. Rubal JJ, Guerra FM, Moreno-Dorado FJ et al (2004) Sulfur-containing sesquiterpenes from *Thapsia villosa*. Tetrahedron 60:159–164

11. Lemmich E, Smitt UW, Jensen JS et al (1991) Guaiane esters from *Thapsia villosa*. Phytochemistry 30:2987–2990
12. Avato P, Trabace G, Smitt UW (1996) Essential oils from fruits of three types of *Thapsia villosa*. Phytochemistry 43:609–612
13. Benahmed M, Akkal S, Louaar S et al (2006) A new furanocoumarin glycoside from *Carum montanum* (Apiaceae). Biochem Syst Ecol 34:645–647
14. Benahmed M, Akkal S, Elomri A et al (2014) Constituents from *Bupleurum montanum* (Coss. & Dur.) (Apiaceae). Arab J Chem 7:1065–1069
15. Djeddi N, Benahmed M, Makhloufi E et al (2015) Study on methylene dichloride and butanolic extracts of *Reutera lutea* (Desf.) Maire (Apiaceae) as effective corrosion inhibitions for carbon steel in HCl solution. Res Chem Intermed 41:4595–4616
16. Raja PB, Qureshi AK, Rahim AA et al (2013) *Neolamarckia cadamb*a alkaloids as eco-friendly corrosion inhibitors for mild steel in 1 M HCl media. Corros Sci 69:292–301
17. Tang Y, Zhang F, Hu S et al (2013) Novel benzimidazole derivatives as corrosion inhibitors of mild steel in the acidic media. Part I: gravimetric, electrochemical, SEM and XPS studies. Corros Sci 74:271–282
18. Muthukrishnan P, Jeyaprabha B, Prakash P (2013) Adsorption and corrosion inhibiting behavior of *Lannea coromandelica* leaf extract on mild steel corrosion. Arab J, Chem
19. Li L, Zhang X, Lei J et al (2012) Adsorption and corrosion inhibition of *Osmanthus fragran* leaves extract on carbon steel. Corros Sci 63:82–90
20. Shuduan D, Li X (2012) Inhibition by Ginkgo leaves extract of the corrosion of steel in HCl and H_2SO_4 solutions. Corros Sci 55:407–415
21. Qian B, Hou B, Zheng M (2013) The inhibition effect of tannic acid on mild steel corrosion in seawater wet/dry cyclic conditions. Corros Sci 72:1–9
22. Lebrini M, Robert F, Lecante A et al (2011) Corrosion inhibition of C38 steel in 1 M hydrochloric acid medium by alkaloids extract from *Oxandra asbeckil* plant. Corros Sci 53:687–695
23. Lecante A, Robert F, Blandinières PA et al (2011) Anti-corrosive properties of *S. tinctoria and G. ouregou* alkaloid extracts on low carbon steel. Curr Appl Phys 11:714–724
24. Bentiss F, Traisnel M, Lagrenee M (2000) The substituted 1,3,4-oxadiazoles:a new class of corrosion inhibitors of mild steel in acidic media. Corros Sci 42:127–146
25. Torres VV, Amado RS, de Sá CF et al (2011) Inhibitory action of aqueous coffee ground extracts on the corrosion of carbon steel in HCl solution. Corros Sci 53:2385–2392
26. Ramya K, Mohan R, Anupama KK et al (2015) Electrochemical and theoretical studies on the synergistic interaction and corrosion inhibition of alkyl benzimidazoles and thiosemicarbazide pair on mild steel in hydrochloric acid. Mater Chem Phys 14(9–150):632–647
27. Nazeer AA, Shalabi K, Fouda AS (2015) Corrosion inhibition of carbon steel by Roselle extract in hydrochloric acid: electrochemical and surface study. Res Chem Intermed 41:4833–4850
28. Odewunmi NA, Umoren SA, Gasem ZM (2015) Watermelon waste products as green corrosion inhibitors for mild steel in HCl solution. J Environ Chem Eng 3:286–296
29. Eduok UM, Khaled M (2015) Corrosion inhibition for low-carbon steel in 1 M H2SO4 solution by phenytoin: evaluation of the inhibition potency of another "anticorrosive drug". Res Chem Intermed 41:6309–6324
30. Rochdi A, Touir R, Bakri ME (2015) Protection of low carbon steel by oxadiazole derivatives and biocide against corrosion in simulated cooling water system. J Environ Chem Eng 3:233–242
31. Mu GN, Li XH, Qu Q et al (2006) Molybdate and tungstate as corrosion inhibitors for cold rolling steel in hydrochloric acid solution. Corros Sci 48:445–459
32. Obot IB, Umoren SA, Gasem ZM et al (2015) Theoretical prediction and electrochemical evaluation of vinylimidazole as corrosion inhibitors for mild steel in 1 M HCl. J Ind Eng Chem 21:1328–1339
33. Vilamiml RFV, Corio P, Rubin JC et al (1999) Effect of sodium dodecylsulfate on copper corrosion in sulfuric acid media in the absence and presence of benzotriazole. J Electroanal Chem 472:112
34. Li X, Deng S, Fu H et al (2014) Synergistic inhibition effects of bamboo leaf extract/major components and iodide ion on the corrosion of steel in H3PO4 solution. Corros Sci 78:29–42
35. Li X, Deng S, Xie X (2014) Inhibition effect of tetradecylpyridinium bromide on the corrosion of cold rolled steel in 7.0 M H_3PO_4. Arab J Chem. doi:10.1016/j.arabjc.2014.05.004
36. Desimone MP, Gordillo G, Simison SN (2011) The effect of temperature and concentration on the corrosion inhibition mechanism of an amphiphilic amido-amine in CO_2 saturated solution. Corros Sci 53:4033–4043
37. El-Haddad MN (2014) Hydroxyethylcellulose used as an eco-friendly inhibitor for 1018 C-steel corrosion in 3.5% NaCl solution. Carbohydr Polym 112:595–602
38. Tebbji K, Faska N, Tounsi A et al (2007) The effect of some lactones as inhibitors for the corrosion of mild steel in 1 M hydrochloric acid. Mater Chem Phys 106:260–267
39. Hamdy A, El-Gendy NS (2013) Thermodynamic, adsorption and electrochemical studies for corrosion inhibition of carbon steel by henna extract in acid medium. Egypt J Pet 22:17–25
40. Ostovari A, Hoseinieh SM, Peikari M et al (2009) Corrosion inhibition of mild steel in 1 M HCl solution by henna extract: a comparative study of the inhibition by henna and its constituents (Lawsone, Gallic acid, Glucose and Tannic acid). Corros Sci 51:1935–1949
41. Moretti G, Guidi E, Fabris F (2013) Corrosion inhibition of the mild steel in 0.5 M HCl by 2-butyl-hexahydropyrrol [1,2-b][1,2] oxazole. Corros Sci 76:206–218

11

Temperature progression in a mixer ball mill

Robert Schmidt[1] · H. Martin Scholze[1] · Achim Stolle[1]

Abstract The influence of the operating frequency, the milling ball and grinding stock filling degree, the material of the milling balls and beakers, the milling ball diameter and the size of the milling beakers on the temperature increase inside the milling beakers in a mixer ball mill was investigated. These parameters influence the temperature progression and the equilibrium temperature of the system. The grinding stock filling degree with regard to the void volume in the milling ball package showed huge influence on the heating rate and the equilibrium temperature. In this context, the behavior of the temperature progression changes if the complete void volume is filled with the grinding stock.

Keywords Ball milling · Temperature measurement · Milling parameters

Introduction

Mechanochemistry using ball mills is a promising technique with applications in organic and inorganic chemistry as well as material sciences [1–4]. During ball milling, up to 80 % of the energy that is generated in the mill is dissipated as heat [5]. This is why measurement and control of the temperature in the ball mills is important, for example, if heat-sensitive products are formed that would be

degraded, or side reactions that would be favored at high temperatures [6, 7]. Furthermore, liquefaction of organic substrates because of a temperature increase can diminish the energy transfer and may disturb the reaction [8].

Ball milling procedures are often referred to as "milling at room temperature" [9, 10]. This term should be used carefully, as even within short milling times a temperature increase can be observed, if there are no precautions for temperature control. For example, it was shown that milling in a mixer ball mill (MBM) for 10 min with two milling balls in a 10 ml beaker raised the temperature from 25 to 30 °C, and Colacino and co-workers measured a temperature increase of approximately 14 K after 30 min milling in an MM200 mixer mill [6, 11]. McKissic et al. observed a temperature of 50 °C after 1 h milling in a Spex mixer mill and Takacs and McHenry reported milling ball temperatures of 66 °C [12, 13]. Comparably higher temperatures can be reached in planetary ball mills (PBMs), where temperatures from 60 to 600 °C can be measured, depending on the type of PBM, the grinding stock, the grinding material and the filling degree [5, 13, 14].

The temperature of the milling beaker is often determined on the surface of the milling beaker, which can be done for example with temperature data loggers [12, 15] or thermocouples [13, 16]. However, the temperature on the surface is not necessarily the temperature inside the milling beaker, and a temperature difference between milling beaker and balls of 25 K was reported [13]. For determination of the milling balls̓ temperature, calorimetric measurements can be performed [5, 13]. The Lamaty group developed a mathematical model for the prediction of the milling walls̓ temperature. Experimental and simulated results (MBM MM200, operated at 30 s^{-1} with 10 ml steel vessels) were in good accordance. The calculated temperature difference of the inner and outer side of the vessel was negligible [11].

✉ Achim Stolle
Achim.Stolle@uni-jena.de

[1] Institute for Technical Chemistry and Environmental Chemistry (ITUC), Friedrich-Schiller University Jena, Lessingstr. 12, 07743 Jena, Germany

Temperature control and more importantly temperature regulation in MBMs and PBMs can be challenging. One option is the integration of milling pauses to the milling cycle that allow a cooling down of the milling beakers, but also increase the total reaction time [10, 17, 18]. Technical solutions for a temperature-controlled milling are: cryogenic milling, where the beakers are cooled with liquid nitrogen [19]; water cooling of the vessels [5] and of the milling beaker holder [20]; (forced) air cooling [21]; use of heating tapes [22]; the application of double-walled milling beakers, which are equipped with an inlet and outlet for a circulating liquid that can be tempered by a thermostat [23–25].

In this study, the influence of milling parameters on the temperature progression in a MBM was investigated.

Methods

All chemicals were purchased from Sigma Aldrich or Alfa Aesar and used as received. The reactions were accomplished in a Retsch MM400 Mixer Mill. If not stated otherwise, milling beakers made of stainless or tempered steel with a volume of 35 mL and steel milling balls with a diameter of 5 mm were used. The temperature was measured with a K-type thermocouple.

The temperature was determined inside of the milling beaker ($T_{\mathrm{milling\ bed}}$) in the milling ball/grinding stock mixture with a thermocouple. The milling beakers were equipped with the respective number of milling balls; the grinding stock was added and milling was accomplished at the respective frequency, v_{osc}, and milling time t. Afterward, the beaker was opened and the thermocouple placed in the middle of the beaker. The time for measurement was <1 min. If not stated otherwise, quartz sand was used as grinding stock in the basic experiments to avoid interference with mechanochemical reactions.

The following formulas were used for the calculation of the milling ball filling degree Φ_{MB} and the grinding stock filling degrees Φ_{GS} and $\Phi_{\mathrm{GS,\ rel}}$; with the milling ball volume V_{MB}, the grinding stock bulk volume V_{GS}, the vessel volume V_{MV}, the diameter of the milling balls d_{MB}, the number of milling balls n_{MB} and the porosity of the milling ball packing ε.

$$\Phi_{\mathrm{MB}} = \frac{\sum V_{\mathrm{MB}}}{V_{\mathrm{MV}}} = \frac{1/6\pi d_{\mathrm{MB}}^3 n_{\mathrm{MB}}}{V_{\mathrm{MV}}}, \tag{1}$$

$$\Phi_{\mathrm{GS}} = \frac{V_{\mathrm{GS}}}{V_{\mathrm{MV}}}, \tag{2}$$

$$\Phi_{\mathrm{GS,rel}} = \frac{V_{\mathrm{GS}}(1 - \varepsilon)}{V_{\mathrm{MV}}\Phi_{\mathrm{MB}}\varepsilon}. \tag{3}$$

Results and discussion

The amount of heat that is dissipated is strongly dependent on the milling parameters that influence the energy input in the milling beaker. These parameters are the frequency v_{osc}, the milling ball diameter d_{MB}, the milling ball filling degree Φ_{MB} (Eq. 1), the milling beaker size, the grinding stock filling degree Φ_{GS} (Eq. 2) and material properties of the grinding stock as well as of the grinding tools, such as Youngs modulus, density, hardness and heat capacity [12, 26].

Temperature measurements at the surface and inside of the milling beaker showed that the temperature difference is negligible at low v_{osc}, but becomes larger for higher v_{osc} (ΔT up to 15 K, Online resource 1). Temperatures recorded with surface measurements for example with temperature data loggers are therefore only an imprecise indicator for the internal temperature conditions [12]. A more precise, alternative method would be the temperature measurement inside the wall with embedded thermocouples [27], but even in this case a temperature gradient form the middle of the milling bed to the wall should occur.

For measurements on the surface, the thermocouple position was reported to be of slight influence [16]. We measured the temperature difference between the surface of the wall and the cap and found differences between 0.2 and 1.4 K (Online resources 1, 2). Thus, for further investigations, we determined the temperature inside the milling beaker ($T_{\mathrm{milling\ bed}}$) in the milling ball/grinding stock mixture with a thermocouple.

Influence of the operating frequency v_{osc}

The first investigated parameter was the operating frequency. As shown in Fig. 1, the temperature increased at $v_{\mathrm{osc}} = 15$ s^{-1} within 90 min from 22 to 30 °C. With higher v_{osc}, the heating rate as well as the final temperature was higher. Thus, at $v_{\mathrm{osc}} = 30$ s^{-1}, a temperature of 87 °C was measured. The energy that is dissipated at higher v_{osc} is raised because of the higher kinetic energy of the milling balls. This led to an enhanced energy input.

Influence of the milling ball filling degree

Figure 2 illustrates the change in the temperature progression, if the milling ball filling degree Φ_{MB} is varied. An increase of Φ_{MB} from 0.06 to 0.36 resulted in a higher end temperature and heating rate, whereas the differences in

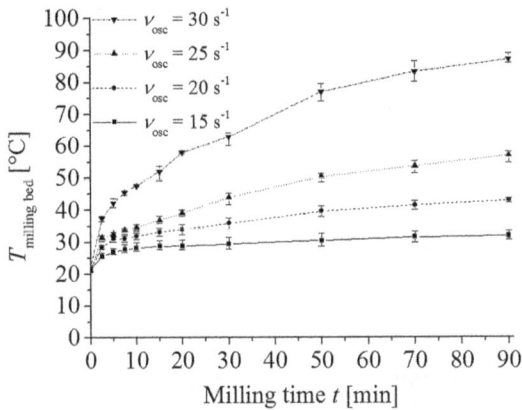

Fig. 1 Influence of the operating frequency on the temperature measured in the milling bed. Conditions: MBM MM400, 35 ml steel beaker, steel balls, $d_{MB} = 5$ mm, $\Phi_{MB} = 0.24$, $m_{quartz\ sand} = 15.36$ g ($\Phi_{GS} = 0.31$)

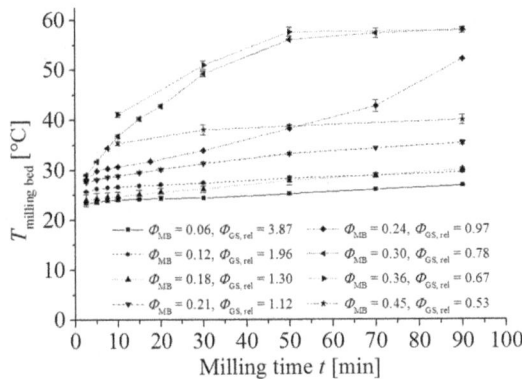

Fig. 2 Influence of the milling ball filling degree on the temperature measured in the milling bed. Conditions: MBM MM400, $v_{osc} = 20$ s^{-1}, 35 ml steel beaker, steel balls, $d_{MB} = 5$ mm, $m_{quartz\ sand} = 7.48$ g ($\Phi_{GS} = 0.15$)

$T_{milling\ bed}$ for $0.06 \leq \Phi_{MB} \leq 0.18$ are small compared to experiments at higher filling degrees (for discussion, see next chapter). At $\Phi_{MB} = 0.45$, the temperature reached after 90 min was considerably reduced. For values of $\Phi_{MB} < 0.36$, the increase of milling balls number resulted in more ball–ball and ball–wall collisions and therefore led to a higher energy input and heat dissipation. The ball movement for $\Phi_{MB} > 0.36$ is hindered because of the reduced space for acceleration and less energy is dissipated, which justifies the lower final temperature [28, 29]. A similar result was reported by Fang et al. for the temperature that was generated in a lysis mill [30]. They observed the highest temperature if 60 % of the beaker was filled with milling balls, which approximately corresponds to Φ_{MB} of 0.4. A higher heating rate was also found with an increased number of milling balls in a Spex mixer mill [13]. The observed dependency of the temperature from Φ_{MB} is also in good agreement with experimental results

for the reaction of vanillin and barbituric acid in an MBM, in which the yield increased for $0.06 < \Phi_{MB} < 0.30$ and strongly decreased at $\Phi_{MB} = 0.45$ [24]. This indicates the influence of the temperature on an organic reaction in a ball mill and the demand for a temperature control for organic syntheses in ball mills.

Interestingly, if Φ_{MB} was changed from 0.21 to 0.3, a huge difference in the measured heating curve was observed. As the amount of grinding stock was kept constant at $\Phi_{GS} = 0.15$, we calculated $\Phi_{GS,\ rel}$ (Eq. 3) defined as the grinding stock filling degree with regard to the void volume in the milling ball packing. The complete void volume is filled with the grinding stock if $\Phi_{GS,\ rel} = 1$. Results show that the change in the heat up curve occurs if $\Phi_{GS,\ rel}$ is approximately at this value. The milling beakers warm up significantly slower for $\Phi_{GS,\ rel} > 1$ (see Online resource 1). The grinding stock overfills the void volume, the velocity of the milling balls is reduced and thus less heat is generated.

Influence of the grinding stock filling degree

Next, we varied Φ_{GS} at a constant Φ_{MB} (Fig. 3). The highest end temperature of 77 °C was measured without any grinding stock. By adding even small amounts of quartz sand, the temperature decreased considerably [13]. The grinding stock influences the elasticity of the collisions; the velocity and motion of the milling balls are reduced and as a consequence less energy is dissipated as heat [29]. Thus, with grinding stock a "damping effect" can be observed. The grinding stock acts as a heat sink as the energy is dissipated in more material and the average temperature of the milling bed is lower. Furthermore, the convection of heat from the milling balls to the milling beaker wall is improved if a grinding stock is loaded, as the main mechanism for heat transfer between a moving ball

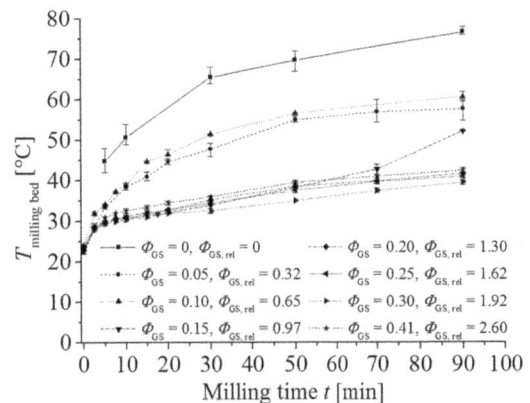

Fig. 3 Influence of the grinding stock filling degree on the temperature measured in the milling bed. Conditions: MBM MM400, $v_{osc} = 20$ s^{-1}, 35 ml steel beaker, steel balls, $d_{MB} = 5$ mm, $\Phi_{MB} = 0.24$, quartz sand

and the wall is forced convection [5]. As observed for the variation of Φ_{MB} (see Fig. 2), a significant change in the heating curve was observed if Φ_{GS} was varied. Temperatures obtained for $\Phi_{GS} \geq 0.20$ are nearly in the same range, independent of Φ_{GS}, but considerably higher temperatures were measured if Φ_{GS} is equal or smaller than 0.15. The point of change is, as observed in the investigation of Φ_{MB} (Fig. 2), if $\Phi_{GS, rel}$ reaches a value of 1.0 (Online resource 1). Thus, the effect of the amount of grinding stock on the temperature is little for values of $\Phi_{GS, rel} > 1$ ($\Phi_{GS} \geq 0.20$). Obviously, a further addition of grinding stock to the milling balls has a negligible effect on the temperature progression in the milling bed. A similar effect was found for changing $\Phi_{MB} \geq 0.20$ as shown in Fig. 2, which indicates a less pronounced temperature increase for $1.3 \leq \Phi_{GS, rel} \leq 3.87$ ($0.18 \leq \Phi_{MB} \leq 0.06$), similar to the results summarized in Fig. 3.

Influence of the milling ball and milling beaker material

The material of the milling balls is an important factor for the energy input in the milling beaker. We performed reactions with milling balls made of steel ($\rho = 7.8$ g cm^{-1}), zirconium oxide ($\rho = 5.9$ g cm^{-1}), sintered corundum ($\rho = 3.8$ g cm^{-1}), silicon nitride ($\rho = 3.25$ g cm^{-1}) and agate ($\rho = 2.65$ g cm^{-1}). From Fig. 4, it becomes clear that the higher the density of the milling balls, the higher is the temperature. A linear increase with the density of the milling balls was found. Milling balls made of a high density material are heavier. Thus, the kinetic energy of the milling balls is elevated. This led to a higher energy that is provided in the collision.

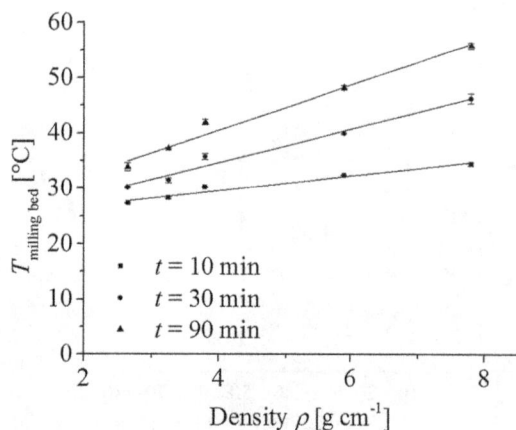

The milling beaker material affects the temperature of the milling beakers as well. The temperature in zirconium oxide milling beakers was approximately 6 K higher as in steel beakers (with zirconium oxide balls in both types of milling beakers) because of the lower thermal conductivity of zirconium oxide.

Influence of the milling ball diameter

As shown in Fig. 5, the final temperature depends on d_{MB}: $T_{milling bed}$ passes through a maximum for 7 mm balls. Milling balls with lower or larger diameter led to a reduced dissipation of heat. Aside from density and number, the diameter of the milling balls can affect the energy dissipation, because larger milling balls correspond to a higher kinetic energy of the single balls. However, the number of milling balls is affected if Φ_{MB} is kept constant, which influences the number of collisions [24]. In addition, d_{MB} affects the friction coefficient and the frictional energy [31]. The results are in consensus with data published by Kwon et al. for mechanical alloying, who reported elevated temperatures in planetary ball mills if d_{MB} was increased from 3 to 9 mm, but a decreased temperature for larger milling balls [5]. In contrast, Takacs reported that the final temperature was relatively independent of the number and size of the milling balls [13]. For milling with a constant number of milling balls, a higher temperature can be assumed for larger milling balls (as long as Φ_{MB} is not too high) because of the higher kinetic energy with higher d_{MB}. For instance, the temperature increase for milling with three balls with a diameter of 10 mm resulted in a temperature increase of 3 K after 30 min, whereas with 15 mm balls ΔT was 20 K.

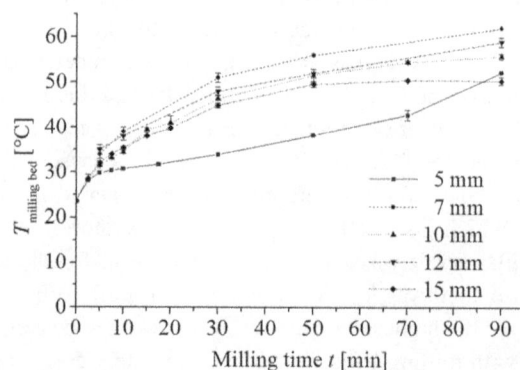

Fig. 4 Influence of the milling ball density on the temperature measured in the milling bed. Conditions: MBM MM400, $v_{osc} = 20$ s^{-1}, 35 ml steel beaker, $d_{MB} = 10$ mm, $\Phi_{MB} = 0.24$, $m_{quartz\ sand} = 7.48$ g ($\Phi_{GS} = 0.15$)

Fig. 5 Influence of the milling ball diameter on the temperature measured in the milling bed. Conditions: MBM MM400, $v_{osc} = 20$ s^{-1}, 35 ml steel beaker, steel balls, $\Phi_{MB} = 0.24$, $m_{quartz\ sand} = 7.48$ g ($\Phi_{GS} = 0.15$)

Variations of the milling beaker dimension

Experiments in milling beakers with varied volume revealed that the size of the milling beaker has a strong influence on the temperature (Online resource 1). While in 10 ml beakers after 90 min a temperature of 35 °C was observed, $T_{\text{milling bed}}$ was 52 and 58 °C in 35 and 50 ml beakers, respectively. On the one hand, the number of milling balls is higher in larger milling beakers, which results in an increased number of collisions and therefore a higher energy input. On the other hand, the larger milling beakers have a lower volume to surface ratio. The ratio is 1.45 times lower for 50 ml beakers as for 10 ml beakers. Thus, the energy dissipation from the beaker to the environment is slower, resulting in a higher $T_{\text{milling bed}}$.

Influence of the grinding stock material

The temperature that was measured after 90 min milling strongly depends on the loaded material. The highest temperature was measured without any grinding stock (compare section grinding stock filling degree, Fig. 3). The addition of powder led to lower temperatures in every case. The temperature ranges between 40 and 63 °C for quartz sand and vanillin, respectively (Online resource 1). The material influence on the temperature seems to be a complicated interaction of material properties like Youngs modulus and hardness [13]. The kind of material influences the elasticity of the collision and the motion of the balls, as shown for planetary ball mills [32]. Thus, it acts on the temperature progression and on the heat transfer from the balls to the wall of the beaker [13]. Youngs modulus for example is insufficient to describe the increase of temperature. For example, with MgF_2 and CaF_2 as grinding stocks, final temperatures of 51 °C were measured, although Young's modulus is considerably different with 139 and 76 GPa, respectively [33]. In addition to the material properties of the grinding stock, physical phenomena like the compaction of the material on the milling wall influence $T_{\text{milling bed}}$. If the grinding stock material adheres to the wall, less material is trapped when the balls collide. Furthermore, the compact layers reduce the free space for acceleration of the balls.

Several grinding stock materials were examined at constant Φ_{GS} or constant mass. The trend of the end temperatures was similar for both conditions (Online resource 1). Material properties seem to have strong influence, balancing the changes in the grinding stock filling degree.

In Table 1 shows the maximal observed temperature difference that was obtained by variation of one parameter, while other parameters were kept constant. In the investigated range, the effect of the operating frequency was highest. A temperature difference of roughly 55 K was detected

Table 1 Maximal measured temperature difference obtained by variation of one parameter

Condition	$\Delta T_{\text{max}-\text{min}}$ (K)
ν_{osc}	55.1
Changed Φ_{GS}	37.3
Changed Φ_{MB}	31.2
Beaker size	23.9
Grinding stock material ($\Phi_{GS} = $ const.)	23.5
Grinding tool material	22
Grinding stock material ($m_{GS} = $ const.)	13
d_{MB}	11.6

between milling at the lowest and highest operating frequency. By variation of the filling degree of the grinding stock as well as of the milling balls, huge values for $\Delta T_{\text{max}-\text{min}}$ could be found. The impacts of the beaker size, grinding stock material and grinding tool material were at the same level with approximately 23 K. The lowest effect was observed for changes in the grinding stock material (at constant m_{GS}) and for the milling ball diameter. Thus, the highest effects were induced by ν_{osc} and the filling degree, which is important with regard to the design of the experiments.

Conclusion

The measurement of the temperature that was generated in a mixer ball mill indicates a strong dependence of $T_{\text{milling bed}}$ on several milling parameters. Higher temperatures were measured with increased operating frequency and in milling beakers with larger volume. The heat dissipation passes through a maximum for the milling ball filling degree and the milling ball diameter. Regarding the milling ball material, a linear correlation to the density of the milling balls was found. The results indicate that $\Phi_{GS, \text{rel}}$ is of great influence on the temperature progression. A changed behavior was observed if $\Phi_{GS, \text{rel}}$ was increased over $\Phi_{GS, \text{rel}} = 1$. These results can be helpful for the experimental design and for performing reactions in ball mills successfully. Furthermore, the results seem to be transferrable to other types of ball mills (for example, planetary ball mills) and are not restricted to the investigated type of mixer ball mills.

Acknowledgments This work was funded by the Deutsche Bundesstiftung Umwelt (DBU; AZ 29622-31).

References

1. Wang GW (2013) Mechanochemical organic synthesis. Chem Soc Rev 42(18):7668–7700. doi:10.1039/C3CS35526H
2. Zhu SE, Li F, Wang GW (2013) Mechanochemistry of fullerenes and related materials. Chem Soc Rev 42(18):7535–7570. doi:10.1039/C3CS35494F

3. Stolle A, Szuppa T, Leonhardt SES, Ondruschka B (2011) Ball milling in organic synthesis: solutions and challenges. Chem Soc Rev 40(5):2317–2329. doi:10.1039/C0cs00195c

4. Stolle A, Ranu B (eds) (2015) Ball milling towards green synthesis: applications, projects, challenges. The Royal Society of Chemistry, Cambridge. doi:10.1039/9781782621980

5. Kwon YS, Gerasimov KB, Yoon SK (2002) Ball temperatures during mechanical alloying in planetary mills. J Alloys Compd 346(1–2):276–281. doi:10.1016/S0925-8388(02)00512-1

6. Naimi-Jamal MR, Mokhtari J, Dekamin MG, Kaupp G (2009) Sodium tetraalkoxyborates: intermediates for the quantitative reduction of aldehydes and ketones to alcohols through ball milling with NaBH4. Eur J Org Chem 21:3567–3572. doi:10.1002/ejoc.200900352

7. Gérard EMC, Sahin H, Encinas A, Bräse S (2008) Systematic study of a solvent-free mechanochemically induced domino oxa-Michael-Aldol reaction in a ball mill. Synlett 17:2702–2704. doi:10.1055/s-0028-1067255

8. Kaupp G (2003) Solid-state molecular syntheses: complete reactions without auxiliaries based on the new solid-state mechanism. CrystEngComm. doi:10.1039/B303432a

9. Tan YJ, Zhang Z, Wang FJ, Wu HH, Li QH (2014) Mechanochemical milling promoted solvent-free imino Diels-Alder reaction catalyzed by FeCl3: diastereoselective synthesis of cis-2,4-diphenyl-1,2,3,4-tetrahydroquinolines. RSC Adv 4(67):35635–35638. doi:10.1039/C4RA05252H

10. Yu J, Li Z, Jia K, Jiang Z, Liu M, Su W (2013) Fast, solvent-free asymmetric alkynylation of prochiral sp3 C–H bonds in a ball mill for the preparation of optically active tetrahydroisoquinoline derivatives. Tetrahedron Lett 54(15):2006–2009. doi:10.1016/j.tetlet.2013.02.007

11. Colacino E, Nun P, Colacino FM, Martinez J, Lamaty F (2008) Solvent-free synthesis of nitrones in a ball-mill. Tetrahedron 64(23):5569–5576. doi:10.1016/j.tet.2008.03.091

12. McKissic KS, Caruso JT, Blair RG, Mack J (2014) Comparison of shaking versus baking: further understanding the energetics of a mechanochemical reaction. Green Chem 16(3):1628–1632. doi:10.1039/C3gc41496e

13. Takacs L, McHenry JS (2006) Temperature of the milling balls in shaker and planetary mills. J Mater Sci 41(16):5246–5249. doi:10.1007/s10853-006-0312-4

14. Tullberg E, Peters D, Frejd T (2004) The Heck reaction under ball-milling conditions. J Organomet Chem 689(23):3778–3781. doi:10.1016/j.jorganchem.2004.06.045

15. Jiang XJ, Trunov MA, Schoenitz M, Dave RN, Dreizin EL (2009) Mechanical alloying and reactive milling in a high energy planetary mill. J Alloys Compd 478(1–2):246–251. doi:10.1016/j.jallcom.2008.12.021

16. Takacs L (2002) Self-sustaining reactions induced by ball milling. Prog Mater Sci 47(4):355–414. doi:10.1016/S0079-6425(01)00002-0

17. Meine N, Rinaldi R, Schüth F (2012) Solvent-free catalytic depolymerization of cellulose to water-soluble oligosaccharides. ChemSusChem 5(8):1449–1454. doi:10.1002/cssc.201100770

18. Rodriguez B, Bruckmann A, Bolm C (2007) A highly efficient asymmetric organocatalytic aldol reaction in a ball mill. Chem Eur J 13(17):4710–4722. doi:10.1002/chem.200700188

19. Retsch CryoMill (2015) http://www.retsch.com/products/milling/ball-mills/mixer-mill-cryomill/function-features. Accessed 22 June 2015

20. Retsch Emax (2015) http://www.retsch.com/products/milling/ball-mills/emax/function-features. Accessed 22 June 2015

21. He S, Qin Y, Walid E, Li L, Cui J, Ma Y (2014) Effect of ball-milling on the physicochemical properties of maize starch. Biotechnol Rep 3:54–59. doi:10.1016/j.btre.2014.06.004

22. Immohr S, Felderhoff M, Weidenthaler C, Schüth F (2013) An orders-of-magnitude increase in the rate of the solid-catalyzed CO oxidation by in situ ball milling. Angew Chem Int Ed 52(48):12688–12691. doi:10.1002/anie.201305992

23. Kaupp G, Naimi-Jamal MR, Schmeyers J (2002) Quantitative reaction cascades of ninhydrin in the solid state. Chem Eur J 8(3):594–600. doi:10.1002/1521-3765(20020201)8:3<594:Aid-Chem594>3.0.Co;2-5

24. Schmidt R, Burmeister CF, Baláž M, Kwade A, Stolle A (2015) Effect of reaction parameters on the synthesis of 5-arylidene barbituric acid derivatives in ball mills. Org Process Res Dev 19(3):427–436. doi:10.1021/op5003787

25. Etman HA, Metwally HM, Elkasaby MM, Khalil AM, Metwally MA (2011) Green, two components highly efficient reaction of ninhydrin with aromatic amines, and malononitrile using ball-milling technique. Am J Org Chem 1:10–13. doi:10.5923/j.ajoc.20110101.03

26. Yazdani A, Hadianfard MJ, Salahinejad E (2013) A system dynamics model to estimate energy, temperature, and particle size in planetary ball milling. J Alloys Compd 555:108–111. doi:10.1016/j.jallcom.2012.12.035

27. Strukil V, Fabian L, Reid DG, Duer MJ, Jackson GJ, Eckert-Maksic M, Friscic T (2010) Towards an environmentally-friendly laboratory: dimensionality and reactivity in the mechanosynthesis of metal-organic compounds. Chem Commun 46(48):9191–9193. doi:10.1039/C0cc03822a

28. Sato A, Kano J, Saito F (2010) Analysis of abrasion mechanism of grinding media in a planetary mill with DEM simulation. Adv Powder Technol 21(2):212–216. doi:10.1016/j.apt.2010.01.005

29. Burmeister CF, Kwade A (2013) Process engineering with planetary ball mills. Chem Soc Rev 42(18):7660–7667. doi:10.1039/C3cs35455e

30. Fang Y, Salame N, Woo S, Bohle DS, Friscic T, Cuccia LA (2014) Rapid and facile solvent-free mechanosynthesis in a cell lysis mill: preparation and mechanochemical complexation of aminobenzoquinones. CrystEngComm 16(31):7180–7185. doi:10.1039/C4CE00328D

31. Köster A, Scherge M, Teipel U (2014) Energy distributions of grinding balls in frictional contact. Chem Ing Tech 86(3):361–364. doi:10.1002/cite.201300127

32. Rosenkranz S, Breitung-Faes S, Kwade A (2011) Experimental investigations and modelling of the ball motion in planetary ball mills. Powder Technol 212(1):224–230. doi:10.1016/j.powtec.2011.05.021

33. Korth Kristalle. http://www.korth.de/index.php/material-161.html. Accessed 2015/03/16

12

Stabilization of γ-sterilized low-density polyethylene by synergistic mixtures of food-contact approval stabilizers

Sameh A. S. Alariqi[1] · Niyazi A. S. Al-Areqi[1] · Elyas Sadeq Alaghbari[1] ·
R. P. Singh[2]

Abstract In our previous studies, we have found the synergistic combinations of stabilizers which follow different mechanisms of stabilization and are approved for food contact applications. The present attempt is to test the potentials of those systems in stabilizing γ-sterilized low-density polyethylene (LDPE). The results were discussed by comparing the stabilizing efficiency of mixtures with and without phenol systems as well as with their counterparts of isotactic polypropylene (iPP) and ethylene-propylene copolymers (EP) matrices. LDPE has been melt-mixed with tertiary hindered amine stabilizer (tert-HAS), oligomeric HAS stabilizer, phenolic and organo-phosphite antioxidants and subjected to γ-sterilization. Stabilization in terms of changes in oxidation products, tensile properties, yellowing and surface morphology was evaluated by FT-IR spectroscopy, Instron, colorimetry, and scanning electron microscopy (SEM), respectively. The results of the present study confirm the validity of those systems for protecting various polyolefins against γ-sterilization. The results showed that the synergism, antagonism and the trend in stabilization efficiency of the binary, ternary and quaternary stabilizer systems were almost similar in LDPE, iPP and EP matrices. The binary system of oligomeric HAS and tert-HAS has shown the antagonistic effect of

stabilization, whereas their combination with organo-phosphite has exhibited synergistic effect even at higher doses of γ-sterilization. The combination of oligomeric HAS, tert-HAS, organo-phosphite and hindered phenol exhibited improved stabilization efficiency than single or binary additive systems. The phenol systems have shown long term of stability than that of phenol-free systems. It was found that the consumption of oligomeric stabilizer significantly depends on the components of stabilization mixture. It was concluded that the stability of polyolefins (LDPE, iPP and EP) against γ-sterilization can be achieved by blends of different stabilizers which are approved for food contact applications.

Keywords LDPE · γ-Sterilization · Stabilization · Oligomeric HAS stabilizer · Synergistic mixtures

Introduction

LDPE is one of the most popular polymers in the manufacturing of food packaging and medical disposables, because it exhibits high transparency, good mechanical properties, low cost, good sealability and chemical resistance, and can be employed over a wide temperature range. Treatment with gamma radiation is becoming a common process for the sterilization of food packaging and medical plastics. The most commonly validated dose used for sterilization is 25 kGy [1]. However, using γ-radiation for sterilization of packaging and medical plastics is known to result in physical changes, including embrittlement, stiffening, softening, discoloration, odor generation and a decrease in molecular weight [2–5]. The degradation of sterilized plastics continues for a long time during their shelf life and service, which is called post-degradation or

✉ Sameh A. S. Alariqi
samehalariqi@yahoo.com

R. P. Singh
rp.singh@ncl.res.in

[1] Department of Chemistry, Faculty of Applied Science, University of Taiz, P. O. Box: 4007, Taiz, Yemen

[2] Division of Polymer Science and Engineering, National Chemical Laboratory, Dr. Homi Bhabha Road, Pune 411008, India

post-sterilization. Radiation-induced changes in the physical properties of a packaging material and medical plastics should not impair its function and the degradation products as well as the utilized additives should be non-toxic [6]. The radiation stability of polyolefins can be done at different stages of degradation process by adding very small amounts of additives (0.05–0.5% w/w) called as 'stabilizers'. They are radical scavengers, antioxidants and hydroperoxide decomposers which follow different action mechanism. Antioxidants are incorporated in the polymer formulation to inhibit the attack of oxygen during the processing and γ-sterilization of the polymer [7]. Phenols, phosphites, or amine compounds are used as antioxidants depending on the free radicals expected to form. Phenolic antioxidants (Primary antioxidants) are generally radical scavengers or H-donors such as Irganox-1010. They are extremely effective at preserving physical properties of polymer during and after γ-sterilization, but at the expense of yellow color formation [8]. Consequently, antioxidants such as hindered phenols are unacceptable medically and for the food packaging because of the intense yellow discoloration which results from the formation of compounds such as stibenequinones upon γ-sterilization. Secondary antioxidants (organo-phosphites) are typically hydroperoxide decomposers (i.e. Irgafos-168) inhibiting oxidation by decomposing the hydroperoxides to form stable products. Unlike primary antioxidants, secondary antioxidants are inadequate if they used alone, so they are usually used in combination with primary antioxidants to get synergistic effects [9]. An organo-phosphite may be used as a short-term antioxidant to protect the polymer during processing, while phenolic antioxidants are used for long-term protection. Hindered amine stabilizers (HAS) are widely used radical scavengers having multifunctional capabilities for scavenging radicals. A substituted piperidine was found to give good protection (little yellowing or embrittlement) against γ-irradiation as well as post-irradiation storage under accelerated test conditions (60 °C in air) [10]. The efficiency of stabilizers is very much dependent upon the type and the grade of polymer in which they are compounded; thus the judicious selection of stabilizers is very important in the formulation of plastic [11]. These additives are not chemically bound to the polymer matrix and migrate or leach out under the influence of physicochemical factors such as temperature, sterilization and type of solvents and pH of the packaged product [12]. Unfortunately, the toxicological data on most of the stabilizers are either not available or incomplete and for many antioxidants are available from feeding studies only [7]. Polymer stabilization is a dynamic process resulting in many transformed and degradation products which are potentially leachable and extractable [12]. Many antioxidants and stabilizers act sacrificially and are converted to oxidation products during the process of stabilization [13]. In fact, there is a little knowledge regarding the toxicity of antioxidant transformation products; thus there is a doubt that they may be more toxic than the antioxidants from which they are derived [13]. It is urged that when more than one stabilizer is utilized, toxicity must be estimated by considering the combination rather than each agent alone, since a different synergistic effect could be the result very often [14]. Consequently, the toxicology of food packaging and medical plastics depends on many factors such as the effect of stabilizer loss, toxicity of migrated or leached stabilizers, degradation process and the degradation products and the effect of sterilization methods on the plastics or its constituents [7]. Migration and leachability of non-polymeric components to its environment (esp. into drugs, body fluids and foods) gives rise to major concerns in case of food packaging plastics, packaging materials for pharmaceuticals and other medical applications. This migration is associated with health hazards and has become a major factor in regulations regarding the safety and quality of packaged food. Thus, stabilization should be done with stabilizers which are approved for food contact and biomedical applications.

It is well known that the efficiency of stabilizers is disturbed by a loss of the active form of stabilizers. This loss can be either chemical consumption or physical loss. The consumption of the stabilizers occurs during chemical reactions in the presence of light, heat and radiation. However, the physical loss of the stabilizers occurs by diffusion toward the polymer surface by evaporation, volatilization, poor solubility, leachability and migration into the material in contact with the polymer [15, 16]. The consumption and loss of the stabilizers accelerate the aging of the polymer more than thermal- or radio- or photo-oxidation [17, 18]. The long-term protection was observed with oligomeric HAS stabilizer, whereas very short-time protection was found with low-molecular weight HAS [19–22]. Polymeric hindered amine light stabilizers (HALS) shows a much higher thermal stability and better extraction resistance than that of low molecular weight; thus, the tendency for developing amine stabilizers in the form of oligomeric/polymeric macromolecules recently established [23, 24]. HALS was developed from low-molecular-weight stabilizers to high molecular weight to counteract the effects of volatilization and extraction from the polymer matrix during outdoor application [25, 26]. The low-molecular-weight stabilizers mostly are liquid, volatile and easily decompose in thermal processing temperatures; thus the effective concentration of such admixture in the polymer is reduced. On the other hand, the low-molecular-weight stabilizer has good mobility and usually it can be dispersed more homogeneously in the polymeric materials than the high-molecular-weight stabilizer [23]. It was concluded that the stabilizer mobility played an important role in the overall mechanism of stabilization of HAS in PP [27]. It was reported that approximately 95% molecules of oligomeric stabilizer were translationally immobile in the polymer matrix and it was explained that decreased efficiency of oligomeric stabilizers with increased molecular weight was a result of reduced stabilizer mobility [28]. Gugumus [29] determined the optimum molecular weight (MW) to be about 2700 for poly (1,2,2,6,6-pentamethyl-4-piperidyl acrylate) for the light stabilization of PP. Thus, there have been always some disadvantages in using single-additive system such as compatibility, migration with low-molecular-weight stabilizers (especially HAS), immobility with high-molecular-weight stabilizers, yellowing with phenolic antioxidants and reduced efficiency of organo-phosphites by fast consumption.

Considering the above said aspects, we have found the combinations of synergistic mixtures of oligomeric stabilizers in our previous works [20–22] where we have used mixtures of oligomeric HAS and tert-HAS, primary and secondary antioxidants, which are approved for food contact applications [30], and their selection has been based on different molecular weights and protecting mechanisms in iPP and EP matrices. The previous work aimed to comprise heterosynergistic combinations of these stabilizers to be an alternative stabilization system of the phenolic antioxidants as well as to improve the discoloration of the phenolic system. Two groups of stabilization systems were prepared and tested on iPP and EP matrices upon γ-sterilization. The first group was phenol-free system where phenol is excluded due to its discoloration disadvantage and the second group was a phenolic system where the phenol is blended with other stabilizers to improve its discoloration. The binary (1:1) phenol-free system (oligomeric HAS–organo-phosphite) and binary (1:1) phenolic system (oligomeric HAS–phenol antioxidant), ternary (1:1:1) phenol-free system (oligomeric HAS–tertiary HAS–organo-phosphite) and ternary (1:1:1) phenolic system (oligomeric HAS–phenol antioxidant–organo-phosphite) and quaternary (1:1:1:1) system of all four stabilizers has shown improved stability against γ-sterilization. However, the antagonistic effect of stabilization was found in the binary (1:1) phenol-free system (oligomeric HAS–tert-HAS). Stabilization and synergistic mechanisms as well as the reactivity of the products resulted due to the combination of those stabilizers have been discussed in our previous work [20–22]. Since the stabilizer blend systems we tested are still new (especially with γ-irradiation), there are no clear data on the toxicity of the formed products. The aim

of our research series was to improve the stability of polyolefins against γ-sterilization by preparing synergistic mixtures based on oligomeric stabilizer using food-approved stabilizers. Since we tested those stabilization systems on iPP and EP matrices it will be beneficial to test their effectiveness on PE matrix and to study the validity of those systems for protecting different polyolefins against γ-sterilization. Thus, it will be worthwhile for making a vast array of data on stabilization of food packaging and biomedical plastics, to study stabilizing efficiency of those systems with LDPE, which are widely desired as food packaging and medical plastic. The major objective of this work was to conduct a comparative study on the effect of those stabilizing systems on LDPE with other polyolefin matrices (iPP and EP copolymers).

Experimental method

Materials and chemicals

A commercial sample of LDPE (density: 0.92 g/cm^3, Melt flow index (MFI): 1.2 g/10 min) was obtained from Indian Petrochemicals Corp Ltd., India, under the trade name INDOTHENE. The LDPE was purified as follows: pellets of LDPE were dissolved in xylene by gentle heating under reflux under a nitrogen atmosphere for 1 h. Addition of cold methanol caused a precipitate to form. This was filtered and then dried at 50 °C in a vacuum oven until constant weight. The sample was assumed to be "additive free" and designated as purified sample. Solvents were obtained from M/s. SD. Fine Chemicals Ltd, Mumbai, India. Four different stabilizers supplied by M/s. Ciba-Geigy, Switzerland, which is approved for food contact applications [30], were used in this study and they are as follows:

1. Tinuvin 765 (CAS No.: 41556-26-7 and 82919-37-7), a low molecular weight (MW = 508 g/mol), tertiary HAS, yellow liquid, designated as (T),
2. Chimassorb 944 (CAS No.:71878-19-8), an oligomeric HAS, high molecular weight (MW = 2790 g/mol, $M_n \approx 3000$), secondary HAS, white powder, mp 115–125 °C, designated as (C),
3. Irganox 1010 (CAS No.: 6683-19-8), a hindered phenol (MW = 1178 g/mol) designated as (X) and
4. Irgafos 168 (CAS No.: 31570-04-4), an organo-phosphite, IV (MW = 649.9 g/mol) designated as (S).

The chemical structure and IUPAC name of the stabilizers are given below:

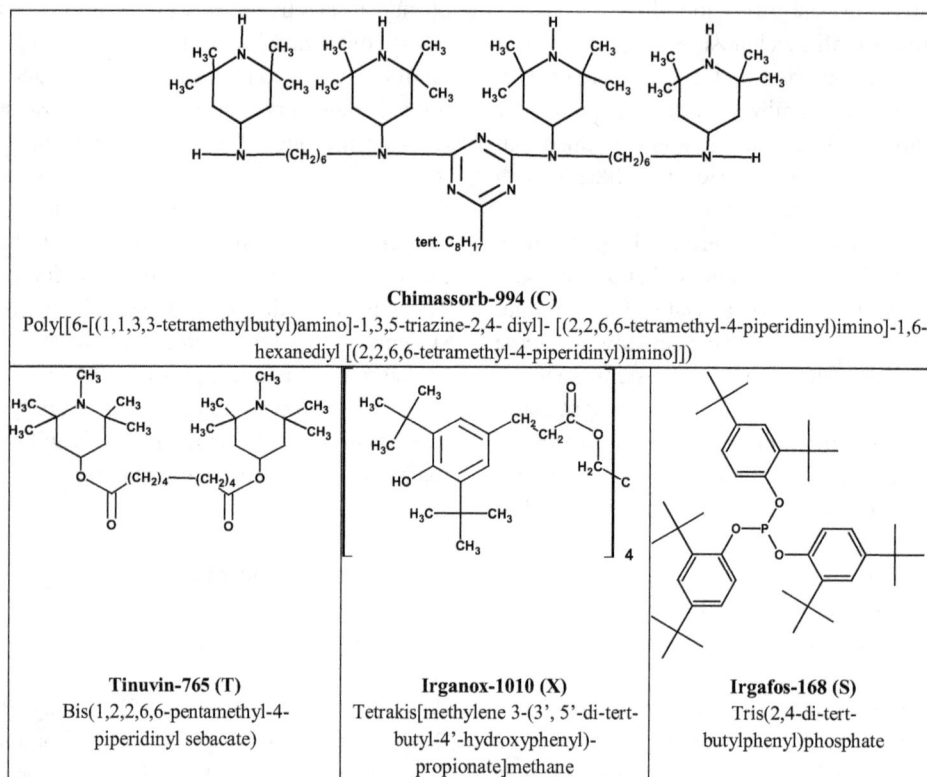

Chimassorb-994 (C)
Poly[[6-[(1,1,3,3-tetramethylbutyl)amino]-1,3,5-triazine-2,4- diyl]- [(2,2,6,6-tetramethyl-4-piperidinyl)imino]-1,6-hexanediyl [(2,2,6,6-tetramethyl-4-piperidinyl)imino]])

Tinuvin-765 (T)
Bis(1,2,2,6,6-pentamethyl-4-piperidinyl sebacate)

Irganox-1010 (X)
Tetrakis[methylene 3-(3', 5'-di-tert-butyl-4'-hydroxyphenyl)-propionate]methane

Irgafos-168 (S)
Tris(2,4-di-tert-butylphenyl)phosphate

Mixing of stabilizers and preparation of specimens

The weighed amount of stabilizers was dissolved in chloroform and mixed with the required amount of dried polymer powder (LDPE) for better distribution and chloroform was evaporated and dried at 50 °C in vacuum oven. After drying, this polymer was melt-mixed in a microcompounder (DSM, The Netherlands) for 5 min at 160 °C. Keeping the ratio between polymer and stabilizer for each blend system as 99.6 polymer: 0.4% (w/w) stabilizers (9.6 g polymer: 0.4 g stabilizer), and the ratio between the stabilizers as 1:1 for binary (9.6 g polymer: 0.2 g for each stabilizer), 1:1:1 for ternary (9.6 g polymer: 0.133 g for each stabilizer) and 1:1:1:1 for quaternary blend systems (9.6 g polymer: 0.1 g for each stabilizer) as tabulated in Tables 1 and 2,

the samples were compounded. Then, they were molded as films in aluminum foil between two plates by heating up to 160 °C and holding for 3–5 min and then increasing the molding pressure to 15,000 lb. The pressure was allowed to fall, and the molds were then immediately quenched into a large bath filled with water at 20 °C. Their thickness was found to be about 100 ± 10 µm.

γ-Irradiation

The films were kept in a cylindrical well-type ^{60}Co γ-irradiation chamber (Made by Bhabha Atomic Research Centre, Bombay, India) in the position which allows a uniform irradiation for all films. The samples were irradiated at different doses: 25 (sterilization dose), 50, 75 and

Table 1 Phenol-free system

S. no	Sample code	Polymer % (w/w)	G	Stabilizers % (w/w) or g Irganox-1010	Chimassorb-944	Tinuvin-765	Irgafos-168
1	C	99.6	9.6	–	0.4	–	–
2	T			–	–	0.4	–
3	S			–	–	–	0.4
4	CT			–	0.2	0.2	–
5	CS			–	0.2		0.2
6	CTS			–	0.133	0.133	0.133

Table 2 Phenol System

S. no.	Sample code	Polymer		Stabilizers % (w/w) or g			
		% (w/w)	G	Irganox-1010	Chimassorb-944	Tinuvin-765	Irgafos-168
7	X	99.6	9.6	0.4	–	–	–
8	CX			0.2	0.2	–	–
9	CTX			0.133	0.133	0.133	–
10	CXS			0.133	0.133	–	0.133
11	CTXS			0.1	0.1	0.1	0.1

100 KGy (dose rate 0.4 kGy h^{-1}) at room temperature in air. Irradiation experiments were performed at the Nuclear Chemistry Department, University of Pune, India.

Characterization

FT-IR spectroscopy

FT-IR spectroscopy (Perkin Elmer 16 PC FT-IR spectrophotometer) was used to characterize the chemical changes caused by γ-radiation in the polymer specimens. Oxidation products were identified and quantified and our interest was mainly focused on the changes in the carbonyl region (1600–1800 cm^{-1}) to follow γ-induced oxidation. The IR spectrometer was used to measure the concentration of carbonyl compounds in the polymer specimens at 1720 cm^{-1}. A value of 220 L mol^{-1} cm^{-1} was used for absorption coefficient [31]. The spectrometer was operated at a resolution of 4 cm^{-1}. The oxidized specimens were analyzed immediately to minimize the post γ-effect.

Universal testing machine

The changes in the mechanical properties were measured by a universal testing machine (Instron model 4201, Instron, MA, USA). Elongation at break was determined from stress–strain curves. The cross speed used was 10 mm min^{-1}. The specimens were cut according to IS: 2808–1984:A4, (100 mm length, 10 mm width and the gauge space 50 mm). The results of each sample were taken as the average of five specimens.

Color measurements

Yellowness index (YI) was determined in accordance with ASTM D1925 [32] by reflectance measurements using a Color Mate HDS Colorimeter (Milton Roy, USA) with integrating sphere. The samples were placed in the reflectance part of a sphere using a standard white ceramic as reference tile. The instrument is designed to give direct yellowness index value on the basis of CIE standard illumination C (CIE 1931) 2° standard observer viewing [33]. It was obtained from the tristimulus values X_{CIE}, Y_{CIE} and Z_{CIE} relative to source C using the equation YI = [100 (1.28 X_{CIE} − 1.06 Z_{CIE})]/Y_{CIE}. Several values of YI obtained from different parts of the samples were generally used to obtain an average value of the yellowness index. The yellowness index represented in terms of delta yellowness index (d YI):

Delta yellowness index (d YI)
= Yellow index after γ-irradiation
− Yellow index before γ-irradiation.

Scanning electron microscopy

Fracture surface produced by subjecting specimens to γ-radiation was determined by scanning electron microscopy. The stained samples were dried under vacuum for 24 h at 50 °C. These gold-coated samples were scanned under electronmicroscope (Leica Cambridge Stereoscan 440 model).

Results and discussion

The incorporation of single, binary, ternary and quaternary systems of oligomeric HAS stabilizer, tertiary hindered amine, hindered phenol and organo-phosphite into LDPE was discussed by comparing the stabilizing efficiency of mixtures with and without phenol systems as well as with their counterparts of iPP and EP matrices.

Tensile properties

Figures 1 and 2 illustrate the changes in mechanical properties of neat and stabilized samples of LDPE before and after γ-sterilization with a dose of 25 kGy where the stabilizing efficiency of phenol and phenol-free systems can be seen, in terms of elongation at break (%), respectively. It is clearly seen that the elongation at break (%) of neat sample was the lowest, while the samples with single, binary, ternary and quaternary stabilizer systems have shown higher values of elongation at break (%). It can be understood from Figs. 1 and 2 that the order of efficiency of the stabilization in terms of tensile properties (i.e. protection against embrittlement) is as

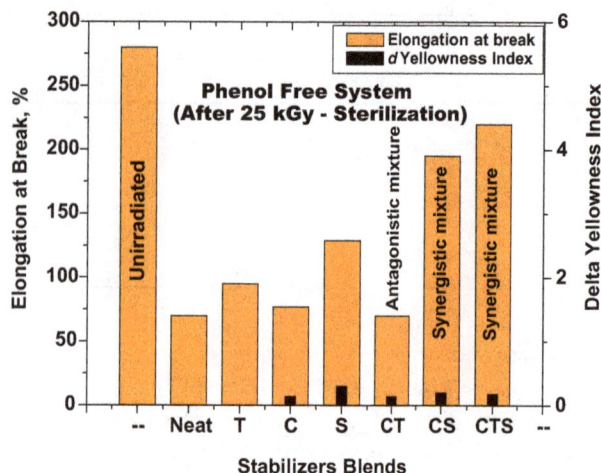

Fig. 1 Changes in tensile properties and yellow index of phenol-free system after γ-irradiation (25 kGy)

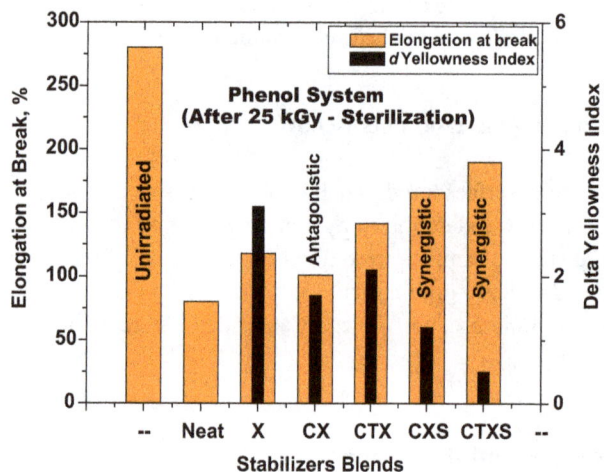

Fig. 2 Changes in tensile properties and yellow index of phenolic system after γ-irradiation (25 kGy)

follows: CTS > CS > CTXS > CXS > CTX > S > X > CX > T > C > CT. This indicates that the combination of various stabilizers with different molecular weights and protecting mechanisms significantly enhanced the efficiency of stabilization. In comparison with iPP & EP matrices, the trend in stabilizing efficiency in terms of tensile properties is almost similar but the difference was observed only in the magnitude of the elongation at break (%).

Single-stabilizers systems

In comparison of single-stabilizer systems, the samples containing organo-phosphite 'S' (Irgafos-168) have shown higher stabilization against γ-sterilization, while the samples stabilized with oligomeric HAS stabilizer 'C' (Chimassorb-944) were the lowest and became the liable of embrittlement. The lower stabilization efficiency of the

oligomeric stabilizer 'C' may be ascribable to its high molecular weight and immobility in the polymer matrix [28]. Among the individual stabilizer system, HAS stabilizers [i.e. sec-HAS (Chimassorb-944) 'C' and tert-HAS (Tinuvin-765) 'T'] have shown the lower protection against γ-degradation compared to hindered phenol (Irganox-1010) 'X' and organo-phosphite (Irgafos-168) 'S'. The order of stabilization efficiency in the individual stabilizer systems in terms of tensile properties can be S > X > T > C.

Phenol-free systems

The changes in tensile properties of stabilized samples with phenol-free systems are revealed in Fig. 1. Among all stabilized samples, the reduction in elongation at break (%) of 'CTS' samples was lesser than other samples. In case of binary systems, 'CS' have shown higher value of elongation than that of 'CT'. As observed in iPP and EP matrices, the sample CT in LDPE has shown to be more susceptible to breakage indicating that 'CT' is an antagonistic mixture in the three polymer matrices. It is obvious that the combination of oligomeric HAS stabilizer 'C' and tert-HAS 'T' exhibits an antagonistic effect, while the combination of oligomeric HAS stabilizer and organo-phosphite exhibits synergistic effect. Similarly was found that the addition of organo-phosphite 'S' to the antagonistic mixture (CT) has improved elongation at break (%) considerably. The order of stabilization efficiency in the phenol-free systems is as follows: CTS > CS > S > T > C > CT.

Phenol systems

The changes in tensile properties of stabilized samples with phenol systems are illustrated in Fig. 2. Among the phenol systems, the samples 'CTXS' and 'CXS' were observed to be highly stabilized, while the sample 'CTX' have shown the higher decrease in elongation at break (%). It is also obvious that the combination of 'X' with the oligomeric HAS stabilizer 'C' [i.e. CX] shows the antagonistic effect of stabilization, while addition of 'S' to 'CX' and 'CTX' improves the polymer resistance against degradation. The phenol and phenol-free systems have shown increased stability against tensile breakage after γ-sterilization and it is supporting the trend observed in iPP and EP matrices [20–22]. For example, the elongation at break (%) of antagonistic mixtures (i.e. CT and CX), 'CTX' and C was significantly improved stabilization efficiency after addition of organo-phosphite 'S'. However, the combination of hindered phenol 'X' with 'CT' increases the elongation at break (%) as can be seen for the sample 'CTX'. It is clearly seen that the addition of 'S' in the phenol and phenol-free systems exhibited synergistic effects. The results of tensile test has clearly shown that the stabilization efficiency of oligomeric HAS stabilizer [i.e. Chimassorb-944] can be improved by combinations with other stabilizers having different stabilization mechanisms and molecular weights.

Yellowness index (YI)

The color formation in neat and stabilized samples after γ-serialization is also represented in terms of delta yellowness index (d YI) in Figs. 1 and 2. The effect addition of oligomeric HAS, tert-HAS and organo-phosphite to hindered phenol can be seen in Fig. 2 where yellowness index reduced. It can be seen that the combination of hindered phenol with other stabilizers highly prevent the discoloration. For example, the combination of hindered phenol with oligomeric HAS, tert-HAS and organo-phosphite (i.e. CTXS) drastically reduces the yellowness index from 3.1 to 0.5. It can be seen that the addition of organo-phosphite (S) to (CX) reduced the YI values from 1.9 to 1.1, while addition of tert-HAS (T) to (CX) did not reduce the yellowness index. The significant reduction in yellowing after addition organo-phosphite (S) to CX and CTX can be attributed to the fact that organo-phosphite decomposes the hydroperoxides to reduce further oxidation and their reaction with oxidized products of hindered phenol, i.e. highly colored quinonoids which are transformed into colorless benzenoid forms [20–22, 34]. In contrast to tensile property, where the combination of oligomeric HAS stabilizer 'C' and hindered phenol 'X' (i.e. CX) exhibited the antagonistic effect, the incorporation of oligomeric HAS stabilizer into hindered phenol 'X' significantly diminished the YI from 3.1 to 1.8. As explained [20–22], the reduction in YI after combination oligomeric HAS with hindered phenol may be due to the radical scavenging effect of oligomeric HAS stabilizer. The efficiency of phenolic stabilizer systems for protection against yellowing is as follows: CTXS > CXS > CX > CTX > X. This trend is similar to that observed for iPP and EP, but the magnitude of discoloration was lower in all samples of LDPE than that of iPP and EP due to the high durability of LDPE against γ-radiation. The results clearly indicate that the combination of hindered phenol with different stabilizers highly prevent discoloration. This could be explained by the presence of various stabilizers which prevent the polymer degradation via different protection mechanisms at different stages synergistically.

Morphological aspects

SEM is a reliable tool to monitor the surface changes during degradation of polymers. Figure 3 shows the scanning electron micrographs of neat and stabilized samples after 25 kGy γ-sterilization. It is evident from these micrographs that under γ-sterilization, neat sample was observed to show deformed/cracked surface (Fig. 3a). The crack formation on the surface of stabilized samples is lesser and samples 'S' (Fig. 3b) and 'CTS' (Fig. 3c) have shown mostly smooth surface. The higher surface erosion was observed in sample 'CT' (Fig. 3d) which shows the antagonistic effect confirming the result of elongation at break (%). As seen in Figs. 1 and 2, the decrease in tensile properties of 'CT' sample also can be explained through the crack formation on the surface. Since eroded surface or cracks on the surface can act as 'defects' where failure mechanism is initiated, the tensile properties of the surface eroded samples can be lowered. The synergistic effect of organo-phosphite 'S' can be appreciated (by adding 'S' to 'CT') in 'CTS' which exhibited the higher stability against surface crack. The sample 'CTS' has shown stability against the crack formation than other samples. In comparison with iPP and EP matrices, the trend in stabilization efficiency is almost similar, but the surface of LDPE matrix was not much cracked as it was observed in iPP and EP matrix.

IR spectroscopic analysis

Kinetics by carbonyl group evolution

The order of stabilization efficiency can be confirmed further by determining the concentration of carbonyl group at the dose of γ-sterilization (25 kGy), while the stability period can be studied by monitoring the concentration of carbonyl group upon γ-irradiation with higher doses (above 25 kGy). The evolution of carbonyl group concentration (mmol L^{-1}) upon γ-irradiation with different doses (0–100 kGy) is plotted in Figs. 4 and 5. The concentration of carbonyl group increases with increasing the dose of γ-irradiation. It can be found in Fig. 4 that the rate of increase in concentration of carbonyl group of neat sample is higher than that of single-stabilizer systems with increasing irradiation dose. Up to sterilization dose, the order of efficiency of the stabilization in terms of preventing oxidation is as follows: S > X > T > C, i.e. the sample containing organo-phosphite 'S' is more stabilized against oxidation than others. With increasing doses of γ-irradiation above sterilization dose (25 kGy), this stabilization efficiency order is affected and for 100 kGy γ-irradiated samples, this order is changed as X > T > C > S. The changes in the efficiency order can be attributed to the fact that there may be the consumption of stabilizers during γ-irradiation up to higher doses. Figure 5 shows the carbonyl evolution in the phenol and phenol-free systems. It is obvious that the stabilization efficiency in the phenol and phenol-free systems is higher than that in single-stabilizer systems. In phenol-free systems, the higher stabilization efficiency in terms of

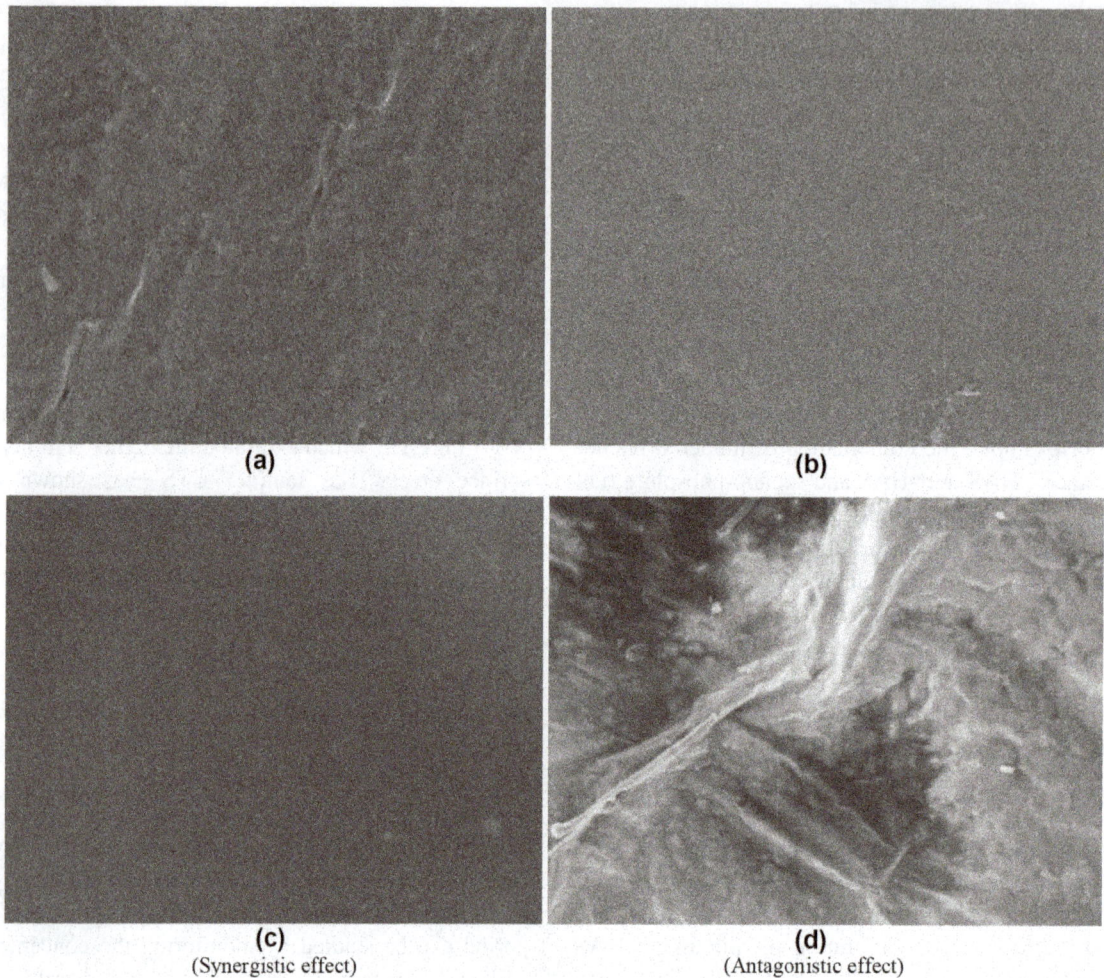

(a)   **(b)**

(c)   **(d)**
(Synergistic effect)   (Antagonistic effect)

Fig. 3 Scanning electron micrographs of 25 kGy γ-irradiated LDPE samples

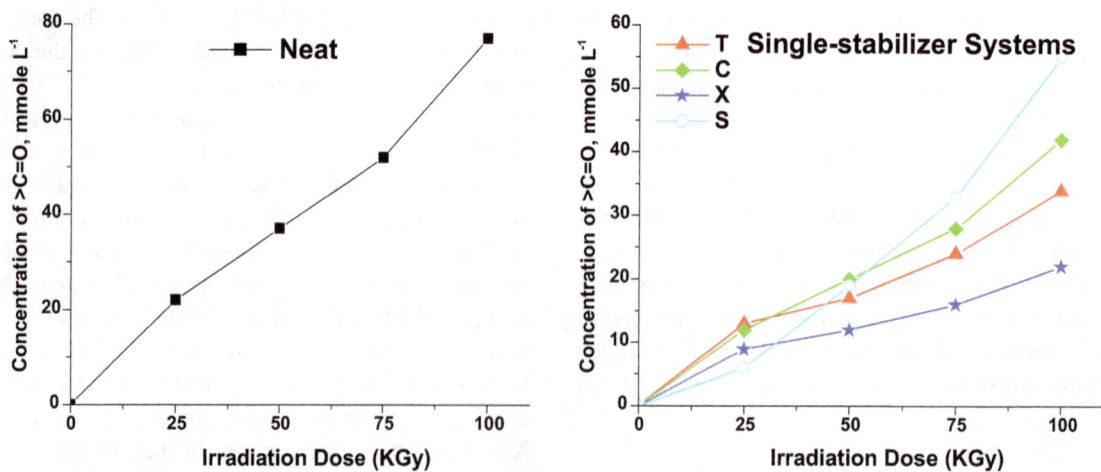

Fig. 4 Carbonyl group increase in neat and single-stabilizer systems

reduction in carbonyl group evolution was observed in CTS. However, the CT mixture has shown the antagonistic effects on oxidation with a linear increase in carbonyl group concentration (up to 100 kGy). An important observation is that the kinetic accumulation of carbonyl group of CT mixture was a linear during irradiation up to 100 kGy, in accordance with the results of iPP and EP [20–22]. Similar to iPP and EP, it was also found that the

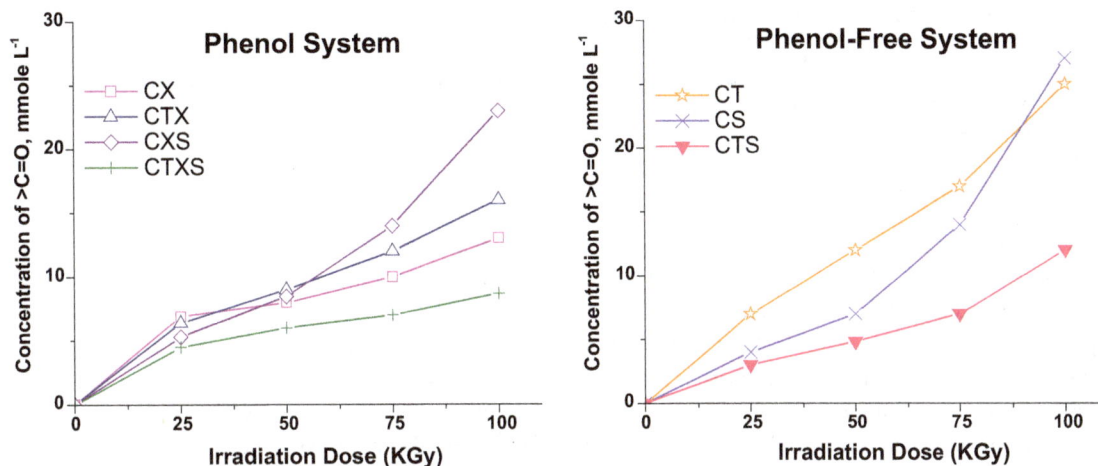

Fig. 5 Change in carbonyl group concentration of phenol and phenol-free systems upon γ-irradiation

stabilization efficiency in the phenol systems is higher than that in phenol-free systems. Accordingly, the addition of hindered phenol stabilizer (X) to oligomeric mixtures (CT, CTS, CS and C) lead to reduce the rate of oxidation during irradiation up to higher doses. It is evidence that the phenol systems exhibited longer period of stabilization against γ-irradiation for iPP, EP and LDPE. Among the phenol systems, the samples of binary (CX) and quaternary mixtures (CTXS) have shown higher stability. Among all the mixtures, CTXS and CTS have shown higher stability against all the doses of γ-irradiation indicating the longer durability for longer period. Up to 100 kGy γ-irradiation, the samples containing organo-phosphite became lower stabilized against oxidation, while samples containing hindered phenol showed higher stability. The samples containing organo-phosphite (i.e. S, CS and CXS) have shown low rates of carbonyl group concentration below 25 kGy and thereafter suddenly increased indicating the short stabilization period.

The change in the order of stabilization efficiency, in terms of reduction in the rate of carbonyl group evolution, can be attributed to the consumption of stabilizers. This may be in the form of physical loss of stabilizer from the samples and/or transformations of active form stabilizer to inactive form during γ-irradiation. It was argued that if one stabilizer is not affecting the stabilization mechanism of other stabilizer, the rate of carbonyl group evolution should not be changed or the order of stabilization efficiency of mixtures should not be altered. As it was observed in Fig. 4 for the sample of 'S', sudden increase in the carbonyl group above γ-irradiation dose of 25 KGy can be attributed to the fact of the disappearance of Irgafos-168. This observation was already reported by Kawamura et al. [35] that Irgafos-168 was fastest to disappear from the samples. These results are in agreement with those of Stoffer [36] who found that the amount of Irgafos-168 from polyolefins

decreased with higher irradiation doses. Carlsson et al. [37] observed a complete degradation of phosphite to give mainly phosphate, but at quite low γ-irradiation doses (∼5 kGy) during γ-sterilization of HDPE trays. They also detected that any residual phosphite is lost progressively in post-irradiation reactions. Allen et al. [38] also reported that both gamma and electron beam irradiation had similar effects with regard to the extent to which Irgafos-168 was not detected after the dose of 25 kGy. Likewise, it was observed in Figs. 4 and 5 that in the samples containing organo-phosphite (S, CS and CXS) the rate of carbonyl group concentration suddenly increased above 25 kGy. The sudden increase in carbonyl group and the fast consumption of 'S' and its mixtures (i.e. CS and CXS) above 25 kGy were also seen in the iPP and EP matrices. On the other hand, for the hindered phenol 'X' sample, such increase in carbonyl group concentration was not observed, indicating that consumption of hindered phenol 'X' is slower than others and/or lower rate of loss of hindered phenol is observed in the film. In mixture systems also, the samples containing 'X' have shown higher protection ability at higher irradiation doses, especially in 'CTXS' and 'CX' samples, indicating the synergistic interaction of oligomeric HAS and hindered phenol by forming the compounds which are in turn radical scavengers [39, 40]. The retained concentration of oligomeric HAS stabilizer (Chimassorb-994) in all mixtures which determined from the triazine absorption [41] will reveal the contribution of added stabilizers in polymer matrix. Table 3 shows the retained concentration of oligomeric HAS stabilizer in all samples after sterilization with 25 kGy. It can be seen that the consumption of oligomeric HAS stabilizer was higher (43% w/w) when used in combination with 'T' (i.e. CT) indicating the antagonistic effect of the CT mixture. However, the retained concentration was higher (97% w/w) when it is combined with 'S' (i.e. CS). The preservation of

Table 3 The retained concentration of oligomeric HALS (Chimassorb 994) after γ-sterilization (25 kGy)

Components	C	CT	CX	CS	CTX	CTS	CXS	CTXS
Chimassorb 994 (C), retained (%)	80	43	71	97	na	92	87	na

na not available

oligomeric HAS stabilizer was also observed whenever it combined with 'S' (i.e. CS, CTS and CXS), which is an indication of the synergistic effect. Norman Allen [42] has found a synergistic effect in the heat and light degradation of PP and PE films when he examined the reaction between two phosphite stabilizers and universal polymeric hindered piperidine ultraviolet light stabilizers. This is because the phosphite stabilizers could destroy hydroperoxide and remove oxygen, thus protecting the Chimassorb-944, which may reflect why the retained concentration of oligomeric HAS (Chimassorb-994) in CS is higher (97% w/w) after γ-sterilization. In comparison, the retained concentration of Chimassorb-994 was higher in the LDPE matrix than that in iPP and EP matrices.

In comparison with iPP and EP copolymers matrices, it can be found that the trend in stabilizing efficiency against embrittlement, discoloration, surface crack and oxidation is almost similar but the difference was observed only in the magnitude. In case of single-stabilizer systems, it was observed that the samples stabilized with high-molecular-weight stabilizer (i.e. oligomeric HAS stabilizer, MW 2790 g/mol) and with relatively small-molecular-weight stabilizer [i.e. tert-HAS (Tinuvin-765) MW 508 g/mol] have shown the lower protection against γ-degradation. However, the samples stabilized with moderate-molecular-weight stabilizers [Irgafos-168, MW 649.9 g/mol and Irganox-1010, MW 1178 g/mol] have shown higher stability in terms of preventing embrittlement, surface erosion and oxidation. The lower stabilization efficiency of the high-molecular-weight stabilizer (oligomeric HAS) may be attributed to its immobility in the polymer matrix. However, the lower stabilization efficiency of the tert-HAS may be attributed to the physical loss due to its low molecular weight. It is well known that if a stabilizer molecule is too small, not only it will be incompatible with certain polymeric materials, but it probably diffuses and volatilize away from the polymer. This problem has been associated with relatively small-molecular-weight compounds such as Tinuvin-770 (MW 478). It was confirmed that the consumption of organo-phosphite (Irgafos-168) is faster in iPP, EP and LDPE matrices than other stabilizers. In case of binary, ternary and quaternary stabilizer systems, it was observed that the mixtures of oligomeric HAS and tert-HAS (i.e. CT) and hindered phenol (i.e. CX) shows the antagonistic effect. However, the addition of organo-phosphite 'S' in the phenol and phenol-free systems (C,

CT, CX and CTX) exhibited synergistic effects of stabilization against embrittlement, discoloration, surface erosion and oxidation. Likewise, the consumption of oligomeric HAS stabilizer was found to be higher when it is used in combination with 'T' and with 'X' (i.e. CT, CX) indicating the antagonistic effect of CT and CX mixtures. However, the retained concentration of oligomeric HAS stabilizer was highly preserved whenever it combined with 'S' (i.e. CS, CTS and CXS), which is an indication of the synergistic effect. The synergistic effect obtained by combination of organo-phosphite with hindered phenols and hindered amines could be explained by the fact that organo-phosphite preserves the concentration of hindered phenols and replaces the hindered amines during polymer processing [43]. The phenol systems have shown longer period of stabilization than phenol-free systems. The longest period of stabilization was explained through the interaction between hindered phenol and hindered amine [20–22, 44–48]. Our experimental results demonstrated that the synergism and antagonism of the binary, ternary and quaternary stabilizer systems were almost similar in LDPE, iPP and EP matrices indicating that the interaction/mechanisms of stabilizers are same, but LDPE was highly stabilized than iPP and EP matrices. The results of our research series proved that the synergistic mixtures based on oligomeric HAS stabilizer (i.e. CS, CTS, CTX, CXS, CTXS) significantly improve the stability of polyolefins (LDPE, iPP and EP copolymers) against γ-sterilization. The results also demonstrated that the phenol-free systems (i.e. CTS) can be good alternatives for the phenolic systems (i.e. X, CTX, CXS and CTXS).

The main objective of our research series was to find out the various possibilities of combinations of stabilizers which follow the different mechanisms of stabilization and are already approved for food contact and biomedical polyolefin applications. The objective of the present work is to study the synergism and antagonism and the stabilizing efficiency of those systems with LDPE and to conduct a comparative study on their effectiveness with other polyolefins matrices (iPP and EP copolymers). In iPP and EP matrices, possibilities of various binary, ternary and quaternary additive systems have been found [20–22]. Potentials of those systems in LDPE matrix were tested here. Almost the trend in stabilization efficiency is the same. The trend in stabilization efficiency of binary, ternary and quaternary systems of hindered amines, hindered phenol and organo-phosphite was confirmed in terms of tensile properties, discoloration and oxidation products as observed for iPP and EP copolymers. Thus, the mechanisms of stabilization of each stabilizer are not disturbed by various polyolefins. The obtained data may reflect the suitability, selection of different kinds of stabilizers and antioxidants to be combined with oligomeric HAS stabilizer and its effectiveness. The data also reveal the stabilization behavior of oligomeric HAS stabilizer when it was used alone and when it is used in combination with different stabilizers for protecting different polyolefins (LDPE, iPP and EP copolymers) against γ-sterilization. The results of the present study confirm the validity of those systems for protecting various polyolefins (LDPE, iPP and EP copolymer) against γ-sterilization using synergistic mixtures of stabilizers, which follow the different mechanisms of stabilization and are already approved for food contact applications.

Conclusion

The stabilization of γ-sterilized LDPE was tested with the mixtures of different stabilizers, which follow the different stabilization mechanisms and are having various molecular weights and are approved for food contact applications. The major objective of this study ws to conduct a comparative study on the effectiveness of those mixtures on LDPE with other polyolefin matrices (iPP and EP copolymers). In this study, we found that the synergism, antagonism and the trend in stabilization efficiency of the binary, ternary and quaternary stabilizer systems of oligomeric HAS, tertiary hindered amine, hindered phenol and organo-phosphite were almost similar in LDPE, iPP and EP copolymer matrices indicating that the interaction/mechanisms of stabilizers are same. The results show that the polyolefin's durability and yellowing formation due to phenolic antioxidants can be improved significantly by adding oligomeric HAS, tert-HAS and organo-phosphite, leading also to long-term stability. The molecular weight distribution of stabilizers in the mixture plays an important role in the overall stability. It was demonstrated that the phenol-free systems can be suitable alternatives for the phenolic systems. Thus, it can be concluded from the results of our research series that the stability of food packaging and medical polyolefins (LDPE, iPP and EP copolymers) against γ-sterilization can be improved by blends of different stabilizers which are approved for food contact applications.

Acknowledgements The authors are grateful to the director and the scientists in National Chemical Laboratory, Pune, council of Scientific and Industrial Research (CSIR), New Delhi, India and to Prof. B. S. M. Rao, Department of Chemistry, University of Pune, Pune, India for providing the facilities during this work.

References

1. Booth AF (1979) Sterilization of Medical devices. Interpharm Press, Illinois, Buffalo Grove
2. Sturdevant M (1991) Plast Eng 47:27–32
3. Deng M, Shalaby SW (1995) Effects of gamma irradiation, gas environments, and post-irradiation aging on ultrahigh molecular weight polyethylene. J Appl Polym Sci 58:2111–2119
4. Goldman M, Lee M, Grousky R, Pruitt L (1997) Oxidation of ultrahigh molecular weight polyethylene characterized by Fourier Transform Infrared Spectrometry. J Biomed Mater Res 37(1):43–50
5. Fisher J, Reeves EA, Isaac GH, Saum KA, Sanford WA (1997) Comparison of the wear of aged and non-aged ultrahigh molecular weight polyethylene sterilized by gamma irradiation and by gas plasma. J Mater Sci–Mater Med 8(6):375–378
6. Shintani H (2002) Effects of ionizing radiation sterilization treatment on medical use plastic materials. Biocontrol Sci 7(1):1–8
7. Klemchuk PP, Horng PL (1991) Transformation products of hindered phenolic antioxidants and colour development in polyolefins. Polym Degrad Stab 34:333–346
8. Emaldi I, Liauw CM, Potgieter H (2015) Stabilization of polypropylene for rotational molding applications. J Res Updates Polym Sci 4:179–187
9. Carlsson DJ, Chmela S (1983) Polymers and high-energy irradiation: degradation and stabilization. In: Jellinek HHG (ed) Degradation and stabilization of polymers, vol 1, chap 4. Amsterdam, Elsevier 1983
10. Nasibullah M, Ahmad N, Hassan F, Patel DK, Khan AR, Masihur R (2012) Assessment of physicochemical parameters of tubing's of intravenous infusion sets. Res J Pharm Sci 1(4):1–9
11. Nasibullah M, Ahmad N, Patel DK, Masihur R (2012) Assessment of Physicochemical Parameters in Nasal Feeding Tubes. Int J Life Sci Pharma Res 2(3):20–24
12. Vedanarayanan PV, Fernandez AC (1987) Toxicology of Biomedical Polymers. Def Sci J 37(2):173–183
13. Scott G (1995) Antioxidants in food packaging: a risk factor? Biochem SOC Symp 61:235–246 **(printed in Great Britain)**
14. David F (1981) In: Williams (ed) Systemic Aspects of Biocompatibility, vol 11. CRC-Press
15. Billingham N (1990) In: Pospisil J, Klemchuk PP (eds) Oxidation Inhibition in Organic Materials, chap 6:2. CRC Press, Boca Raton
16. Al-Malaika S, Goonetilleka MRJ, Scott G (1991) Migration of 4-substituted 2-hydroxy benzophenones in low density polyethylene: Part I - Diffusion characteristics. Polym Degrad Stab 32:231–247
17. Hassanpour S, Khoylou F (2007) Synergistic effect of combination of Irganox 1010 and zinc stearate on thermal stabilization of electron beam irradiated HDPE/EVA both in hot water and oven. Radiat Phys Chem 76:1671–1675
18. Haidar N, Karlsson S (2001) Loss of Chimassorb 944 from LDPE and identification of additive degradation products after exposure to water, air and compost. Polym Degrad Stab 74:103–112
19. Malik J, Hrivik A, Alexyova D (1992) Polym Degrad Stab 35:25
20. Alariqi SAS, Pratheep Kumar A, Tevtia AK, Rao BSM, Singh RP (2006) Stabilization of γ-sterilized biomedical polyolefins by synergistic mixtures of oligomeric stabilizers. Polym Degrad Stab 91:2451–2464
21. Alariqi SAS, Kumar AP, Rao BSM, Singh RP (2007) Stabilization of γ-sterilized biomedical polyolefins by synergistic mixtures of oligomeric stabilizers. Part II. Polypropylene matrix. Polym Degrad Stab 92:299–309

22. Alariqi SAS (2015) Stabilization of γ-sterilized food-packaging materials by synergistic mixtures of food-contact approval stabilizers. Int J Biol Food Vet Agric Eng 9(2)

23. Przybytniak G, Mirkowski K, Rafalski A, Nowicki A, Legocka I, Zimek Z (2005) Effect of hindered amine light stabilizers on the resistance of polypropylene towards ionizing radiation. NUKLEONIKA 50(4):153–159

24. Sun GJ, Jang HJ, Kaang S, Chae KH (2002) A new polymeric HALS: preparation of an addition polymer of DGEBA-HALS and its photostabilizing effect. Polymer 43:5855–5863

25. Malik J, Hrivik A, Tomova E (1992) Diffusion of hindered amine light stabilizers in low density polyethylene and isotactic polypropylene. Polym Degrad Stab 35:61–66

26. Kikkawa K (1995) New developments in polymer photostabilization. Polym Degrad Stab 49:135–143

27. Chmela S, Hrdlovic P, Manasek Z (1985) Polym Degrad Stab 11:233

28. Clough L, Billingham NC, Gillen KT (1996) Polymer durability degradation, stabilization, and lifetime prediction. American Chemical Society, Washington, DC, pp 455–470

29. Gugumus F (1981) Res Discl 209:357

30. Ciba specialty chemicals additives for polyolefins key products selection guide, Pub. No. 016269.00.040, Switzerland: Ciba Specialty Chemicals Inc. http://cibasc.com/

31. Carlsson DJ, Wiles DM (1969) The photodegradation of polypropylene films. II. photolysis of ketonic oxidation products. Macromolecules 2:587–597

32. Annual Book of ASTM Standard (1988) American Society for Testing and Materials, vol 8.02, Philadelphia

33. Billmeyer FW, Sultzman (1966) Principles of colour Technol. Interscience, New York, p 38

34. Clough L, Billingham NC, Gillen KT (1996) Polymer durability degradation, stabilization, and lifetime prediction, chap 39. American Chemical Society, Washington, DC, pp 375–396

35. Kawamura Y, Sayama K, Yamada T (2000) Shokuhin Shosha 35(1–2):7–14

36. Stoffer NH (2004) Certified reference materials for food packaging specific migration tests: development, validation and modeling. Ph.D. Thesis, University Wageningen, The Netherlands

37. Carlsson DJ, Krzymien ME, Deschenes L, Mercier M, Vachon C (2001) Phosphite additives and their transformation products in polyethylene packaging for γ-irradiation. Food Addit Contam 18(6):581–591

38. Allen DW, Leathard DA, Smith C (1991) The effects of gamma-irradiation on the fate of hindered phenol antioxidants in food contact polymers - analytical and C-14 labeling studies. Radiat Phys Chem 38(5):461–465

39. Pospisil J (1993) Chemical and photochemical behaviour of phenolic antioxidants in polymer stabilization: a state of the art report, part II. Polym Degrad Stab 39:103–115

40. Vyprachticky D, Pospisil J (1990) Possibilities for cooperation in stabilizer systems containing a hindered piperidine and a phenolic antioxidant—a review. Polym Degrad Stab 27(3):227–255

41. McFarlane D (2002) Diffusion controlled investigation of the parameters responsible for the efficiency of hindered amine light stabilizers, Ph.D. Thesis, Department of Applied Chemistry, Faculty of Applied Science, Royal Melbourne Institute of Technology, Melbourne, Australia

42. Allen NS, Ortiz RA, Anderson GJ (1998) Interaction in the thermal and light stabilizing action of novel aromatic phosphites with a 2-hydroxybenzophenone and hindered piperidine stabilizer in polyolefin film. Polym Degrad Stab 61(2):183–199

43. Murayama KJ (1971) Synth Org Chem 29:366

44. Pospisil J (1991) The key role of antioxidant transformation products in the stabilization mechanisms—a critical analysis. Polym Degrad Stab 34:85–109

45. Pospisil J (1992) Exploitation of the current knowledge of antioxidant mechanisms for efficient polymer stabilization. Polym Adv Technol 3:443–455

46. Carloni P, Greci L, Stipa P, Rizzoli C, Sgatabotto P, Ugozzoli F (1993) Antioxidants and light stabilizers. Part 1. Reactions of an indolinone nitroxide and phenoxy radicals. X-ray crystallographic analysis of 1-[O-(3,5-di-tert-butyl-4-hydroxy)-benzyl]-1,2-dihydro-2-methyl-2-phenyl-3-oxo-3H-indole and 3,5,3′5′-tetra-tert-butylstilbene-4,4′-quinone. Polym Degrad Stab 39:73–83

47. Pauquet JR (1992) In: Processing of wire and cable, the 42nd international symposium. Eatontown, NJ, pp 35:105

48. Drake WO, Pauquet JR, Tedesco RV, Zweifel H (1990) Processing stabilization of polyolefins. Angew Makromol Chem 176(177):215–230

13

Kinetics study of platinum and base metals precipitation in gas–liquid chloride system

J. Siame[1] 🄳 · H. Kasaini[2]

Abstract The use of dissolved sulphur from gaseous phase was tested for its ability to precipitate platinum ions in chloride system with alterations in the parameters such as metal ion concentration, hydrochloric acid concentration, pressure and temperature. The precipitation process was analysed on the basis of mass transfer coefficient, diffusivity models and also by absorption kinetic models. The maximum physical and reactive absorption capacity for sulphur dioxide was found to be 0.015 and 0.018 mol/L, respectively, at 298.15 K and 1.125 bars in a short contact time of 10 min. Based on the sulphur dioxide solubility data in gas–liquid system for both physical and reactive absorption conditions, the introduction of sulphur atoms in the chloride system was achieved. The mass transfer coefficients obtained for all metal ions were in the range 0.10–0.26 min^{-1}, while the diffusivity of SO_2 at different temperatures and pressures was found to be in the range $(2.36$–$3.43) \times 10^{-9}$ and $(1.36$–$2.22) \times 10^{-9}$ $m^2 s^{-1}$, respectively. Thermodynamic parameters such as the changes in Gibbs free energy, enthalpy and entropy were also evaluated. The results indicated that the precipitation of platinum and base metals in gas–liquid (G–L) chloride system was an endothermic process.

Keywords Diffusivity · Mass transfer coefficient · Platinum · Precipitation · Sulphur dioxide

✉ J. Siame
john.siame@cbu.ac.zm

[1] Department of Chemical Engineering, School of Mines and Mineral Sciences, Copperbelt University, Kitwe, Zambia

[2] US Metals Refining Group, Inc.,, Colorado, USA

Introduction

Global concerns about the impact of toxic gases and effluents on the environment have led to the platinum industry leaders to adopt new strategies of recycling or capturing mine waste. Carbon dioxide (CO_2), for example, from coal-fired power stations and the environment is currently being captured and stored in coal seams underground [1, 2]. Hesselmann and Hough have reported the development of the technology of converting toxic oxides of nitrogen (NO_x) from well-designed boiler burners using ammonia-based reagent and hydrocarbon into a flue gas containing NO_x and some O_2 [3]. They found that at elevated temperatures the hydrocarbon auto ignites, forming plasma, and creating radicals. These radicals catalysed the NO_x reduction reactions—autocatalysis—and the resulting flue gas was found to contain very low NO_x and small level of ammonia slip. Other researchers have also reported on the concept of capturing SO_2 from fluidized-bed reactors during coal gasification using dolomite [4]. However, there is scanty literature on the application of acidic chloride solutions as a means of capturing SO_2 from flue gases and the same time precipitate the precious metals as sulphides. It is established that dissolved sulphur reacts readily with base and precious metal ions to form metal sulphides [5–7]. In industrial sedimentation processes, reduction of metal ions such as sulphides is accomplished by contacting metal-rich solutions with commercial H_2S or SO_2 gases. To utilize commercial SO_2 gases or other sulphur-bearing gases for recovering metals from solution, it is necessary to put in place effective mechanisms for capturing and solubilizing gases.

The conventional separation and purification of platinum group metals (PGMs) in industry combines liquid–liquid precipitation in alkaline media, distillation (boiling off volatile metals) and ion exchange in resin columns. Distillation is an energy-intensive process and ion exchange process requires huge operational costs due to imported resins. Resins are not only expensive but also toxic. The concept of selective gas–liquid precipitation of PGMs using waste sulphide gases from the smelter is more attractive.

In this study, the concept of capturing SO_2 from industrial flue gases by means of acidic chloride solutions has been assessed in a continuous process. The main postulate is that dissolved sulphides can serve as precipitants for platinum ions in chloride system. The effects of different parameters such as initial metal ion concentration, hydrochloric acid concentration, pressure and temperature have been investigated. The absorption rates and thermodynamic parameters were deduced from the absorption measurements.

Theory of absorption kinetics

The absorption kinetic studies in G–L process were conducted to establish the parameters which affect metal (Fe, Co and Cr) precipitation in G–L chloride systems. The study was predominantly concerned with the precipitation rates of Pt as this was the main target metal in the mixed chloride systems. Furthermore, the mass transfer coefficients (K_L), diffusivities (D_L), absorption rates (N), activation energy (E_a) and the thermodynamic properties (ΔH, ΔG and ΔS) associated with platinum were quantified.

In G–L process using the continuous stirred tank reactor (CSTR), chloride solutions of Pt and base metals are used to absorb SO_2 to form solid products (precipitates). There are several reactions and physical steps which may affect the rate of SO_2 absorption within acidic chloride solutions. These include [8, 9]:

(a) The diffusion of SO_2 through the gas film near the gas–liquid interface.

(b) The dissolution of SO_2 in liquid phase (gas–liquid equilibrium):

$$SO_{2(g)} \leftrightarrows SO_{2(aq)} \tag{1}$$

(c) The first dissociation of $SO_{2(aq)}$:

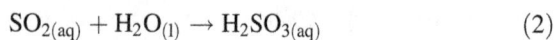

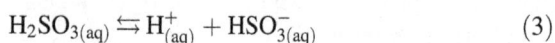

$$SO_{2(aq)} + H_2O_{(l)} \rightarrow H_2SO_{3(aq)} \tag{2}$$

$$H_2SO_{3(aq)} \leftrightarrows H^+_{(aq)} + HSO^-_{3(aq)} \tag{3}$$

(d) The second dissociation of SO_2:

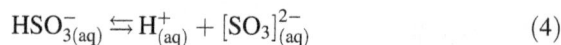

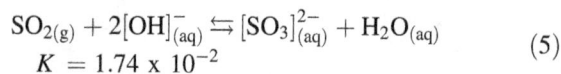

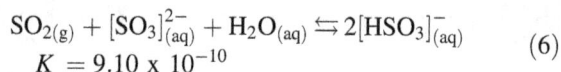

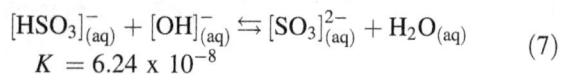

$$HSO^-_{3(aq)} \leftrightarrows H^+_{(aq)} + [SO_3]^{2-}_{(aq)} \tag{4}$$

$$SO_{2(g)} + 2[OH]^-_{(aq)} \leftrightarrows [SO_3]^{2-}_{(aq)} + H_2O_{(aq)} \\ K = 1.74 \times 10^{-2} \tag{5}$$

$$SO_{2(g)} + [SO_3]^{2-}_{(aq)} + H_2O_{(aq)} \leftrightarrows 2[HSO_3]^-_{(aq)} \\ K = 9.10 \times 10^{-10} \tag{6}$$

$$[HSO_3]^-_{(aq)} + [OH]^-_{(aq)} \leftrightarrows [SO_3]^{2-}_{(aq)} + H_2O_{(aq)} \\ K = 6.24 \times 10^{-8} \tag{7}$$

where K is the equilibrium constant. The equilibrium constant is important because it gives where the equilibrium lies. The larger the equilibrium constant, the further the equilibrium lies towards the products. Therefore, on the basis of reactions (5) and (7), $[SO_3]^{2-}$ ions are the main product when SO_2 reacts with water $[OH]^-$ molecules in a few stages. This suggests that at equilibrium, most of the chemical species present ($[SO_3]^{2-}$) are products. From this view, it is possible to correlate the absorption rate (N) of SO_2 with SO_3^{2-} in the bulk solution as shown in Eq. (8):

$$N = C^*_{SO_2} \times a \times \sqrt{D_{SO_2} \times k \times C^{bulk}_{SO_3^{2-}}} \tag{8}$$

where N is the absorption rate of SO_2 (mol m^{-3} s^{-1}); $C^*_{SO_2}$ and D_{SO_2} the solubility and diffusivity of SO_2 in the aqueous phase, respectively; a the interfacial area, k the rate constant for the reaction between SO_2 and SO_3^{2-} ions at the interface and $C^{bulk}_{SO_3^{2-}}$ is the SO_3^{2-} concentration in the bulk of aqueous phase.

Two-film model

In this study, to describe this process in terms of the reaction kinetics, two-film model was used and the reaction rates, mass transfer and diffusion coefficients have been evaluated accordingly. The focus of the gas absorption experiments is based on understanding the mass transfer processes in reactive and non-reactive solvent systems. To model the mass transfer of the gases, it is necessary to generate the absorption kinetics data with or without accompanying chemical reaction. In the two-film model, gaseous molecules in the bulk phase diffuse through a gas–liquid film and then are transferred into the bulk solution; therefore, the total resistance to mass transfer consists of diffusion resistance at the gas–liquid interface and mass transfer resistance in liquid phase.

The overall mass transfer of sulphur atoms from bulk gas phase into chloride solution (mol s^{-1}) may be expressed by Eq. (9):

$$\frac{dm}{dt} = \frac{1}{\left(\frac{1}{k_L} + \frac{1}{k_G}\right)} \times a \times \left(C_{A,i} - C_{A,L}\right)$$

$$= K \times a \times \left(C_{A,i} - C_{A,L}\right) \tag{9}$$

where $\left(\frac{1}{k_L} + \frac{1}{k_G}\right)$ is the overall mass transfer resistance; K and $\frac{1}{k_G} \approx 0$ since the gas phase is pure with only sulphur atoms. Rate of mass transfer may be rewritten to take into account the constant volume of the solution (mol m^{-3} s^{-1}) and interfacial area (m^2 m^{-3}) exposed to a pure gas phase as shown by Eqs. (10) and (11):

$$d\frac{\left(\frac{m}{V}\right)}{dt} = \frac{dC_{A,L}}{dt} = K_L \times a \times \left(C_{A,i} - C_{A,L}\right) \tag{10}$$

and then reduces to Eq. (8)

$$\frac{dC_L}{dt} = K_L \times a \times \left(C_i - C_L\right) \tag{11}$$

where K_L is the mass transfer resistance in the liquid phase. The concentration gradient can be correlated to the pressure drop of the gas during absorption [10].

Methods

Reagents and instruments

All reagents used were pure analytical standard grade solutions purchased from Merck Company (Johannesburg, South Africa). Pure analytical grade SO_2 (99.9 %) gas in pressurized gas cylinders purchased from AFROX (Pty) Ltd, South Africa was used. All standard solutions were prepared in equal concentrations of 100 mg/L using double-distilled water with zero levels of dissolved oxygen. Different concentrations of HCl acid solutions (1.0, 3.0 and 4.0 M) were prepared by diluting HCl acid solution (10 M, 32 % HCl) with known quantities of distilled water. 99.9 % pure analytical grade SO_2 gas in pressurized gas cylinders was used in G–L precipitation tests at moderately high pressure (0.5, 1.0, 1.5 and 2.0 bars). The concentration (99.9 %) and flow rate (177 ml/min) of SO_2 gas were kept constant throughout the tests. The stirring speed (500 rpm) was also kept constant throughout the tests. Barren solution samples which remained after precipitation were analysed for platinum, iron, cobalt and chromium by means of an Inductively Coupled Plasma–Optical Emission Spectrophotometer (ICP–OES) instrument (*Shimadzu model ICPE–9000*). Solid samples (dry precipitates) were analysed using a Scanning Electron Microscope (SEM), X-ray Diffraction (XRD) and Energy Dispersive X-Ray

Spectroscopy (EDS). The experiments were carried out using a CSTR, (2L capacity, Büchiglasuster pressure glass vessel autoclave).

Experimental setup

The experiments were carried out in a thermostatted 2-L Büchiglasuster glass reactor as shown in Fig. 1. The reactor is heated by water through a jacketed wall around it. The reactor dimensions are shown in Table 1. The reactor was operated batch-wise with respect to the liquid phase but continuous with respect to the gas phase. The liquid was stirred mildly with a six-bladed Rushton turbine, diameter 50 mm, height 15 mm, located centrally in the liquid at a height above the reactor bottom equal to half the reactor diameter. The gas phase (on top of the liquid phase) was stirred with a three-bladed impeller having a diameter of 50 mm. Both impellers were mounted on a single axis with a diameter of 10 mm. The speed of stirrer was kept constant at 500 rpm throughout the tests. The reactor was equipped with four symmetrically mounted glass baffles with a height of 104 mm and a width of 8 mm. The glass baffles increased the effectiveness of stirring and prevented the formation of a vortex.

Isothermal conditions in the reactor were maintained by circulating hot water from a thermostatted water bath (Tamson T-1000) through the jacket of the reactor. The temperature of the reactor contents was measured by means

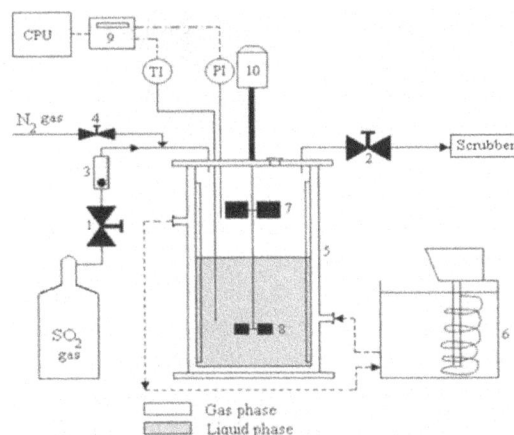

Fig. 1 Stirred cell reactor: SO_2 gas inlet and outlet valves (*1, 2*); SO_2 gas flowmeter (*3*); N_2 gas inlet valve (*4*); stirred cell reactor (*5*); thermostat water bath (*6*); gas stirrer (*7*); liquid stirrer (*8*); data logger (*9*); magnetic coupling mixer (*10*)

Table 1 Reactor dimensions: continuously stirred tank reactor

Reactor diameter	0.0825 m
Reactor volume	2.0×10^{-3} m^3
Gas–liquid contact area	5.28×10^{-3} m^2 m^{-3}
Liquid impeller type	Six-bladed turbine, 0.05 m diameter
Gas impeller type	Three-bladed turbine, 0.05 m diameter

of a K-type thermocouple. The pressure in the reactor was determined with a Druck PDCR 910 pressure transducer. The pressure and temperature transducers were connected to a Squirrel SQ 1000 series Data Logger and Mercer Premium computer, enabling data collection and programmed reactor operation. A desired mass flow rate of pure SO_2 from the cylinder to the reactor was controlled using a mass flow meter (Brooks 5150 T). The gas leaving the reactor was passed through a caustic scrubber before discharging it to the atmosphere. A pressure regulator was employed to control gas pressure in the reactor (Tescom series 1700 back pressure regulator).

For SO_2 absorption experiments, the gas–liquid reactor was loaded with specific solvent (with or without metal ions in a chloride solution). The concentration of each metal ion was kept constant at 100 mg/L. Acid strength was varied in the range 1–4 M HCl. The reactor content was initially degassed by purging with pure nitrogen gas. The starting solution was allowed to equilibrate with its own saturated vapour at room temperature. The temperature of the metal ion solutions was varied in the range 298 K (25 °C)–313 K (40 °C), after which, at $t = 0$, a feed gas (SO_2) was introduced into the reactor from a high-pressure cylinder possessing a regulator at a constants flowrate of 177 Nml/min. The pressure in the reactor was varied in the range 0.5–2.0 bars (50–200 kPa). Table 2 summarizes the test conditions in the CSTR. After filling the reactor with a SO_2 gas, the stirrer was started and the pressure drop in the reactor was recorded over time. At equilibrium, the pressure drop levelled off and a solution sample was taken from the bulk solution to determine the maximum absorption capacity of SO_2 and amount of precipitate formed. The amount of sulphur sequestered in metal-rich and metal-free chloride solvents (>1 M HCl acid) was recorded. The rate of metal precipitation was not quantified due to difficulties (lack of a sampling gun) associated with solution sampling at high pressure. After 24 h, the composite precipitate from the batch glass vessel was filtered and dried at room temperature. The composite clear filtrate solution remaining was analysed for metal ions using ICP-OES. Dry precipitates were analysed for metal ions and sulphur by means of XRD, SEM and EDS. The flow of the gas was continuous but the liquid was stationary in the vessel. From scale-up point of view, this

can be a huge constraint. However, the objective was not to develop parameters for scaling up the process but rather to optimize SO_2 solubility and Pt selectivity at this stage.

Results and discussion

Effect of initial metal ion concentration

The feasibility and efficiency of a precipitation process depends not only on the properties of the precipitants, but also on the concentration of the metal ion solution. In this study, the effect of initial concentrations of metal ions (single- and multi-component system) was investigated and the results are illustrated in Figs. 2, 3, and 4.

Figure 2 shows the effect of varying initial Pt and base metal (Fe, Co and Cr) ion concentrations from 50 to 120 mg/L. The data are plotted as SO_2 concentration in the bulk liquid phase versus contact time. The concentration of SO_2 in the bulk liquid phase of chloride solution increased with an increase in initial metal ion concentration. Diffusion is a passive transport process driven by the concentration gradient at the G–L interface. At higher initial metal ion concentration, the driving force for precipitation, which is the difference between the gas phase film concentration and the liquid phase film concentration, is higher. This leads to higher absorption rate in short contact time, for higher initial metal ion concentration. Furthermore, Fig. 2 shows that the absorption capacity sharply increased in the first 10 min (over 80 % gas absorbed) followed by a slower subsequent absorption capacity that gradually approached an equilibrium condition. The average absorption of SO_2

Table 2 Experimental conditions: gas–liquid precipitation

Temperature	25–40 °C
Initial pressure	0.5×10^5 Pa
Liquid volume	1.5×10^{-3} m³
Gas	N₂, purity >99.5 %; SO_2, purity >99.5 %
	SO_2, purity >99.5 %
Stirrer speed	500 rpm

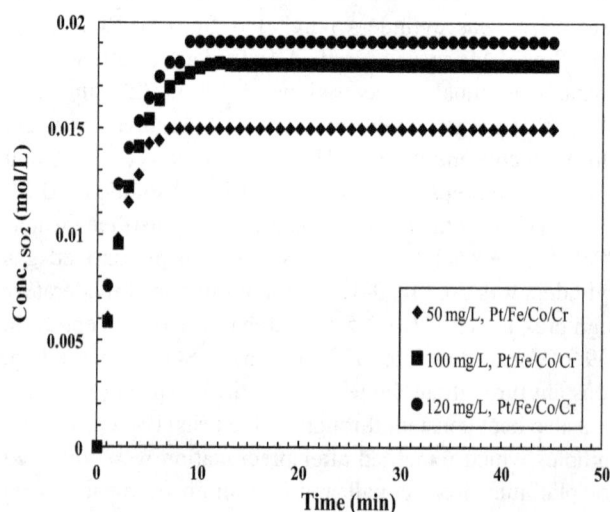

Fig. 2 Effect of initial concentration of multi-metal system (Pt/Fe/Co/Cr) on the absorption of SO_2 in a CSTR. Conditions: $P = 1.125$ bar; $T = 25$ °C; stirring speed = 500 rpm; inlet SO_2 gas conc. = 99.9 vol %; [HCl] = 4 M; [Me^{n+}] = 100 ppm; contact time = 40 min

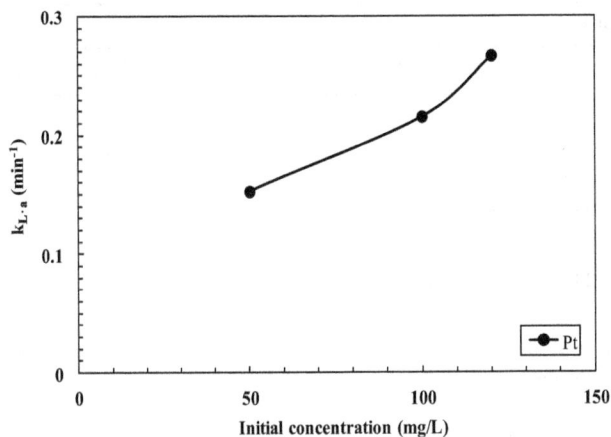

Fig. 3 Mass transfer coefficients vs. initial concentration for Pt ions in chloride system. Conditions: $P = 1.125$ bar; $T = 25$ °C; stirring speed = 500 rpm; inlet SO_2 gas conc. = 99.9 vol %; [HCl] = 4 M; $[Pt^{2+}] = 100$ ppm; contact time = 40 min

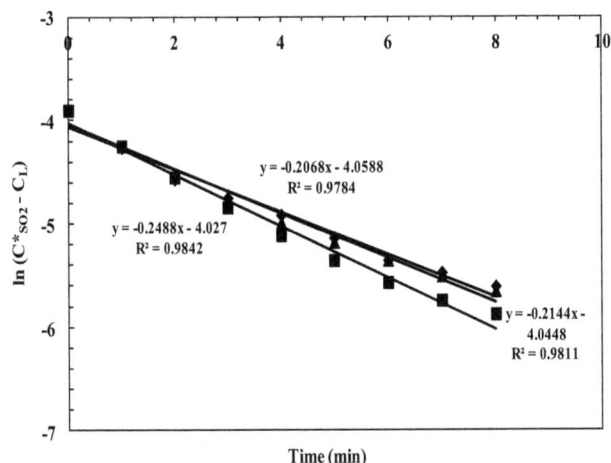

Fig. 4 A plot of $\ln\left(C^*_{SO_2} - C_L\right)$ vs. contact time. Determination of mass transfer coefficients for absorption of SO_2 into 4 M HCl chloride solution containing Pt/Fe/Co/Cr ions. Conditions: $P = 1.125$ bar; $T = 25$ °C; stirring speed = 500 rpm; inlet SO_2 gas conc. = 99.9 vol %; [HCl] = 4 M; $[Me^{n+}] = 100$ ppm; contact time = 40 min

gas after 10 min was 0.0165 mol/L and recovery of Pt from all chloride solutions averaged above 99 % in less than 10 min. It is evident that the rate of increase of SO_2 concentration in the bulk liquid phase was dependent on the initial metal ion concentration.

The absorption kinetics of SO_2 in chloride solution (4 M HCl) containing Pt/Fe/Co/Cr ions are shown in Figs. 3 and 4, respectively. The rate of mass transfer $K_L \times a$ may be limited by resistance in the gas phase film or liquid phase film. Thus, the $K_L \times a$ values were evaluated using Eq. (11), (Fig. 4); the results are summarized in Table 3. As the initial concentrations were raised, the mass transfer coefficients increased in a non-linear fashion as shown in Fig. 3.

The experimental data in this study indicate that the rate of increase of SO_2 concentration in the bulk liquid phase increased by approximately 18 % when the initial concentration of Pt and base metal ions concentration was doubled from 50 to 100 mg/L, and also increased by approximately 27 % when the initial concentration of Pt and base metal ions was increased from 50 mg/L to 120 mg/L. The high rate of increase of SO_2 concentration in bulk liquid phase was achieved at the initial concentration of 120 mg/L in both cases (Fig. 2). This indicates that metal precipitation with SO_2 was found to follow the two-film theory. The mass transfer coefficients obtained for all metals increased with an increase in metal ion concentrations.

Determination of mass transfer coefficient

Figure 5 illustrates the absorption kinetics of SO_2 in 4 M HCl acid solution at 25 °C and also the gas pressure of SO_2 changing with time. The gas pressure dropped from 1.125 to 0.22 bars within approximately 20 min of contact time (Fig. 5). The corresponding increase in sulphur concentration is shown as a mirror image of the pressure drop. The average absorption of SO_2 after 20 min was 0.018 mol/L. This data imply that physical absorption of SO_2 in concentrated HCl solution is possible at room temperature. The reason why we carried out absorption tests at high HCl acid strength was because the final solution from the elution circuit of Pt recovery comes at high acid concentrations in the range of 3.5–4.0 M, HCl. From the literature, it is known that Pt can only be precipitated out of the solutions at high HCl concentration (>3.5 M) using sodium thiosulphate [11, 12]. The mass transfer coefficient of SO_2 in HCl acid solution was evaluated using Eq. (11) and data points from Fig. 5. Therefore, $\ln\left(C^*_{SO_2} - C_{SO_2,i}\right)$ was plotted against contact time according to Eq. (11) and (Fig. 6). The mass transfer coefficient and corresponding concentration values of dissolved sulphur atoms are summarized in Table 4.

From the gas absorption point of view, the mass transfer characteristics of SO_2 from the bulk gas phase into mixed chloride solvents were established. According to the results, it was found out that the presence of metal ions in the solvent accelerated the uptake of sulphur atoms from the gas phase into a liquid phase through a chemical reaction. Dissolved Sulphur (DS) atoms react covalently with base metal ions to form insoluble sulphides; however, they interact with platinum chloro-complex anions by exchanging Cl^- and S^{2-} ions. The latter reaction is characterized by a very small activation energy which makes it possible for platinum sulphide to nucleate preferentially. Resistance to diffusion of a species in a pure gas phase was

Table 3 Mass transfer coefficients with metal ions species in chloride system at different initial metal ion concentrations

Metal ions initial conc. (mg/L)	Pt ions		Pt/Fe/Co/Cr ions	
	$K_L \times a$ (min^{-1})	$[S^0]_{equiv.}$ (mol/L)	$K_L \times a$ (min^{-1})	$[S^0]_{equiv.}$ (mol/L)
50	0.1523	0.014	0.2068	0.015
100	0.2155	0.017	0.2144	0.018
120	0.2667	0.018	0.2488	0.019

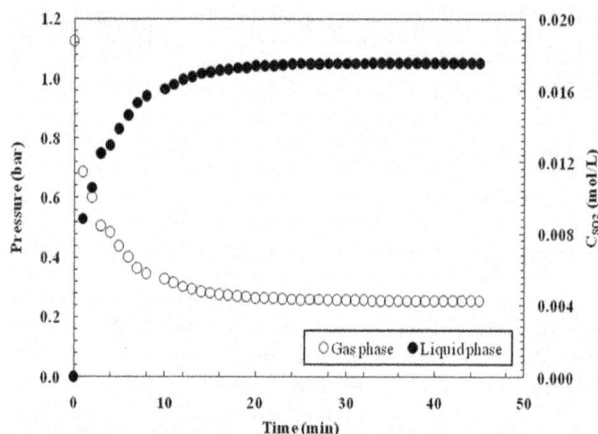

Fig. 5 A plot of pressure drop of SO_2 vs. contact time. Physical absorption of SO_2 into chloride solution at 25 °C. [HCl] = 4 M. In gas phase, the concentration of SO_2 gas was calculated using an ideal gas law. Conditions: P = 1.125 bar; T = 25 °C; stirring speed = 500 rpm; inlet SO_2 gas conc. = 99.9 vol %; [HCl] = 4 M; [Me^{n+}] = 100 ppm; contact time = 40 min

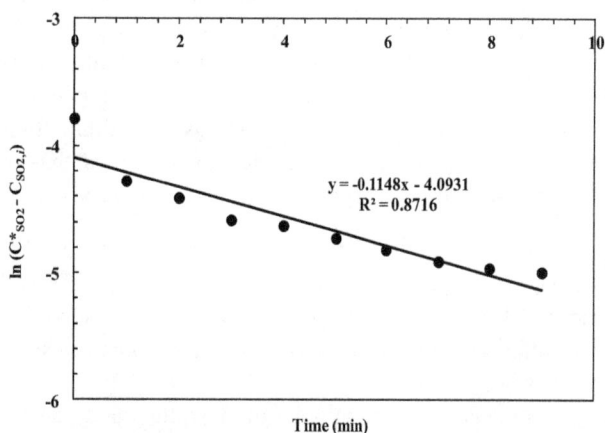

Fig. 6 A plot of $\ln\left(C^*_{SO_2} - C_{SO_2,i}\right)$ vs. contact time. Evaluation of mass transfer coefficient for physical absorption of SO_2 into 4 M (HCl) chloride solution at 25 °C and a gas pressure of 1.125 bars. Conditions: P = 1.125 bar; T = 25 °C; stirring speed = 500 rpm; inlet SO_2 gas conc. = 99.9 vol %; [HCl] = 4 M; [Me^{n+}] = 100 ppm; contact time = 40 min

Table 4 Mass transfer coefficients with and without metals in solution

System	$K_L \times a$ (min^{-1})	$[S^0]_{equiv.}$ (mol/L)
Physical absorption	0.1148	0.0161
Absorption with *Pt*	0.1575	0.0164
Absorption with *Pt/Fe/Co/Cr*	0.2495	0.0182

bulk liquid increases as the absorption process proceeds. Therefore, the mass transfer model of SO_2 was modelled through the absorption kinetics data with or without an accompanying chemical reaction.

Effect of hydrochloric acid concentration on physical absorption of SO_2 gas

Figure 7 shows that the increase in HCl concentration adversely affects the physical absorption of SO_2. In this study, HCl was varied in the range 1–4 M because Pt exists as anionic chloro-complexes in this range. As the concentrations of HCl were increased, the mass transfer coefficient decreased in non-linear form as shown in Fig. 7. The results in Fig. 7 were attributed to the salting-out effect of gaseous molecules in the acidic solution and also probably due to an increase in the interfacial tension at the liquid surface. Platinum ions exist

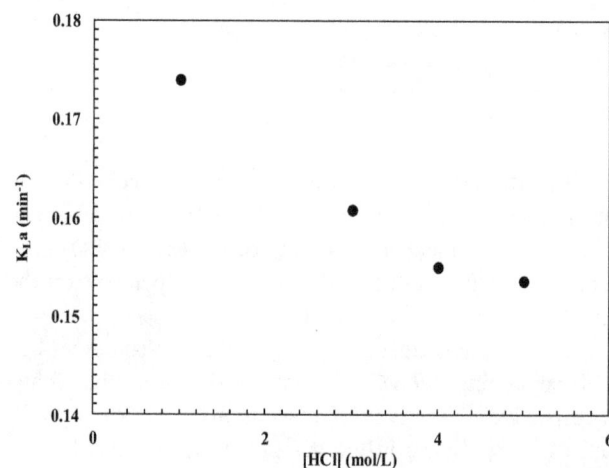

Fig. 7 A plot of mass transfer coefficients in liquid phase vs. [HCl]. Conditions: P = 1.125 bar; T = 25 °C; stirring speed = 500 rpm; inlet SO_2 gas conc. = 99.9 vol %; [HCl] = 1–5 M; [Me^{n+}] = 100 - ppm; contact time = 40 min

considered to be negligible because the process of pure SO_2 absorption with no suspended particles, allows the gas to dissolve faster and directly in the bulk liquid and dissolved SO_2 can be consumed by reactions in Eq. (5) and Eq. (6) simultaneously. The concentration of SO_3^{2-} in the as stable chloro-complex anions in highly acidic chloride solutions [13]. This implies that the degree of Pt ionization is strongly dependent on the [Cl] concentration.

Thermodynamics parameters

The effect of chloride solution temperature on SO_2 absorption, at constant acid strength (4 M HCl), pressure (1.125 bar), stirring speed (500 rpm) and metal concentration doses (100 mg/L) was studied and the results are shown in Fig. 8. The kinetic experiments were conducted at 298.15, 303.15 and 313.15 K. Figure 8 illustrates that SO_2 solubility was slightly affected by temperature in the range 298.15–313.15 K. It is well known that physical absorption of gases is affected by increase in temperature due to the shift in the vapour liquid equilibrium (VLE). According to the Le Chatelier's principle, when the temperature to a system at equilibrium is increased, the system should move in a direction so that the added heat is absorbed. Thus, an increase in the temperature of a chemical system at equilibrium favours an endothermic reaction. Furthermore, the average kinetic energy of molecules in a gas depends on the temperature. Therefore, when the temperature is increased the gas molecules will move more rapidly and the average kinetic energy of the gas molecules increases. Thermodynamically, gas molecules prefer to exist in the gas phase at high temperatures.

Table 5 illustrates the calculated values of diffusion coefficient, D_L, from Eq. (8) and rate constant, k_R at different temperatures. The diffusion coefficient (D_{SO_2}) values of SO_2 into metal solution in chloride systems increased with temperature as expected and the relationship

between diffusion coefficients of SO_2 in metal solution and the temperature is linear as shown in Fig. 9 [14, 15].

The k_R values decrease with temperature due to the fact that k_R is a "phenomenal" constant which corresponds not only to the reaction between SO_2 and metal ions in the solution, but also the total phenomenon where the uptake of SO_2 by metal ions in the solution takes place first [16].

Thermodynamic behaviour of SO_2 absorption was further investigated by considering the dependence of diffusion coefficient on solution temperature (Fig. 9; Arrhenius plot). A change in D_{SO_2} for activation diffusion can be correlated with temperature by use of the Arrhenius-type equation given by:

$$D_{SO_2} = Ae^{-\frac{E_a}{RT}} \tag{12}$$

where A is the pre-exponential factor, R the universal gas constant, T the absolute temperature and E_a the activation energy, representing the minimum energy that the reacting system must attain for the absorption to proceed. The value of activation energy E_a depends on the controlling regime in the absorption process. According to researchers [17], absorption process is said to be film diffusion-controlled when $E_a < 16$ kJ/mol and chemical reaction-controlled when $E_a > 50$ kJ/mol. Linearization of Eq. (12) gives

$$\ln D_{SO_2} = \ln A - \frac{E_a}{RT} \tag{13}$$

The logarithmic diffusional time constant, $\ln D_{SO_2}$, was plotted against the reciprocal of absolute temperature, $1/T$, as shown in Fig. 9. The experimental activation energy was determined from the slope of the plot. An activation energy value of 18.6 kJ/mol was obtained confirming that film diffusion might have been the essential rate-limiting step in the absorption process.

The enthalpy, free energy and entropy changes were also determined. The enthalpy change, ΔH, is given by

$$\Delta H = E_a - RT \tag{14}$$

The enthalpy changes determined according to Eq. (14) were 16.12, 16.08 and 15.99 kJ/mol for chloride solution temperatures of 298.15, 303.15 and 313.15 K, respectively. The positive values of ΔH confirm the endothermic nature of absorption/precipitation process. When absorption is endothermic, the precipitation or recovery of the target ions increases with an increase in solution temperature. On the other hand, the values for the free energy change (ΔG) were calculated at different solution temperature using the following expression:

$$\Delta G = -RT \ln\left(C_{SO_3^{2-}}/C_{SO_2}^*\right) \tag{15}$$

where $C_{SO_3^{2-}}$ is the gas concentration of SO_2 with SO_3^{2-} in the bulk liquid phase of chloride solution at equilibrium

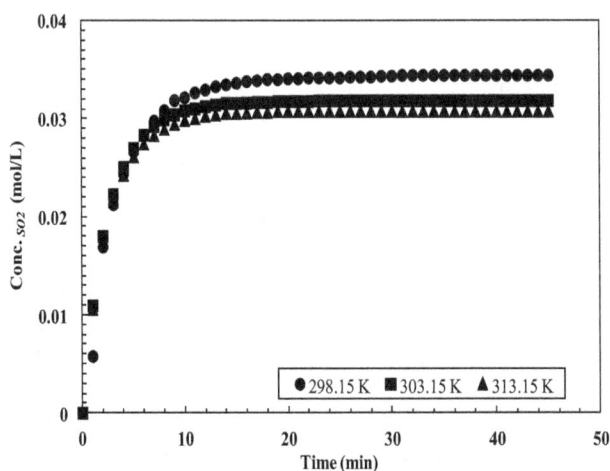

Fig. 8 Dependence of concentration of SO_2 on contact time at different temperatures. The effect of temperature on physical absorption of SO_2 in chloride solution at 298.15, 303.15 and 313.15 K. P = 1.125 bar; stirring speed = 500 rpm; inlet SO_2 gas conc. = 99.9 vol %; [HCl] = 4 M; [Me^{n+}] = 100 ppm; contact time = 40 min

Table 5 Summary of thermodynamic properties of SO_2 absorption in G–L chloride system as a function of temperature

Temperature (K)	$D_{SO_2} (10^{-9}\ m^2/s)$	ΔH (kJ/mol)	ΔG (kJ/mol)	ΔS [J/(mol K)]
298.15	2.36	16.12	−3.14	64.59
303.15	2.89	16.08	−3.16	63.47
313.15	3.43	15.99	−3.24	61.41

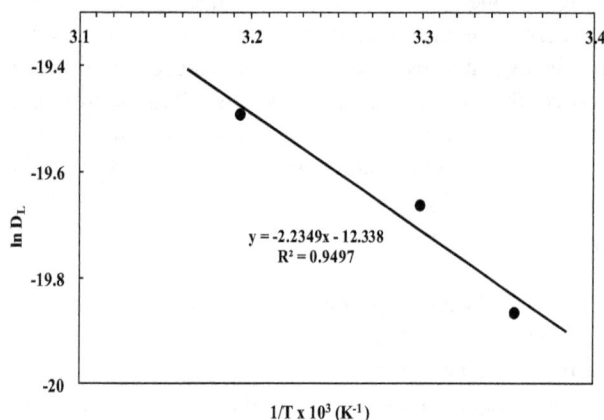

Fig. 9 Dependence of effective diffusion coefficient on temperature: Arrhenius plot. Conditions: $P = 1.125$ bar; T = 298–313 K; stirring speed = 500 rpm; inlet SO_2 gas conc. = 99.9 vol %; [HCl] = 4 M; [Me^{n+}] = 100 ppm; contact time = 40 min

(mol/L), $C^*_{SO_2}$ interfacial concentration of SO_2 (mol/L) at equilibrium. The ΔG values were −3.14, −3.16 and −3.24 kJ/mol for chloride solution temperatures of 298.15, 303.15 and 313.15 K, respectively. The ΔG values indicate the spontaneous nature of absorption of SO_2 into chloride solution. On the other hand, the entropy change (ΔS) was determined according to the expression:

$$\Delta G = \Delta H - T\Delta S \tag{16}$$

ΔS values of 64.59, 63.47 and 61.41 J/(mol·K) for chloride solution temperature of 298.15, 303.15 and 313.15 K, respectively, were obtained. The positive values of the entropy change show the increased mobility or randomness at the gas–liquid interface and an affinity of the absorbent (chloride solution) towards SO_2. The summary of thermodynamic parameters of SO_2 absorption in gas–liquid chloride system as function of temperature is given in Table 5.

Evaluation of absorption rate and diffusivity

The absorption rate (N) and diffusivity (D) values of SO_2 in the chloride system of the continuous stirred tank reactor were evaluated using Eq. (8) considering gas–liquid interfacial area as 0.00528 m^2/m^3 liquid. The absorption rate and diffusivity of SO_2 were in the order of 10^{-5} mol/m^3·s and 10^{-9} m^2/s, respectively.

The absorption rate and diffusivity values of SO_2 were evaluated at different temperatures of the solution in the

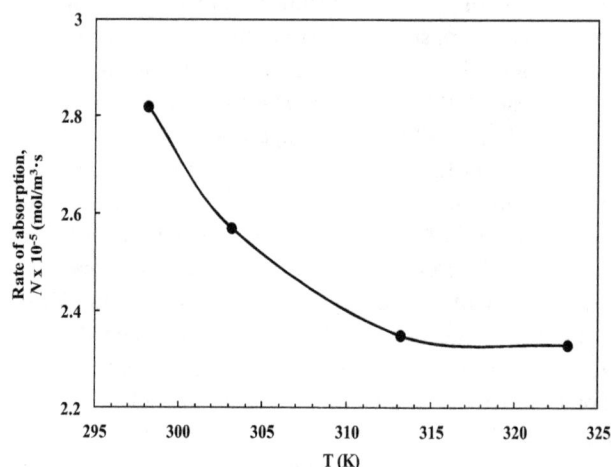

Fig. 10 Effect of temperature on average specific absorption rate. The effect of temperature on physical absorption of SO_2 in chloride media at 298, 303 and 313 K. $P = 1.125$ bar; stirring speed = 500 rpm; inlet SO_2 gas conc. = 99.9 vol %; [HCl] = 4 M; [Me^{n+}] = 100 ppm; contact time = 40 min

Table 6 Effect of temperature on the absorption rate (N), diffusivity (D_{SO_2}) and rate constant (k_R)

T (K)	N (10^{-5} mol/m^3·s)	D_{SO_2} (10^{-9} m^2/s)	k_R (m^3/mol s)
298.15	2.82	2.36	8.81
303.15	2.57	2.89	7.46
313.15	2.35	3.43	6.10

CSTR (298–313 K). By combining information from Eq. (8), Figs. 8 and 10, absorption rate, diffusivity and rate constant values of SO_2 were evaluated. The results are summarized in Table 6.

As shown in Fig. 10, the absorption rate is found to decrease considerably with increasing temperature leading to equilibrium state. In the case of gas absorption with accompanying chemical reaction, the absorption rate is influenced by temperature primarily because of two opposing factors: decrease in gas solubility and increase in reaction rate (for irreversible reaction) as the temperature is increased. The experimental data in this study indicate that the rate of absorption decreases by about 17 % for an increase of temperature from 298 to 313 K. For the same change in temperature, the physical solubility and reaction equilibrium constant decreases by 27 % and 10 %, respectively. The fall in physical absorption rate is also around 27 %. The experimental data further show that the fall in absorption rate in multi-component system (Pt/Fe/Co/Cr) is more sensitive to temperature rise from 303 to 313 K than from 298 to 303 K, compared to the physical absorption rate. This indicates that the reverse reaction rate is more sensitive to temperature than the forward reaction rate. The temperature variation across the liquid film was, however, neglected in the interpretation of data in consideration of the low heat of reaction.

Reactive absorption of SO_2 gas in the presence of metal ion species

In the presence of Pt ions (Fig. 11), the solubility of SO_2 was enhanced slightly due to the chemical reaction with Pt to form Pt–S bond formation. The increase was not significant because the stoichiometric amount of sulphur atoms required to react with the traces of Pt (5.0×10^{-4} mol/L) in solution was small. Initial concentration of Pt was the limiting factor in the reactive solubility of SO_2.

As it can be seen from Tables 3 and 4, the presence of metal ions in the solution has an effect on mass transfer coefficient. The resistance to mass transfer slightly increases as more metal ions are added into a solution. This could be attributed to the fact that interfacial area decreases due to the presence of solid particles in the liquid phase. The formation of solid particles could also influence the effective interfacial area.

Effect of pressure

According to the literature [18], the initial pressures (P_A) are converted to concentrations (C_A) because the rate equations developed in terms of pressures and the

calculated values of activation energy are incorrect when these units are used. Therefore, in this study, the initial pressure values of SO_2 were converted into concentration values using the ideal gas law equation given in Eq. (17).

$$P_A V = n_A RT \tag{17}$$

$$P_A = n_A / VRT; \; P_A = C_A RT$$

Therefore,

$$C_A = P_A / RT \tag{18}$$

where $C_A = n_A / V$ is the concentration of component A (mol/L); P_A is partial pressure of component A (bar); V is volume (L); n_A is moles of the component A; R is the ideal gas law constant (L·bar/mol·K) and T is the absolute temperature (K).

Figures 12 and 13 show the plot of SO_2 concentration vs. time and the effect of initial pressure of SO_2 gas on its absorption at ambient conditions, respectively. It is evident from Figs. 12 and 13 that the initial pressure has an effect on gas solubility. It is obvious that SO_2 absorption rate increases as the SO_2 initial pressure increases from 0.5 to 2.0 bars. For instance, the SO_2 absorption rate increases 72.3 % (2.3×10^{-5} to 8.3×10^{-5} mol/m^3·s) at 25 °C as SO_2 initial pressure increases from 0.5 to 2.0 bar. This is due to the fact that SO_2 absorption process under the tested experimental conditions is a gas phase mass transfer control process, as a result, SO_2 initial pressure has a great impact on its absorption rate.

The results in Table 7 show that as initial gas pressure value of SO_2 was increased and the diffusion coefficient values of SO_2 into metal ions solution in chloride systems were increased. Furthermore, it was observed that

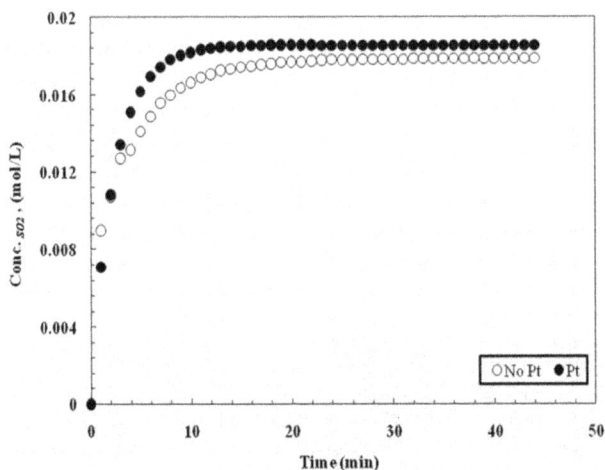

Fig. 11 A plot of SO_2 concentration vs. contact time. Reactive absorption of SO_2 in solution with or without Pt ions. [HCl] = 4 M. $P = 1.125$ bar; $T = 25$ °C; stirring speed = 500 rpm; inlet SO_2 gas conc. = 99.9 vol %; [Me^{n+}] = 100 ppm; contact time = 40 min

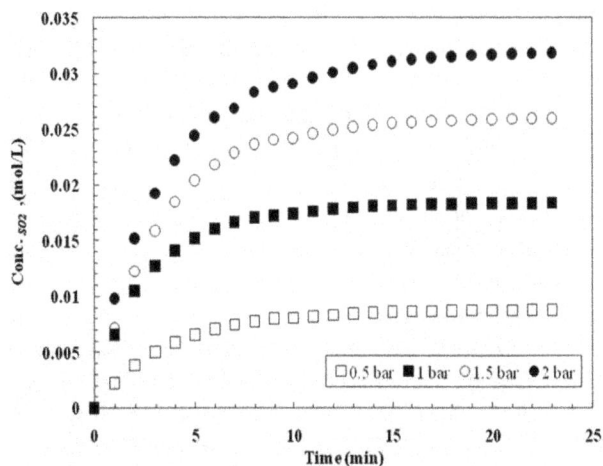

Fig. 12 A plot of SO_2 concentration vs. time. The effect of initial pressure of SO_2 at ambient conditions. Conditions: $P = (0.5–2$ bar); $T = 25$ °C; stirring speed = 500 rpm; inlet SO_2 gas conc. = 99.9 vol %; [HCl] = 4 M; [Me^{n+}] = 100 ppm; contact time = 40 min

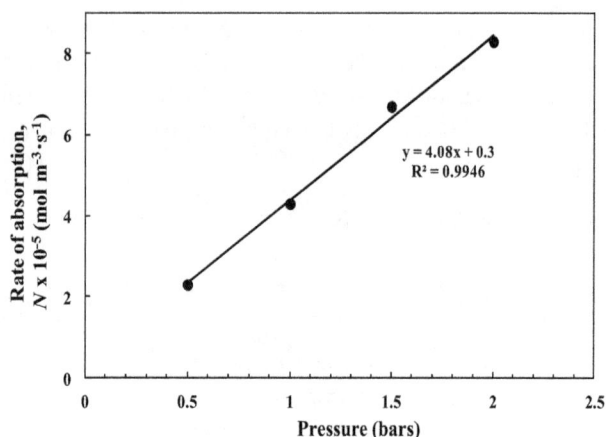

Fig. 13 Effect of SO_2 pressure on its absorption rate at ambient conditions. Conditions: $P = (0.5-2 \text{ bar})$; $T = 25 \text{ °C}$; stirring speed = 500 rpm; inlet SO_2 gas conc. = 99.9 vol %; [HCl] = 4 M; $[Me^{n+}]$ = 100 ppm; contact time = 40 min

Table 7 Effect of pressure on diffusivity (D_{SO_2}), rate constant, (k_R) in G–L chloride system

Pressure (bar)	D_L (10^{-9} m²/s)	k_R (m³/mol s)
0.5	1.36	16.60
1.0	1.62	9.82
1.5	1.99	8.52
2.0	2.22	5.20

increasing SO_2 gas concentration (initial gas pressure) increased the rate of metal depletion in solution.

Conclusions

The kinetics study of platinum and base metals precipitation in gas–liquid chloride system using SO_2 was performed in Büchiglasuster glass reactor. Experimental data in G–L chloride system for both physical and reactive absorption conditions indicated the possibility of introducing significant amounts of sulphur atoms in HCl solutions. Sulphur dioxide gas absorption in chloride system was achieved in short contact time while the formation (appearance) of precipitates was achieved after 24 h due to low metal ion concentration. The rate constants were evaluated, which could be used to predict the correct size of industrial contactors. Sulphur atoms exhibited a higher affinity for Pt. The model that best described the diffusion of SO_2 in chloride system was the two-film Model. The mass transfer model of SO_2 was modelled through the absorption kinetics data with or without an accompanying chemical reaction. The absorption rate and diffusivity values in chloride system were evaluated and were found to be in the order of 10^{-5}

mol/d³ s and 10^{-9} m² s⁻¹, respectively. The changes in enthalpy (ΔH), entropy (ΔS) and Gibbs' free energy (ΔG) of Pt precipitation were evaluated at optimum temperature of 313.15 K. The positive (ΔS) values and low (ΔG) suggested that some energy is required albeit small for the reactive absorption reaction to proceed. The results from this study confirmed that it is possible to replace liquid precipitants such as $Na_2S_2O_3$ with SO_2 gas as a cost-effective precipitant for Pt.

Acknowledgments The authors would like to thank the Copperbelt University (Zambia) and Tshwane University of Technology (South Africa) for the permission to publish the paper, as well as financial and technical support.

References

1. Star Energy Group (2008) CO₂ futures. http://www.co2crc.com.au/dls/co2futures/CO2FUTURES_18.pdf. Accessed 20 July 2015
2. Rasul MG, Brown J (2007) Coal seam methane power generation. Power Energy Syst 539:413
3. Hesselmann G, Hough D (2002) Selective auto-catalytic NO_x reduction (SACR). Mitsui Babcock Energy Limited Technical Review. http://www.netl.doe.gov/publications/proceedings/02/scr-sncr/hesselmannsummary.pdf. Accessed 14 June 2015
4. US Department of Energy (2005) Conference proceedings. http://www.fischer-tropsch.org/DOE. Accessed 20 Jul 2015
5. Bernardis FL, Grant RA, Sherrington DC (2005) A review of method of separation of the platinum-group metals through their chloro-complexes. React Funct Polym 65:205–217
6. Danckwerts PV (1970) Gas–liquid reactions. McGraw-Hill, New York
7. Gomez-Diaz D, Navaza JM, Sanjurjo B (2006) Analysis of mass transfer in the precipitation processes of calcium carbonate using a gas/liquid reaction. Chem Eng J 116:203–209
8. Dagaonkar VM, Beenackers AACM, Pangarkar VG (2001) Enhancement of gas–liquid mass transfer by small reactive particles at realistically high mass transfer coefficients: absorption of sulphur dioxide into aqueous slurries of $Ca(OH)_2$ and $Mg(OH)_2$ particles. Chem Eng J 81:203–212
9. Rodriguez-Sevilla L, Alvarez M, Liminana G, Diaz MC (2002) Dilute SO_2 absorption equilibrium in aqueous HCl and HCl solution at 298.15 K. J Chem Eng Data 47(6):1339–1345
10. Kening YE, Wiesner U, Gorak A (1997) Modelling of reactive absorption using the Maxwell–Stefan equations. Ind Eng Chem Res 36:4325–4334
11. Kasaini H, Masahiro G, Furusaki S (2001) Adsorption performance of activated carbon pellets immobilized with organo phosphorus extractants and amines: a case study for the separation of Pt(IV), Pd(II), and Rh(III) ions in chloride media. Sep Sci Technol 36(13):2845–2861
12. Siame J, Kasaini H (2010) Sulphur-mediated precipitation of Pt/Fe/Co/Cr ions in liquid–liquid and gas–liquid chloride systems. World Acad Sci, Eng Technol 70:302–311
13. Inoue K, Yoshizuka K, Baba Y, Wada F, Matsuda T (1990) Solvent extraction of palladium (II) and platinum (IV) from aqueous chloride media with N, N-dioctylglycine. Hydrometallurgy 25(2):271–279

14. Koliadima A, Kapolos J, Farmakis L (2009) Diffusion coefficients of SO_2 in water and partition coefficients of SO_2 in water-air interface at different temperature and pH values. Instrum Sci Technol 37:274–283

15. Leaist DG (1984) Diffusion coefficient of aqueous sulphur dioxide at 25 °C. J Chem Eng Data 29:281–282

16. Boniface J, Shi YQ, Li JL, Cheung OV, Rattigan P, Davidovits DR, Jayne WST, Kolb CE (2000) Uptake of gas-phase SO_2, H_2S and CO_2 by aqueous solutions. J Phys Chem A 104(32):7502–7510

17. Juang RS, Chen ML (1997) Application of the elovich equation to the kinetics of metal sorption with solvent-impregnated resins. Ind Eng Chem Res 36:813–820

18. Levenspiel O (1999) Chemical reaction engineering. Wiley, New York

14

Chemical, structural and energy properties of hydrochars from microwave-assisted hydrothermal carbonization of glucose

Sunday E. Elaigwu[1,2] · Gillian M. Greenway[1]

Abstract Hydrothermal carbonization has been used as a green and effective technique for the preparation of hydrochars from simple carbohydrates, such as glucose. The chemical and structural properties of hydrochars prepared from glucose have been studied. However, the energy properties of hydrochars prepared from microwave-assisted hydrothermal carbonization of glucose have not been studied. Thus, in this study, microwave-assisted hydrothermal carbonization of glucose, and the energy properties of the prepared hydrochars are reported. The preparation involved heating glucose in de-ionized water at 200 °C for 5–60 min in a microwave oven. The prepared hydrochars were characterized using scanning electron microscope, nitrogen sorption measurement, Fourier transform infrared spectroscopy, elemental (CHN) analyzer, and nuclear magnetic resonance, and their energy properties studied. The result indicated that, in comparison with previous studies using the classical hydrothermal carbonization process, this approach reduced the processing time greatly from hours to 45 min, while the increase in higher heating value of the hydrochar when compared to that of the starting material was higher in this study than some value previously reported.

Keywords Microwave · Hydrothermal · Carbonization · Glucose · Energy properties · Hydrochar

Introduction

The use of cheap, fast and different environmentally friendly strategies for the preparation of carbon materials from renewable resources has been on the increase in recent times in the areas of environmental science and technology. This is because of the vital roles these materials play in different applications, such as adsorbents, catalyst supports, energy storage materials, electrode materials, and stationary phases in liquid chromatography [1, 2]. As a result, a number of approaches such as pyrolysis, arc discharge, chemical vapor decomposition, and hydrothermal treatment have been used in the preparation of these carbon materials [3]. Hydrothermal carbonization is a process of decomposing an organic material in hot water under high pressure to produce solid carbon material (hydrochar) and water soluble organics. It is a green and efficient approach for treating organic materials because of its comparatively low emission, and generation of nontoxic waste [4]. Due to its simple operation, mild reaction conditions, and ability to exploit renewable biomass with minimal pre-treatment, it is of a particular environmental advantage when compared to other techniques of carbonization [5–7]. Among potential precursors used in the preparation of carbonaceous materials, glucose is very promising and its hydrothermal carbonization process has been studied several times using the conventional method of oven heating [6, 8–10]. However, the conventional hydrothermal process requires special systems that support pressure and temperature, usually an autoclave with pressure safety device is used. Also, the reaction times are

✉ Sunday E. Elaigwu
S.E.Elaigwu@2009.hull.ac.uk; sunnietrinex@hotmail.com

[1] Department of Chemistry, University of Hull, Cottingham Road, Hull HU6 7RX, UK

[2] Department of Chemistry, University of Ilorin, PMB 1515, Ilorin, Kwara, Nigeria

usually in hours, which makes the process expensive and time consuming.

In many applications, the use of microwave heating as an attractive alternative to conventional method of heating has shown to be more energy efficient, because it provides selective, fast and homogenous heating, which reduces processing time and costs significantly [11, 12]. It has also been established that irradiation with microwave produced effective internal heating by direct coupling of microwave energy with solvents, reagents and catalysts, which increased the reactions greatly [13]. The use of microwave heating in hydrothermal carbonization process, and in the preparation of hydrochar from biomasses, glucose and other materials, such as human waste, cellulose and starch, has been reported [4, 14–19]. Some of these processes usually proceed via the degradation of starch or cellulose to form glucose and then further carbonization of the formed glucose. However, the energy properties of hydrochars obtained using glucose as starting material have not been previously reported to the best of our knowledge. Therefore, in this study the energy properties of hydrochars from microwave-assisted hydrothermal carbonization process using glucose as starting material are reported.

Materials and methods

Anhydrous D-glucose was purchased from Fisher Scientific, UK and was used throughout without further purification.

Microwave-assisted hydrothermal carbonization of glucose

5 g each of glucose was dissolved in 5 mL of de-ionized water in 100 ml microwave reaction vessels made of Teflon to form supersaturated solutions. The vessels were sealed and placed in a 2.45 GHz magnetron frequency microwave oven (MARS, CEM, Milton Keynes, UK equipped with XP1500 digestion vessels, and 1600 W at maximum power), and were hydrothermally carbonized at 200 °C in the microwave oven which was set to ramp to a given temperature in 5 min, and then held at the temperature for 5–60 min. The reaction system was allowed to cool to room temperature, and the carbonized materials were filtered off using Whatman filter paper number 3, ashless 11 cm. The solid chars (hydrochars) obtained were washed gently with de-ionized water to remove any left over of the liquid phase of the reaction, and were dried in a conventional oven at 80 °C for 16 h.

Mass yield

In each case, the dry mass of the carbonized material was measured and the mass yield (dry mass percentage of starting material) was calculated as follows:

$$\text{Mass yield } (\%) = \frac{\text{Mass of carbonized material (g)}}{\text{Mass of starting material (g)}} \times 100.$$

(1)

Energy properties of the hydrochars

In place of a calorimeter to experimentally determine the energy content, the higher heating value (HHV) was calculated using Eq. 2 (Dulong equation) as previously reported [20]. This formula is one of the first correlations to estimate the HHV of coals, which is still in used by many researchers today and has also been applied to oils [21].

$$\text{HHV} = 0.3383C + 1.422(H - 0/8).$$

(2)

The energy densification ratios of the hydrochars were calculated using the following equation:

$$\text{Energy densification ratio } = \frac{\text{HHV of dried hydrochar}}{\text{HHV of dried starting material}}.$$

(3)

All experiments were carried out in triplicate, and the results obtained are presented in Table 1.

Characterization techniques

The prepared hydrochars were characterized using the following instruments: Fisons instruments EA 1108 CHN analyser (Fison Instrument, Crawley, UK) was used for CHN analysis; samples were ground into fine powder, weighed into tin capsules and placed on the autosampler for analysis. Surface area and pore size distribution were measured using a Micromeritics Tristar BET-N_2 surface area analyser (Micromeritics, Hexton, UK). Before the analysis, samples were degassed under nitrogen atmosphere at 120 °C for 3 h. FT-IR spectra were recorded on Thermo Scientific Nicolet 380 FT-IR (Themo Scientific, Hemel Hempstead, UK), equipped with attenuated total reflectance (ATR). The samples were in direct contact with ATR diamond crystal, and each sample was investigated in wavenumber range of 4000–525 cm^{-1} using 16 scans at a spectral resolution wavenumber of 4 cm^{-1}. The morphologies and particle sizes were visualized with a ZEISS EVO 60 SEM (Carl Zeiss, Cambridge, UK); samples were coated with gold and platinum alloy and impregnated on a sticky disc before analysis. ^{13}C solid state magic angle

Table 1 Elemental composition, mass yield, O/C and H/C atomic ratios, and energy properties of glucose and hydrochars, with data from previous studies using the conventional approach for comparison

Temperature and time	C (%)	H (%)	N (%)	O (%)[a]	Mass yield (%)	O/C[b]	H/C[b]	HHV (MJ/kg)	Energy densification ratio	Energy yield (%)
Glucose	39.84 ± 0.06	6.84 ± 0.09	0.00	53.32 ± 0.15	–	1.00	2.04	13.73	–	–
200 °C for 60 min	61.98 ± 0.56	4.43 ± 0.22	0.00	33.59 ± 0.32	37.28 ± 1.16	0.41	0.85	21.30	1.55	57.84
200 °C for 45 min	62.32 ± 1.80	4.31 ± 0.70	0.00	33.37 ± 1.10	40.03 ± 0.98	0.40	0.82	21.30	1.55	62.06
200 °C for 30 min	60.66 ± 2.15	4.64 ± 1.58	0.00	34.70 ± 0.57	32.67 ± 1.23	0.43	0.91	20.95	1.53	49.87
200 °C for 20 min	58.64 ± 0.60	4.90 ± 0.55	0.00	36.46 ± 1.15	25.01 ± 0.56	0.47	1.00	20.33	1.48	37.03
200 °C for 15 min	52.81 ± 0.12	4.61 ± 0.10	0.00	42.58 ± 0.02	20.69 ± 1.48	0.61	1.04	16.85	1.23	25.40
Glucose (210 °C for 4.5 h) [9]	66.29	4.15	–	29.56	28.00	0.334	0.752	–	–	–
Prosopis africana shell (200 °C for 4 h) [24]	57.86	6.37	1.46	34.31	30.85	0.44	1.31	22.53	1.28	–
Anaerobically digested maize silage (190 °C for 10 h) [25]	62.13	6.87	–	27.44	65.0	0.33	1.33	27.0	–	–

[a] Oxygen content was determined by difference [100 % − (C % + H % + N %)]

[b] Atomic ratios

spinning NMR experiment was carried out on a Bruker Avance II 500 MHz (11.74T) spectrometer (Bruker, Coventry, UK); samples were packed without further treatment into a 4 mm zirconia rotor sample holder spinning at MAS rate $v_{MAS} = 8$ kHz. Carbon sensitivity was enhanced by Proton-to-carbon CP MAS: recycle delay for all CP experiments was 3 s and TPPM decoupling was used during signal acquisition. Cross polarization transfer was carried out under adiabatic tangential ramps to enhance the signal with respect to other known methods. CP time $t_{CP} = 500$ ms. The number of transients for all carbon samples was 200.

Results and discussion

Effect of time on the microwave-assisted hydrothermal carbonization of glucose

The effect of time on the process is presented in Table 1. As the processing time was increased from 15 to 60 min, different mass yields were obtained for the hydrochars. The mass yield increased between 15 and 45 min, and decreased afterwards. Hence, processing time above 60 min was not considered. Maximum mass yield for the hydrothermal carbonization of glucose has been reported to be obtained at 200 °C [6], while further increase in temperature will lead to a gradual decrease in mass yield; reason being that increase in temperature favors gasification reactions, which results in part of the hydrothermal carbon being lost in form of volatile compounds [22, 23].

Hence, temperatures above 200 °C were not considered in this study. Falco et al. [6], using conventional hydrothermal carbonization process, reported a maximum mass yield of about 40 % for glucose hydrochar at 200 °C for 24 h, which is consistent with the maximum mass yield obtained in this study under microwave heating, despite the shorter time (45 min) in the microwave oven.

Elemental composition of the prepared hydrochars

The elemental compositions (C, O, and H) of the starting material (glucose) and different hydrochar samples are listed in Table 1. It was observed that carbon contents increased from 39.84 % in the starting material to about 53–62 % in the hydrochar samples. The oxygen and hydrogen contents of the hydrochars reduced at the same time. These variations, which increased with reaction time, are consistent with hydrothermal carbonization processes. The gradual increase in carbon content, and the decrease in hydrogen and oxygen contents of the hydrochars, with increase in processing time is due to loss of hydrogen and oxygen in deoxygenating, dehydration and decarboxylation reactions that occurred during the microwave-assisted hydrothermal carbonization process [6, 22, 26, 27]. Despite the shorter time used in this study under microwave heating, the result is similar to previous studies [6, 9].

The changes in elemental compositions were further analyzed using the van Krevelen diagram (Fig. 1), by plotting the atomic H/C against the atomic O/C. The diagram provided further evidence about the transformation (dehydration, decarboxylation, and demethanation

Fig. 1 van Krevelen diagram of raw glucose and hydrochars prepared at 200 °C and different processing time

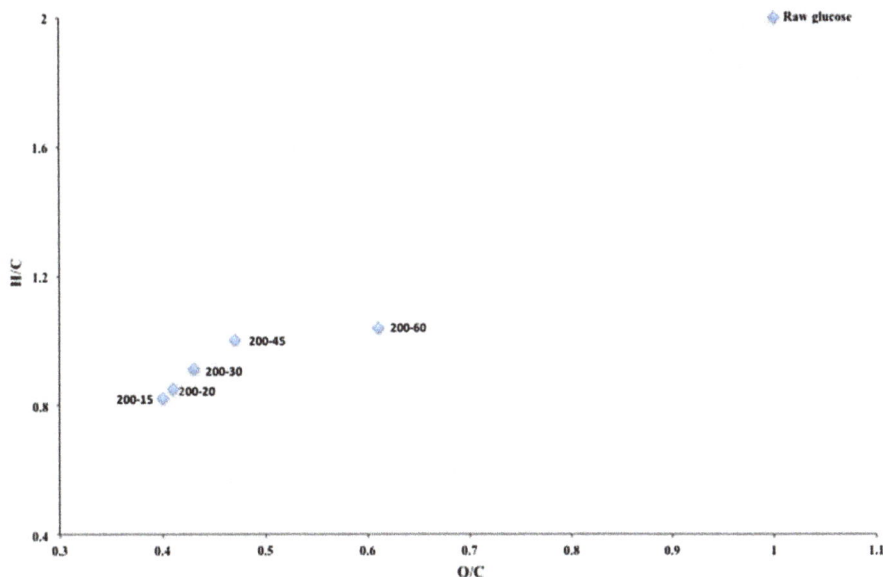

reactions) that took place in the chemical structure of glucose during the microwave-assisted hydrothermal carbonization process. The transformation from the starting material (glucose) to the hydrochar samples follows a diagonal line due to decrease in the O/C and H/C ratios, suggesting dehydration reactions as prevalent reaction during the process, which is consistent with previous reports [9, 28].

Energy properties of the prepared hydrochars

An important parameter to test the quality of hydrochars is their higher heating value (HHV). It provides information about the quantity of energy present in the hydrochar. The calculated HHV increased with increase in residence time from 16.90 MJ/kg in 15 min to 21.30 MJ/kg in 45 min, and remained stable afterwards (Table 1). This increase in higher heating value of the hydrochars with increase in processing time is consistent with previous report [29]. The highest HHV in this study showed an increase of 55.5 % when compared to that of the starting material, against 39 % previously reported for loblolly pine [29], 45.20 % for bamboo [30], and 21 % for dry leaves [31].

As stated earlier, dehydration, decarboxylation, and condensation reactions are associated with hydrothermal carbonization process. This leads to the carbonization of the starting material and consequently results in energy densification, which is used to measure the effectiveness of the hydrothermal carbonization processes [32, 33]. In this study, energy densification ratios of the hydrochars (Table 1) increased with increase in reaction time and ranged from 1.23 to 1.55, which is consistent with previous reports [29, 31].

The energy yield is defined as the mass yield of the hydrochar multiply by its energy densification ratio. The result of the energy yield for the hydrochars is presented in Table 1. The lowest energy yield of 25.40 % was obtained when the reaction time was 15 min, and increased steadily until the maximum hydrochar energy yield of 62.06 % was obtained after 45 min. Thus, the hydrochar prepared at 200 °C for 45 min has the highest mass yield, HHV, energy densification, and energy yield. Therefore, further characterization will be focused on this hydrochar.

FT-IR analysis

The FT-IR spectra (Fig. 2) gave further insight into changes in the chemical composition of glucose during the process. Comparing the FT-IR results of glucose (Fig. 2a) and that of the hydrochar prepared at 200 °C for 45 min (Fig. 2b) indicates that dehydration and aromatization of glucose occurred during the process. The allocation of the functional groups was based on previous reports [9, 34]. The broad peak between 3600 and 3000 cm^{-1} (section 1) corresponds to stretching vibration of aliphatic O–H (hydroxyl and carboxyl), while the peaks between 1500 and 1000 cm^{-1} (section 5) correspond to C–O stretching vibration from esters, ether, phenols, and aliphatic alcohols. The weak intensity of these peaks (sections 1 and 5) is an indication that dehydration and decarboxylation reactions occurred during the process. The peaks between 1800 and 1650 cm^{-1} (section 3) present only in the hydrochar result from C=O vibration of esters, quinone, pyrone, carboxylic acids or aldehydes, while the appearance of peaks between 1650 and 1500 cm^{-1} (section 4) due to C=C vibrations, and well-

Fig. 2 FT-IR spectra of **a** glucose, **b** hydrochar prepared at 200 °C 45 min

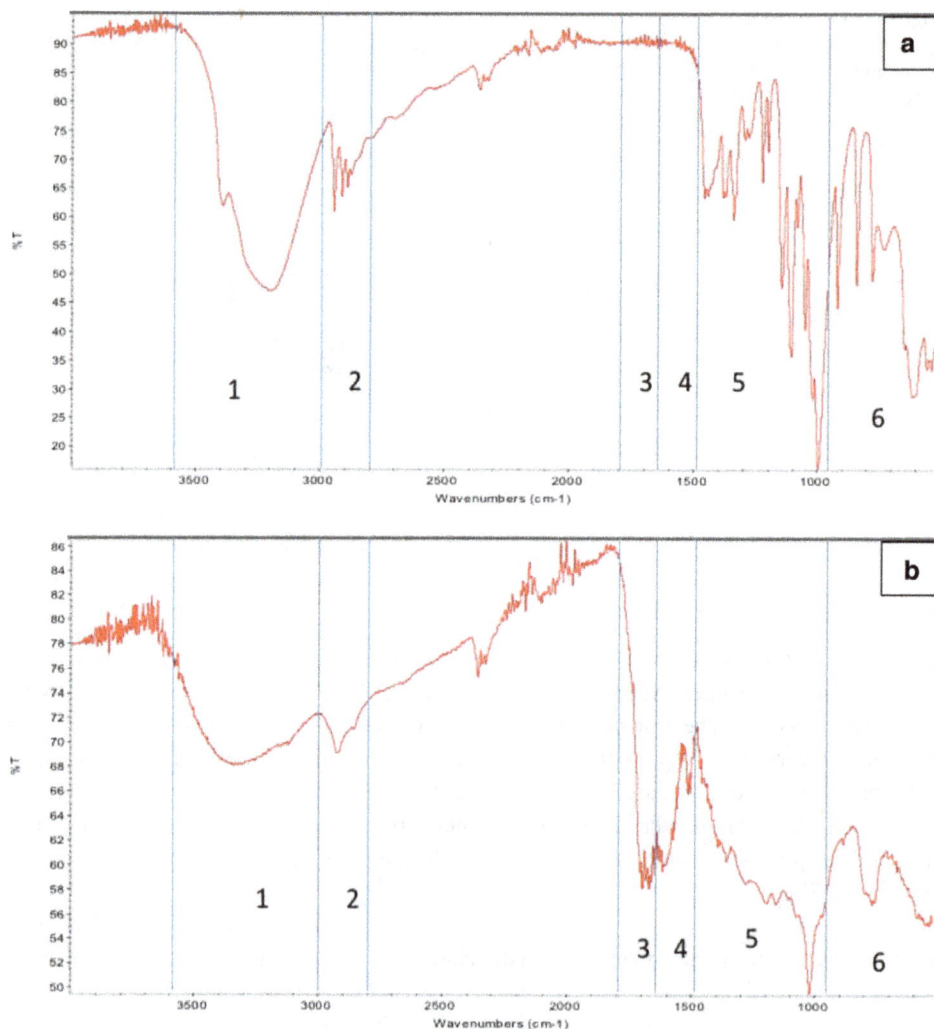

defined peak below 1000 cm^{-1} (section 6) from deformation of C–H out of plane bending vibrations in aromatic compounds, showed the aromatic nature of the hydrochar. The peaks between 3000 and 2800 cm^{-1} (section 2) due to stretching vibration of aliphatic C–H bonds, indicate the presence of aliphatic structures.

NMR analysis

^{13}C solid state NMR analysis is usually used as a complementary technique to FT-IR analysis in describing the level of conversion during hydrothermal carbonization process [24]. The ^{13}C solid state NMR spectrum of the hydrochar prepared at 200 °C for 45 min shown in Fig. S1 (Supplementary information) provided information about its chemical composition, and also confirmed the results from the FT-IR. The peaks between 14 and 60 ppm are due to the presence of aliphatic carbons [8, 35], while those between 100 and 160 ppm usually referred to as the aromatic region are all due to C=C bond, but between 140 and 160 ppm are specifically due to the oxygen bound O–C=C (O-aryl) [35]. The peaks between 170 and 200 ppm are due to the presence of carboxylic acid, aldehydes or ketones moieties [36]. The spectrum also shows carbohydrate resonances (specifically CH$_2$OH groups around 62 ppm, CHOH groups around 72 ppm, and anomeric O–C–O carbons around 90 ppm) in the O-alkyl region between 60 and 100 ppm [37, 38]. These functional groups have already been indicated by the FT-IR result. Thus, both results are in good agreement with each other, and clearly show the formation of the hydrochar structure under the microwave-assisted process.

SEM analysis

The scanning electron microscope (SEM) analysis (Fig. 3) provided information about the structural morphologies of the starting material (glucose) and the hydrochar prepared at 200 °C for 45 min. The morphological features of

Fig. 3 SEM images of
a glucose, **b** hydrochar prepared
at 200 °C 45 min

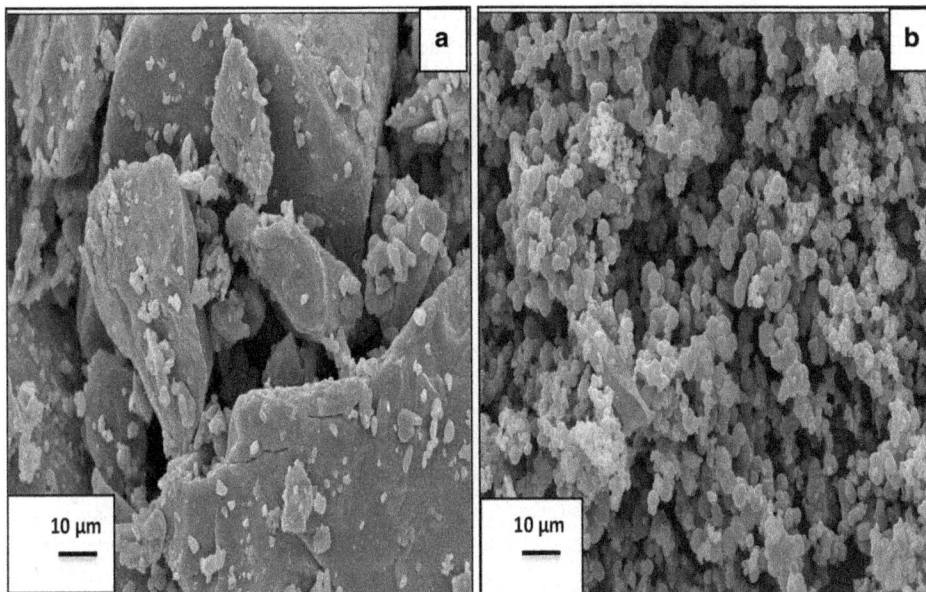

glucose (Fig. 3a) as visualized are completely different from those of the hydrochar (Fig. 3b). The sphere-like microparticles of different sizes (1–10 μm) seen on the SEM image of the hydrochar are in contrast to the block of materials observed in glucose. The mechanism in which these microparticles are generated is based on hydrolysis, dehydration and polymerization of glucose as stated in the mechanism, while the spherical nuclei are formed to minimize the energy of the interface [6, 14].

Surface area analysis

The hydrochar prepared at 200 °C for 45 min showed a Type II isotherm based on the IUPAC system of classification from the nitrogen sorption measurement in Fig. S2 (Supplementary information). This is a typical isotherm obtained with non-porous materials. It is, therefore, not surprising that the hydrochar has a very small BET surface area of 3.3 ± 0.7 m^2g^{-1}, due to poor porosity. The hydrothermal carbonization process involves carbonization and solubilization of organics leading to the formation of tarry substances, which usually contaminate the hydrochars, plugging their pores, and making the apparent BET surface area to be small [39]. Similar results have been reported using the conventional hydrothermal method [9, 40]. Thus, surface area and porosity improvement are usually carried out for the hydrochars to fit into specific applications, such as hydrogen or electrical energy storage [33].

Mechanism for the hydrochar formation

Similar mechanism to that in conventional hydrothermal carbonization process is expected to occur under the microwave-assisted process [41]. However, under the electromagnetic irradiation during the microwave-assisted process, temperature and pressure will play a very vital role by enhancing the different reactions (dehydration, polymerization and aromatization), thereby accelerating the process [12]. Therefore, in this study, glucose fragments by hydrolysis and dehydrates under microwave heating to form different soluble furfural-like compounds. The dehydration reaction is indicated in the hydrochar by the reduction in the OH peak between 3600 and 3000 cm^{-1}. The furfural-like compounds undergo further decomposition to form acids, aldehydes and phenols [18, 42], indicated in the hydrochar by the appearance of the peaks between 1800 and 1650 cm^{-1} (C=O vibration). The dehydration of water from the equatorial hydroxyl groups could be the mechanism responsible for the appearance of C=O groups [43]. The glucose and its decomposition products then undergo polymerization or condensation reactions, which could be caused by intermolecular dehydration or through aldol condensation leading to the formation of soluble polymers [9]. The aromatization of the polymers also occurs simultaneously with the formation of C=C linkages (indicated by the appearance of peaks between 1650–1500 cm^{-1}, and 100–160 ppm in the FTIR and NMR spectra of the hydrochar, respectively), which could result from keto-enol tautomerism of the dehydrated species, intermolecular dehydration or via the condensation of the aromatized molecules formed during the decomposition or dehydration of the glucose [9, 43]. A burst nucleation process occurs once the amount of aromatic clusters in aqueous phase reaches the critical super saturation point. The formed nuclei then grow through diffusion to the surface of the chemical species that are present in solution (as seen on the SEM image of the hydrochar) to minimize the energy of the interfaces [14]. The reactive oxygen functionalities (as indicated by the FTIR and NMR spectra), such as hydroxyl, carbonyl, and carboxylic, present on the outer surface of the microspheres and in the reactive species helps to quickly link the species to the microspheres [43]. Due to this linkage, stable oxygen groups, such as ether, pyrone or quinone, usually found at the center of the resulting microsphere are formed, and as a result, when the growth process ends, the outer surface of the hydrochar particles will have a high concentration of reactive oxygen groups, while the core will have less reactive oxygen groups [9, 43]. Thus, two types of products are formed in the reaction medium at the end of the reaction, namely the insoluble solid residue (hydrochar), and the aqueous soluble organic phase (consisting of furfural-like compounds, aldehydes and acids). Thus, as the reaction time is increased in this study, it is expected that the aromatization and polymerization processes will also be favoured, which will increase the yield of the hydrochar.

Conclusion

Microwave-assisted hydrothermal carbonization was successfully used to prepare hydrochar from glucose. The chemical and structural characterization of the prepared hydrochar showed that it is similar to those previously prepared using the conventional oven method of hydrothermal carbonization. This implies that the microwave-assisted hydrothermal approach is a fast and simple approach for preparing the hydrochar as it reduced the processing time from hours to few minutes. The energy properties of the prepared hydrochars in relationship to some published results showed a higher increase in HHV of the hydrochars when compared to that of the starting material.

Acknowledgments The authors are grateful to Petroleum Technology Development Fund (PTDF), Nigeria for providing the PhD studentship for Dr. Sunday E. Elaigwu, and Bob Knight for his assistant with the CEM microwave oven.

References

1. Elaigwu SE, Greenway GM (2014) Biomass derived mesoporous carbon monoliths via an evaporation-induced self-assembly. Mater Lett 115:117–120
2. Liu S, Tian J, Wang L, Zhang Y, Qin X, Luo Y, Asiri AM, Al-Youbi AO, Sun X (2012) Hydrothermal treatment of grass: a low-cost, green route to nitrogen-doped, carbon-rich, photoluminescent polymer nanodots as an effective fluorescent sensing platform for label-free detection of Cu (II) ions. Adv Mater 24:2037–2041
3. Latham KG, Jambu G, Joseph SD, Donne SW (2014) Nitrogen doping of hydrochars produced hydrothermal treatment of sucrose in H_2O, H_2SO_4, and NaOH. ACS Sustain Chem Eng 2:755–764
4. Elaigwu SE, Rocher V, Kyriakou G, Greenway GM (2014) Removal of Pb^{2+} and Cd^{2+} from aqueous solution using chars from pyrolysis and microwave-assisted hydrothermal carbonization of *Prosopis africana* shell. J Ind Eng Chem 20:3467–3473
5. Kambo HS, Dutta A (2015) A comparative review of biochar and hydrochar in terms of production, physico-chemical properties and applications. Renew Sustain Energy Rev 45:359–378
6. Falco C, Baccile N, Titirici MM (2011) Morphological and structural differences between glucose, cellulose and lignocellulosic biomass derived hydrothermal carbons. Green Chem 13:3273–3281
7. Reza MT, Mumme J, Ebert A (2015) Characterization of hydrochar obtained from hydrothermal carbonization of wheat straw digestate. Biomass Conv Bioref. doi:10.1007/s13399-015-0163-9
8. Demir-Cakan R, Baccile N, Antonietti M, Titirici MM (2009) Carboxylate-rich carbonaceous materials via one-step hydrothermal carbonization of glucose in the presence of acrylic acid. Chem Mater 21:484–490
9. Sevilla M, Fuertes AB (2009) Chemical and structural properties of carbonaceous products obtained by hydrothermal carbonization of saccharides. Chem Eur J 15:4195–4203
10. Yao C, Shin Y, Wang LQ, Windish CF, Samuels WD, Arey BW, Wang C, Risen WM, Exarhos GJ (2007) Hydrothermal dehydration of aqueous fructose solutions in a closed system. J Phys Chem C 111:15141–15145
11. Nüchter M, Ondruschka B, Bonrath W, Gum A (2004) Microwave assisted synthesis—a critical technology overview. Green Chem 6:128–141
12. Elaigwu SE, Kyriakou G, Prior TJ, Greenway GM (2014) Microwave-assisted hydrothermal synthesis of carbon monolith via a soft-template method using resorcinol and formaldehyde as carbon precursor and pluronic F127 as template. Mater Lett 123:198–201
13. Guo F, Fang Z, Zhou TJ (2012) Conversion of fructose and glucose into 5-hydroxymethylfurfural with lignin-derived carbonaceous catalyst under microwave irradiation in dimethyl sulfoxide–ionic liquid mixtures. Bioresour Technol 112:313–318
14. Guiotoku M, Rambo CR, Hansel FA, Magalhaes WLE, Hotza D (2009) Microwave-assisted hydrothermal carbonization of lignocellulosic materials. Mater Lett 63:2707–2709
15. Guiotoku M, Rambo CR, Hotza D (2014) Charcoal produced from cellulosic raw materials by microwave-assisted hydrothermal carbonization. J Therm Anal Calorim 117:269–275
16. Afolabi OOD, Sohail M, Thomas CPL (2015) Microwave hydrothermal carbonization of human biowastes. Waste Biomass Valor 6:147–157
17. Hassanzadeh S, Aminlashgari N, Hakkarainen M (2014) Chemoselective high yield microwave assisted reaction turns cellulose to green chemicals. Carbohyd Polym 112:448–457
18. Wu D, Hakkarainen M (2014) A closed-loop process from microwave-sssisted hydrothermal degradation of starch to utilization of the obtained degradation products as starch plasticizers. ACS Sustain Chem Eng 2:2172–2181
19. Möller M, Harnisch F, Schröder U (2012) Microwave-assisted hydrothermal degradation of fructose and glucose in subcritical water. Biomass Bioenerg 39:389–398
20. Gao Y, Wang X, Wang J, Li X, Cheng J, Yang H, Chen H (2013) Effect of residence time on chemical and structural properties of hydrochar obtained by hydrothermal carbonization of water hyacinth. Energy 58:376–383
21. Kieseler S, Neubauer Y, Zobel N (2013) Ultimate and proximate correlations for estimating the higher heating value of hydrothermal solids. Energy Fuels 27:908–918

22. Berge ND, Ro KS, Mao J, Flora JRV, Chappell MA, Bae S (2011) Hydrothermal carbonization of municipal waste streams. Environ Sci Technol 45:5696–5703

23. Kruse A, Gawlik A (2003) Biomass conversion in water at 330–410 °C and 30–50 MPa. Identification of key compounds for indicating different chemical reaction pathways. Ind Eng Chem Res 42:267–279

24. Elaigwu SE, Greenway GM (2016) Microwave-assisted and conventional hydrothermal carbonization of lignocellulosic waste material: comparison of the chemical and structural properties of the hydrochars. J Anal Appl Pyrol. doi:10.1016/j.jaap.2015.12.013

25. Mumme J, Eckervogt L, Pielert J, Diakité M, Rupp F, Kern J (2011) Hydrothermal carbonization of anaerobically digested maize silage. Bioresour Technol 102:9255–9260

26. Sevilla M, Maciá-Agulló JA, Fuertes AB (2011) Hydrothermal carbonization of biomass as a route for the sequestration of CO_2: chemical and structural properties of the carbonized products. Biomass Bioenerg 35:3152–3159

27. Kang S, Li X, Fan J, Chang J (2012) Characterization of hydrochars produced by hydrothermal carbonization of lignin, cellulose, D-xylose, and wood meal. Ind Eng Chem Res 51:9023–9031

28. Erdogan E, Atila B, Mumme J, Reza MT, Toptas A, Elibol M, Yanik J (2015) Characterization of products from hydrothermal carbonization of orange pomace including anaerobic digestibility of process liquor. Bioresour Technol 196:35–42

29. Hoekman SK, Broch A, Robbins C (2011) Hydrothermal carbonization (HTC) of lignocellulosic biomass. Energy Fuels 25:1802–1810

30. Li MF, Shen Y, Sun JK, Bian J, Chen CZ, Sun RC (2015) Wet torrefaction of bamboo in hydrochloric acid solution by microwave heating. ACS Substain Chem Eng 3:2022–2029

31. Saqib NU, Oh M, Jo W, Park SK, Lee JY (2015) Conversion of dry leaves into hydrochar through hydrothermal carbonization (HTC). J Mater Cycles Waste Manag. doi:10.1007/s10163-015-0371-1

32. Xu Q, Qian Q, Quek A, Ai N, Zeng G, Wang J (2013) Hydrothermal carbonization of macroalgae and the effects of experimental parameters on the properties of hydrochars. ACS Sustain Chem Eng 1:1092–1101

33. Parshetti GK, Kent HS, Balasubramanian R (2013) Chemical, structural and combustion characteristics of carbonaceous products obtained by hydrothermal carbonization of palm empty fruit bunches. Bioresour Technol 135:683–689

34. Tekin K, Pileidis FD, Akalin MK, Karagöz S (2015) Cellulose-derived carbon spheres produced under supercritical ethanol conditions. Clean Technol Environ Policy. doi:10.1007/s10098-015-1014-x

35. Baccile N, Laurent G, Babonneau F, Fayon F, Titirici MM, Antonietti M (2009) Structural characterization of hydrothermal carbon spheres by advanced solid-state MAS C-13 NMR investigations. J Phys Chem C 113:9644–9654

36. Sun K, Ro KS, Guo M, Novak JM, Mashayekhi H, Xing B (2011) Sorption of bisphenol A, 17a-ethinyl estradiol and phenanthrene on thermally and hydrothermally produced biochars. Bioresour Technol 102:5757–5763

37. Cao X, Ro KS, Chappell M, Li Y, Mao J (2011) Chemical structures of swine-manure chars produced under different carbonization conditions investigated by advanced solid-state ^{13}C nuclear magnetic resonance (NMR) spectroscopy. Energy Fuels 25:388–397

38. Mao J, Holtman KM, Scott JT, Kadla J, Schmidt-Rohr K (2006) Differences between lignin in unprocessed wood, milled wood, mutant wood, and extracted lignin detected by ^{13}C solid-state NMR. J Agric Food Chem 54:9677–9686

39. Mochidzuki K, Sato N, Sakoda A (2005) Production and characterization of carbonaceous adsorbents from biomass wastes by aqueous phase carbonization. Adsorption 11:669–673

40. Titirici MM, Antonietti M, Baccile N (2008) Hydrothermal carbon from biomass: a comparison of the local structure from poly- to monosaccharides and pentoses/hexoses. Green Chem 10:1204–1212

41. Guiotoku M, Maia CMBF, Rambo CR, Hotza D (2011) Synthesis of carbon-based materials by microwave hydrothermal processing. In: Chandra U (ed) Microwave heating. InTech, New York

42. Kabyemela BM, Adschiri T, Malaluan RM, Arai K (1999) Glucose and fructose decomposition in subcritical and supercritical water: detailed reaction pathway, mechanisms, and kinetics. Ind Eng Chem Res 38:2888–2895

43. Sevilla M, Fuertes AB (2009) The production of carbon materials by hydrothermal carbonization of cellulose. Carbon 47:2281–2289

15

Chemical modification at the surface and corrosion inhibition response of two semicarbazones on carbon steel in HCl medium

Vinod Raphael Palayoor[1] · Joby Thomas Kakkassery[2] · Shaju Shanmughan Kanimangalath[1] · Sini Varghese[2]

Abstract Two similarly structured heterocyclic semicarbazones (E)-4-(5-((2-carbamoylhydrazano)methyl)furan-2-yl)benzoic acid (CPFASC) and (E)-2-((5-(4-nitrophenyl)-furan-2-yl)methylene)hydrazinecarboxamide (NPFASC) were synthesized, characterized and tested for their corrosion protection capacity on carbon steel (CS) in 1 M HCl solution. Contrary to expectation the non-planar molecule NPFASC showed better inhibition efficiency on CS surface than CPFASC. At a concentration of 0.5 mM, NPFASC displayed 93.4 % while CPFASC showed 89.9 % inhibition efficiency according to impedance studies. This unusual behavior can be explained by the conversion of the nitro group present in the NPFASC molecule into amino group on approaching the metal surface in the corrosive medium. This transformation obviously changes the geometry of the molecule which is more conducive for corrosion inhibition. Analysis of the corrosion product deposited on the surface of the metal revealed the mechanism behind the inhibitive power of molecules. Adsorption studies showed that CPFASC and NPFASC follow Freundlich and El-Awady isotherms, respectively, on the carbon steel surface. Adsorption equilibrium constant and free energy of adsorption were also evaluated. The corrosion investigations were done by gravimetric, EIS and polarization studies and the surface analysis of the metal specimens was performed by SEM, AFM and IR spectral spectroscopy.

Keywords Inhibitor · EIS · Polarization · Freundlich · El-Awady

Introduction

The use of corrosion inhibitors is the most practical method to control the metallic dissolution in acidic media and is widely employed in industrial processes like de-scaling, acid pickling, oil well acidizing, etc. Molecules equipped with active corrosion inhibition sites such as hetero atoms like O, S, N, aromatic rings and Schiff bases are found to act as efficient corrosion inhibitors in aggressive media [1–7]. A few semicarbazones and hydrazones exhibited good corrosion protection ability due to the presence of C=N group [11, 16]. Electronic effects such as inductive, mesomeric, etc., affect the corrosion inhibition power of molecules. Scientists and industrialists are ever in search of soluble, durable and economic corrosion inhibitors. Many researchers are trying to substantiate the mechanism behind corrosion inhibition of organic molecules on various metals using experimental as well as quantum chemical methods. The corrosion inhibition response of a molecule is closely related to its structure and geometry [8, 9]. The reactions of the inhibitor molecules with aggressive medium may also influence the behavior of the molecules. This type of response of the molecules has to be taken into account when one explains the mechanism of corrosion inhibition. For instance, the behavior of the Schiff base N,N'-bis(salicylidene)-1,2-ethylenediamine (Salen) and a mixture of its parent molecules, ethylenediamine and salicylaldehyde, as carbon steel corrosion inhibitors in 1 M HCl solution was studied by da Silva et al. [10] using corrosion potential measurements, potentiodynamic polarization curves, electrochemical impedance spectroscopy and spectrophotometric measurements. They reported that results

✉ Joby Thomas Kakkassery
drjobythomask@gmail.com

[1] Department of Chemistry, Government Engineering College, Thrissur 680009, Kerala, India

[2] Research Division, Department of Chemistry, St. Thomas' College (Autonomous), Thrissur 680001, Kerala, India

obtained in the presence of Salen were similar to those obtained in the presence of the salicylaldehyde and ethylenediamine mixture, showing that in acid medium the Salen molecule undergoes hydrolysis, regenerating its precursor molecules. In one of previous studies, we could establish the corrosion antagonistic behavior of 3-acetylphenylhyrazone (3APPH) on carbon steel in 0.5 M H_2SO_4. The corrosion rate of CS increased with the concentration of 3APPH. Intense hydrolysis to this molecule occurred in sulphuric acid, which was confirmed by UV–visible spectroscopy. The same compound showed very high inhibition efficiency (>99 % at 1.0 mM as per electrochemical studies) on CS surface in 0.5 M H_2SO_4 in the presence of trace amount of KI [11]. It is quite sure that the corrosion inhibiting capacity of a molecule is a combined effect of its molecular structure, geometry and the way in which it behaves in the aggressive medium in the presence of a metal.

In the present course of study, we synthesized two furan-2-aldehyde derived semicarbazones and investigated their corrosion inhibition response on carbon steel in 1 M HCl using gravimetric, surface and electrochemical analytical techniques. The major aim was to relate the structural behavior of the molecules with its corrosion inhibition capacity and to explore the combined influence of the aggressive medium and metal in modifying the performance of the molecules.

Experimental

Synthesis and characterization of semicarbazones

All chemicals for synthesis were purchased from Merck Millipore. Furan-2-aldehyde (98 %), p-aminobenzoic acid (>99 %), $NaNO_2$ (EMSURE®), p-nitroaniline (>99 %), $CuCl_2·2H_2O$ (EMSURE®) and semicarbazide hydrochloride (>99 %) were used for synthesis.

Semicarbazones were synthesized in two steps, i.e., arylation of furan-2-aldehyde (Meerwein arylation) followed by the condensation reaction with semicarbazide. Meerwein arylation was conducted by standard method, reported elsewhere [12]. For this, 10 mmol of

p-aminobenzoic acid/p-nitroaniline was diazotized at 0–5 °C using $NaNO_2$ and HCl. The reaction mixture was kept for 20 min and equimolar amount of furan-2-aldehyde in acetone followed by 3 mmol of $CuCl_2·2H_2O$ in water were added. The entire reaction mixture was kept for 2 days with occasional shaking. The precipitated yellow colored solid was filtered, washed with warm water and dried.

The arylated furfural derivative was dissolved in ethanol and heated to reflux in a water bath. Equimolar amount of semicarbazide hydrochloride was dissolved in ethanol water mixture (9:1) and added drop-wise into the boiling solution. The mixture was kept for 12 h and the precipitated compound was filtered, washed with ethanol–water mixture (1:1) and dried. Figure 1 shows the reaction pathway for the synthesis of semicarbazones. The products were characterized by elemental (Elementar make Vario EL III-CHN analyzer) analysis and mass (Shimadzu, QP 2010 GCMS), nmr (Bruker Avance III HD—dmso-d_6 solvent), UV–visible (Shimadzu UV–visible-1800 Spectrophotometer—DMSO solvent) and IR (Shimadzu Affinity-1-KBr pellet method) spectroscopic analyses.

Metal specimen and aggressive medium

Carbon steel coupons having 0.58 % Mn, 0.07 % P, 0.02 % S, 0.015 % Si, 0.02 % and the rest Fe (estimated by EDAX technique-SEM, Hitachi SU6600 model) were cut (1.5 × 1.5 × 0.15 cm) and polished with various grades of emery papers (120, 400, 600, 800, 1000 and 1200). The specimens were washed with soap solution, degreased with acetone, dried in air oven and weighed [13]. A stock solution of CPFASC and NPFASC (0.5 mM) was prepared in 1 M HCl. This solution was diluted with 1 M HCl to get the inhibitor solutions in the range 0.1–0.5 mM.

Gravimetric corrosion studies

CS specimens were carefully hanged using polyethylene fishing lines in 1 M HCl in the presence of various concentrations of semicarbazones for 24 h at RT (29 ± 0.1 °C). Blank experiment was also conducted in the

Fig. 1 Reaction pathway for the synthesis of semicarbazones

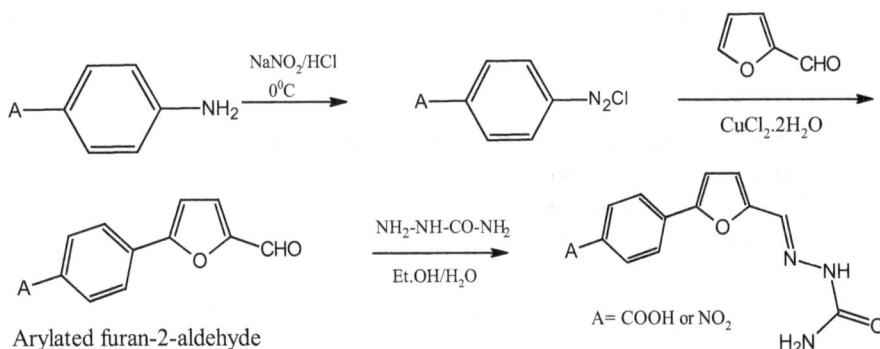

Arylated furan-2-aldehyde

absence of semicarbazones. Total volume of the aggressive medium was 50 ml in each experiment. From the initial and final weight of the metal specimen, the rate of corrosion was determined using the following equation

$$v = \frac{KW}{DSt}, \tag{1}$$

where v corrosion rate (mm year^{-1}), W weight loss (g), S surface area of metal specimen (cm^2), t time of treatment (h), D density of specimen (g cm^{-3}) and K a constant (8.76×10^4). The inhibition efficiency (η_w %) was obtained by the following equation [14–16].

$$\eta_w \% = \frac{v - v'}{v} \times 100, \tag{2}$$

where v and v' are the corrosion rate of the CS specimen in the absence and presence of the inhibitor, respectively. For good reproducibility, all the experiments were conducted in duplicate and the average values were taken.

Electrochemical corrosion investigations

Three-electrode circuitry was used for electrochemical investigations. Saturated calomel electrode (SCE) and platinum electrode with 1 cm^2 surface area acted as reference and counter electrodes, respectively. Polished metal specimen exposed to the aggressive solution with an area of 1 cm^2 acted as working electrode. The working electrode was kept in contact with the aggressive solution to attain steady-state open-circuit potential (OCP). Each metal specimen was immersed in the aggressive medium for a period of 30 min prior to the experiment at 29 °C. The electrochemical studies were carried out by Ivium compactsat-e electrochemical system (Netherlands) controlled by Ivimsoft software. Impedance spectroscopic studies were conducted to evaluate the corrosion inhibition efficiency of the heterocyclic molecules. A frequency range of 1 kHz–100 mHz with an amplitude of 10 mV as excitation signal was used for every experiment [17]. Analysis of impedance plots provided the charge transfer resistance by which one can measure the corrosion inhibition efficiency using following equation [18].

$$\eta_{EIS} \% = \frac{R_{ct} - R'_{ct}}{R_{ct}} \times 100, \tag{3}$$

where R_{ct} and R'_{ct} are the charge transfer resistances of working electrode with and without inhibitor, respectively.

Tafel polarization studies were executed between +100 and −100 mV with a sweep rate of 1 mV/s at RT (29 ± 0.1 °C). Analysis of Tafel lines gave corrosion current densities by which inhibition efficiency was calculated using the following equation [19, 20]:

$$\eta_{pol} \% = \frac{I_{corr} - I'_{corr}}{I_{corr}} \times 100, \tag{4}$$

where I_{corr} and I'_{corr} are the uninhibited and inhibited corrosion current densities, respectively.

Surface analysis

The deposited film of semicarbazone on CS surface was scrapped and analyzed with IR spectroscopy (Shimadzu Affinity-1-KBr pellet method). Morphology of metal surface was studied by SEM (Hitachi SU6600 model). Topography of the CS specimens was monitored by AFM (Park Systems XE-100 model) in the contact mode. Scanning of specimens in the area 5 × 5 μm at a rate of 0.8 Hz provided the topographic images.

Results and discussion

Structure of semicarbazones

Melting point, elemental data and spectral analytical data of CPFASC and NPFASC are shown in the subsequent paragraphs.

CPFASC: M. P. 290 °C, CHN found (calc.): C %; 55.85 (57.14 %), H %; 4.14 (4.03), N %; 14.44 (15.38), Mass spectrum; M + peak m/z = 273, base peak m/z = 230 ($[C_{12}H_{10}N_2O_3]^+$), ^{1}Hnmr spectrum; 10.3 δ (COOH), 7.73 δ(s), (CH=N) 6.38δ (s,br) (NH). ^{13}Cnmr spectrum; 166.87 ppm (COOH), 152.55 ppm (CH=N), IR spectrum; 1666 cm^{-1} (C=N), 3477 cm^{-1} (OH), 3363 cm^{-1} (NH), UV–vis spectrum; 27,777 cm^{-1} (n → π*), 28,490 cm^{-1} (π → π*).

NPFASC: M. P. 221 °C, CHN found (calc.): C %; 52.78 (52.55), H %; 3.99 (3.65), N %; 21.01 (20.43). Mass spectrum; M + peak m/z = 274, base peak m/z = 231 ($[C_{11}H_9O_3N_3]^+$), ^{1}Hnmr spectrum; 8.36 δ(s) (CH=N), 6.41 δ(s,br) (NH), ^{13}Cnmr spectrum; 154.21 ppm (CH=N), IR spectrum; 1693 cm^{-1} (C=O), 1597 cm^{-1} (C=N), UV–vis spectrum; 29,615 cm^{-1} (n → π*), 32,011 cm^{-1} (π → π*).

The structures and optimized geometries of the molecules CPFASC, NPFASC and the reduced form of NPFASC (APFASC—will be discussed later) are given in Fig. 2.

Gravimetric corrosion studies

The corrosion inhibition efficiencies of heterocyclic semicarbazones on CS for a period of 24 h are listed in Table 1. According to gravimetric studies, both molecules appreciably prevented the metallic dissolution in the aggressive medium and the inhibition efficiency increased with the

Fig. 2 Structures and optimized geometries of semicarbazones

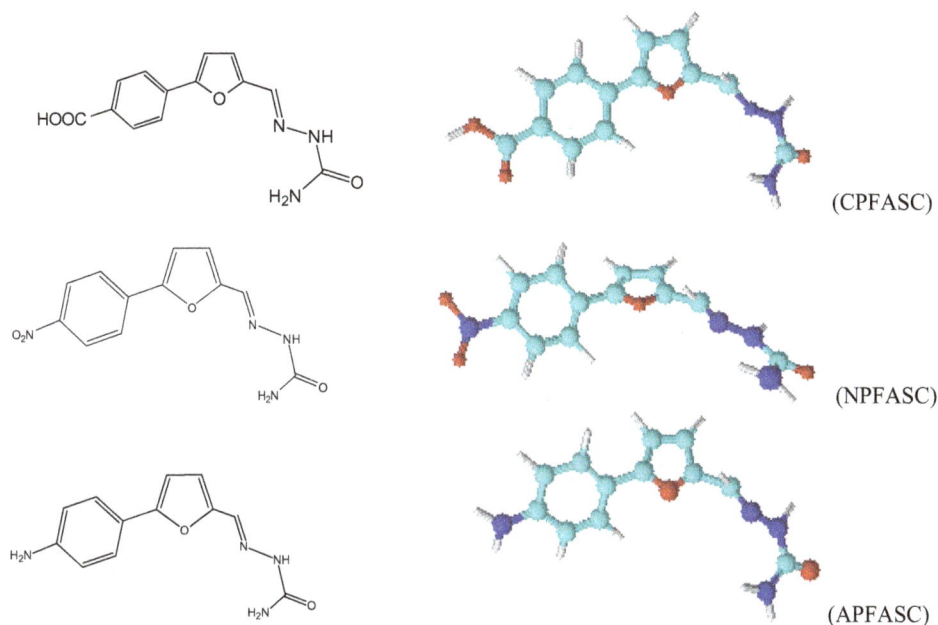

(CPFASC)

(NPFASC)

(APFASC)

Table 1 Corrosion inhibition efficiencies (η_w %) of CPFASC and NPFASC on CS in 1 M HCl for 24 h at 29 °C

C (mM)	CPFASC	NPFASC
0.1	25.88	36.54
0.2	43.24	44.68
0.3	58.72	77.83
0.4	74.54	78.75
0.5	84.37	85.44

concentration. At 0.5 mM, both molecules displayed >80 % inhibition efficiency. Among the semicarbazones CPFASC and NPFASC, former one displayed lesser efficiency than latter on CS surface in acidic medium at all concentrations.

The only difference between the structures of CPFASC and NPFASC is that CPFASC contains a –COOH group attached to the phenyl ring instead, –NO$_2$ group is attached to the phenyl ring in NPFASC. While explaining the corrosion inhibition efficiency of molecules in terms of their structures, it is important to point out two factors: (1) the electron denser sites on the molecule and (2) geometry of the molecule. On comparing the structures of CPFASC and NPFASC, evidently, the electron density of the aromatic rings and azomethine linkage will be poorer in NPFASC than in CPFASC due to the presence of highly electron withdrawing –NO$_2$ group and the NPFASC molecule is expected to display inferior corrosion inhibition efficiency than CPFASC. Moreover, considerable deviation can be observed from co-planarity in optimized geometry of NPFASC (Fig. 2), which is an unfavorable scenario for the firm interaction of molecules on the metal surface. Contrary to expectation, NPFASC molecule showed better corrosion inhibition efficacy than CPFASC at all

concentrations. This unusual behavior can be accounted by Bechamp's reduction, i.e., reduction of nitro group of the molecule into amino group in the presence of Fe and HCl [21], which can be seen from Fig. 3.

The possibility for the reduction of large number of NPFASC into APFASC in HCl medium is high when the molecules approach the metal surface, which helps to improve the corrosion inhibition efficacy of the NPFASC molecule appreciably than CPFASC molecule. Upon reduction, the electron density of the phenyl ring, furan ring and the azomethine linkage increases significantly. This is because, the electron deactivating nitro group is replaced by electron-rich amino group. The amino group of APFASC can donate the lone pair of electron to the vacant molecular orbitals of Fe. The free electron pair can also participate in resonance which will definitely improve the electron richness of the aromatic ring. Furthermore, the reduced molecule gets a complete planar structure (Fig. 2) according to the optimized geometry. Now, these reduced molecules can interact with the metal surface more effectively than before and thus prevent the metal dissolution more efficiently. Very strong evidence regarding the reduction of NPFASC is obtained from the mass spectral analysis of the film formed on CS surface and the spectrum exhibited a clear signal at m/z 244, which is the exact molecular mass of the reduced NPFASC, i.e., APFASC. This peak was not seen in the mass spectrum of NPFASC. The signal due to the molecular ion peak of NPFASC ($m/z = 274$), which was totally absent in the mass spectrum of the corrosion product, indicates the absence of free NPFASC molecules in the corrosion product. Figure 4 shows the mass spectrum of NPFASC and the surface film

Fig. 3 Conversion of NPFASC into APFASC in the presence of Fe and HCl

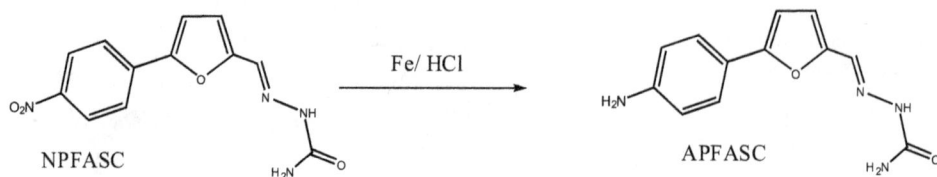

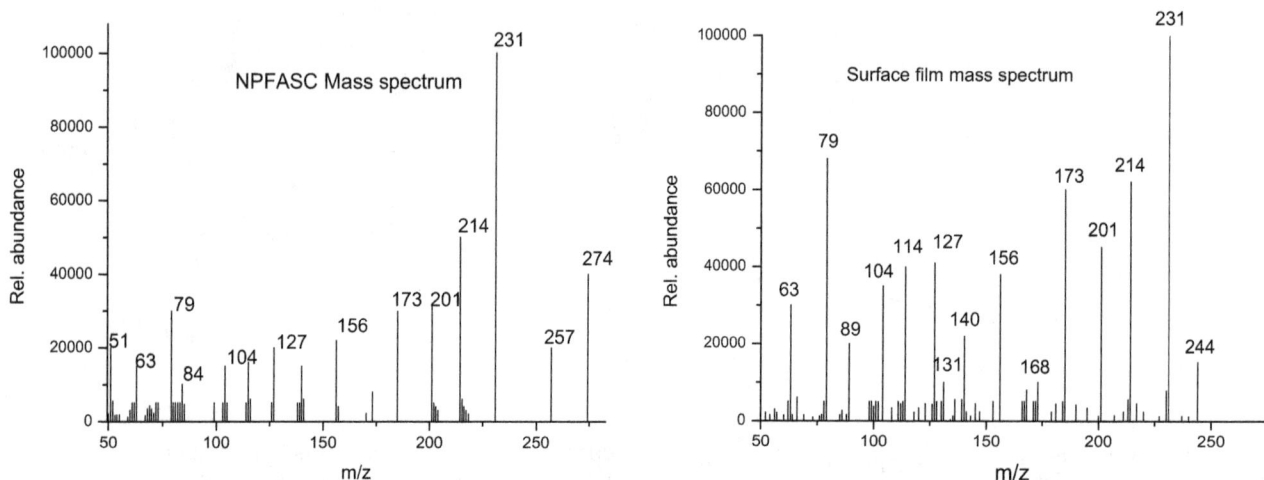

Fig. 4 Mass spectrum of NPFASC and surface film on CS treated with NPFASC (0.5 mM) in 1 M HCl for 24 h

formed on CS after the treatment with NPFASC (0.5 mM) in 1 M HCl for 24 h at 29 °C. The overall mechanism of interaction of NPFASC on CS in 1 M HCl can be visualized from Fig. 5.

Adsorption isotherms

To get more insight into the mechanism of corrosion, adsorption isotherms were plotted. Among the various isotherm models tried, the most suitable one was selected with the help of regression coefficient (Table 2). The most fit isotherm model for CPFASC and NPFASC on CS in 1 M HCl was Freundlich and El-Awady isotherm (Eqs. 5, 6) [22–24], where θ is fractional surface coverage, K_{ads} is the adsorption equilibrium constant and C is the concentration of the inhibitor. In El-Awady isotherm, $K_{ads} = K^{1/y}$. The regression coefficient of El-Awady isotherm for NPFASC/APFASC system on CS surface was 0.899. Considerable deviation from unity of this isotherm may be due to the adsorption of two different molecules (NPFASC and APFASC) at the same time on CS surface. The adsorption equilibrium constant for CPFASC was 1515 while that of NPFASC/APFASC was 6130, indicating the elevated interaction of latter molecules on CS surface. The free energies of adsorption (ΔG_{ads}) calculated by Eq. 7 were −28.6 and −31.9 kJ mol^{-1}, respectively, for CPFASC and NPFASC/APFASC on CS surface. Generally, the values of ΔG_{ads} up to −20 kJ/mol

is an indication of electrostatic interaction (physisorption) of charged molecules and the metal, while those around −40 kJ mol^{-1} stands for chemisorption. Since the ΔG_{ads} values lie between −20 and −40 kJ mol^{-1}, for CPFASC and NPFASC, both physisorption and chemisorption were involved during the interaction of molecules on CS surface [25]. The involvement of chemical force was much higher in the case of NPFASC/APFASC than CPFASC on CS surface as evident from the free energy values. On analyzing the structures of the CPFASC and NPFASC, it may be assumed that the aromatic ring systems, azomethine linkage and the nitrogen atoms of semicarbazide part are the active sites of adsorption. But the transformation of NPFASC into APFASC led to the generation of an additional electron-rich site (amino group) in APFASC. Moreover, the electron density of the aromatic rings and azomethine linkage escalated due to the transformation of NPFASC into APFASC in the presence of Fe and HCl.

Figure 6a, b shows adsorption isotherms for CPFASC and NPFASC on CS surface in 1 M HCl, respectively.

Freundlich adsorption isotherm $\theta = K_{ads}C$ (5)

El − Awady adsorption isotherm

$$\log\left(\frac{\theta}{1-\theta}\right) = \log K + y\log C \quad (6)$$

$$\Delta G_{ads} = -2.303RT\log(55.85K_{ads}) \quad (7)$$

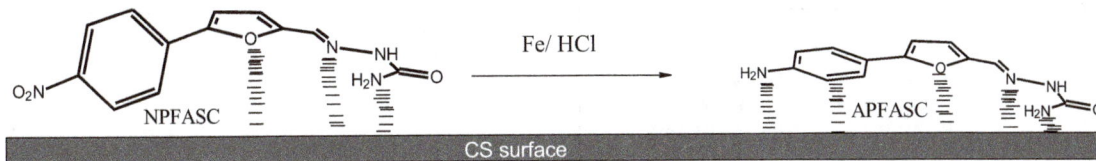

Fig. 5 Illustration of the interaction of NPFASC on CS surface in 1 M HCl

Table 2 Adsorption isotherm models and their regression coefficients for CPFASC and NPFASC on CS in 1 M HCl

Adsorption isotherm	CPFASC	NPFASC
Langmuir	0.973	0.828
Freundlich	0.991	0.872
Temkin	0.973	0.893
Flory–Huggin	0.888	0.555
El-Awady	0.960	0.899

Electrochemical studies

Figure 7a, b represents the Nyquist plots of metal specimens treated with CPFASC and NPFASC, respectively, in acidic medium. The impedance parameters are listed in Table 3. From the figures and table it is evident that the R_{ct} values increased with the concentration of the semicarbazones, which shows the reluctance of metal dissolution with the inhibitor concentration. The solution resistance values did not show appreciable change with the concentration. The simple equivalent circuit (Randle's equivalent circuit) that fits these electrochemical systems is a parallel combination of double-layer capacitance C_{dl} and charge transfer resistance R_{ct}, both in series with the solution resistance R_s [26, 27]. For instance, Fig. 8a, b shows two simulated EIS spectrum for CS in the presence of CPFASC and NPFASC (0.3 mM) using Iviumsoft program. The red square symbols denote the impedance values and the black line represents the fitting curve. The equivalent circuit is also provided in the figures. From the figures it is evident

that the Randle's equivalent circuit is the best circuit that would fit for the measured impedance. Bode plots for various concentrations of CPFASC and NPFASC are given in Fig. 9a, b.

Nyquist plots were not perfect semicircles, which may be attributed to the nonhomogeneity of the CS surface, grain boundaries and impurities [28]. To get an accurate fit, constant phase element (CPE) was introduced in the circuit than pure double-layer capacitor. The impedance of CPE can be represented as [29]

$$Z_{CPE} = \frac{1}{Y_0(j\omega)^n}, \tag{8}$$

where Y_0 is the magnitude of CPE, n is the exponent (phase shift), ω is the angular frequency and j is the imaginary unit. Depending upon the values of n, CPE may be resistive, capacitive and inductive. The capacitive behavior of CPE was established in all analyses since the calculated value of n varied between 0.8 and 0.86 [28]. Since the shapes of Nyquist plots do not change with the concentration of the inhibitor, it can be assumed that the mechanism of corrosion inhibition by the molecules does not alter with the concentration. The charge transfer resistance significantly increased with the concentration of semicarbazones. Even at low concentrations NPFASC exhibited very high corrosion inhibition efficiency and it was greater than that of CPFASC. The double-layer capacitance decreased with the concentration of inhibiting molecules. This is either due to the decrease of local dielectric constant or the increase of thickness of electric double layer (or both) with the adsorption of semicarbazones. It is also

Fig. 6 **a** Freundlich adsorption isotherm for CPFASC and **b** El-Awady adsorption isotherm for NPFASC on CS in 1 M HCl on CS in 1 M HCl

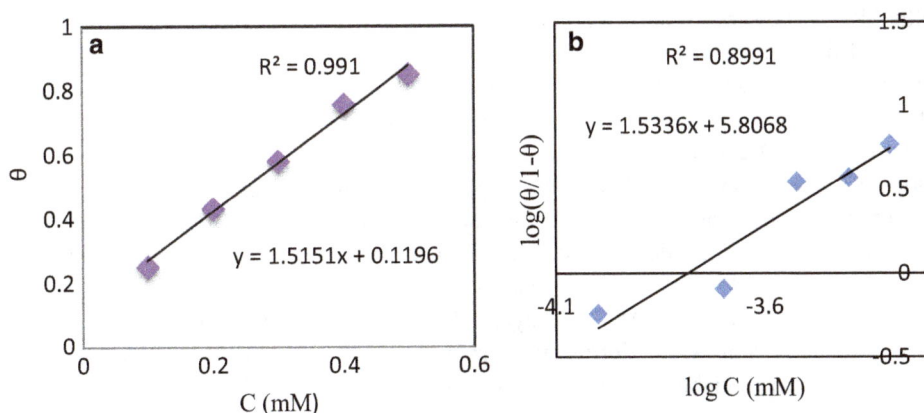

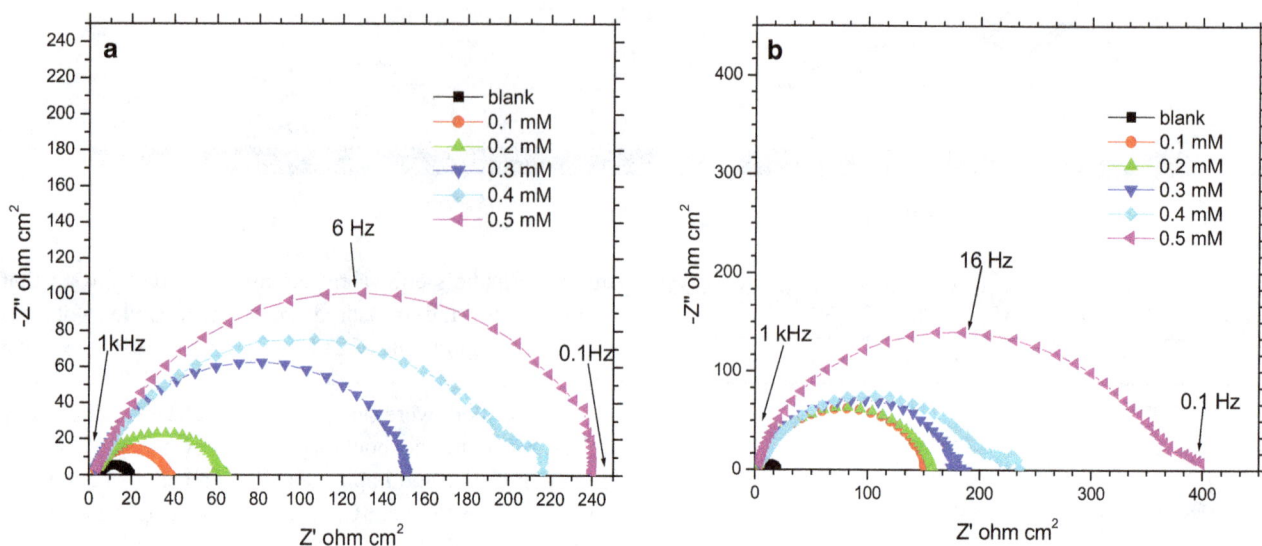

Fig. 7 Nyquist plots of CS in the presence and absence of **a** CPFASC and **b** NPFASC in 1 M HCl at 29 °C

Table 3 Electrochemical impedance and polarization parameters of CS in the presence and absence of semicarbazones CPFASC and NPFASC in 1 M HCl

Semicarbazone	Impedance data					Polarization data				
	C (mM)	C_{dl} (μF cm^{-2})	N	R_{ct} (Ω cm^2)	η_{EIS} %	$-E_{corr}$ (mV)	I_{corr} (μA cm^{-2})	$-b_c$ (mV dec^{-1})	B_a (mV dec^{-1})	η_{pol} %
Blank	0	76.36	0.80	23.08	–	474	499.6	102	77	–
CPFASC	0.1	72.00	0.81	38.96	40.76	474	215.6	54	87	56.85
	0.2	60.69	0.82	55.88	58.70	498	146.0	64	94	70.78
	0.3	58.80	0.82	136.50	83.09	507	102.0	79	93	79.58
	0.4	56.00	0.84	187.80	87.71	493	76.84	77	94	84.62
	0.5	56.10	0.85	227.40	89.85	492	61.9	75	93	87.61
NPFASC	0.1	71.89	0.83	132.80	82.68	489	95.29	82	94	80.93
	0.2	59.37	0.85	134.10	82.85	498	73.12	85	92	85.36
	0.3	54.88	0.83	171.40	86.58	480	66.44	82	86	86.70
	0.4	43.75	0.84	192.40	88.05	487	65.55	78	62	86.88
	0.5	43.67	0.86	350.00	93.40	475	37.36	84	79	92.52

visible from the data that C_{dl} (NPFASC) $<C_{dl}$ (CPFASC) at all concentrations, which implies a better interaction of NPFASC on Fe surface than CPFASC.

Tafel lines of the metal specimens in the presence and absence of the semicarbazones are given in Fig. 10. Polarization parameters like corrosion potential (E_{corr}), corrosion current density (I_{corr}), cathodic slope (b_c), anodic slope (b_a), etc., and inhibition efficiency (η_{pol} %) are represented in the Table 3. From the Tafel data it is evident that NPFASC molecule demonstrated elevated inhibition efficiencies than CPFASC at all concentrations. At a concentration of 0.5 mM, NPFASC displayed 92.5 % inhibition efficiency on CS. Corrosion current densities decreased appreciably with concentration of

semicarbazones suggesting that reluctance of the metal dissolution increased with the concentration. On close examination of polarization data one can assure that CPFASC molecule worked more at cathodic regions of corrosion. This was because cathodic slopes were altered considerably when compared to that of blank. Thus the rate of H$_2$ formation was lowered by CPFASC than the rate of metal dissolution process. The cathodic slopes of Tafel lines in the presence of NPFASC did not change considerably while anodic slopes showed more shifts. This suggests that NPFASC/APFASC molecules were more active at anodic sites of corrosion or metal dissolution process [30, 31]. There was high compatibility between the η_{pol} and η_{EIS} % for the semicarbazones.

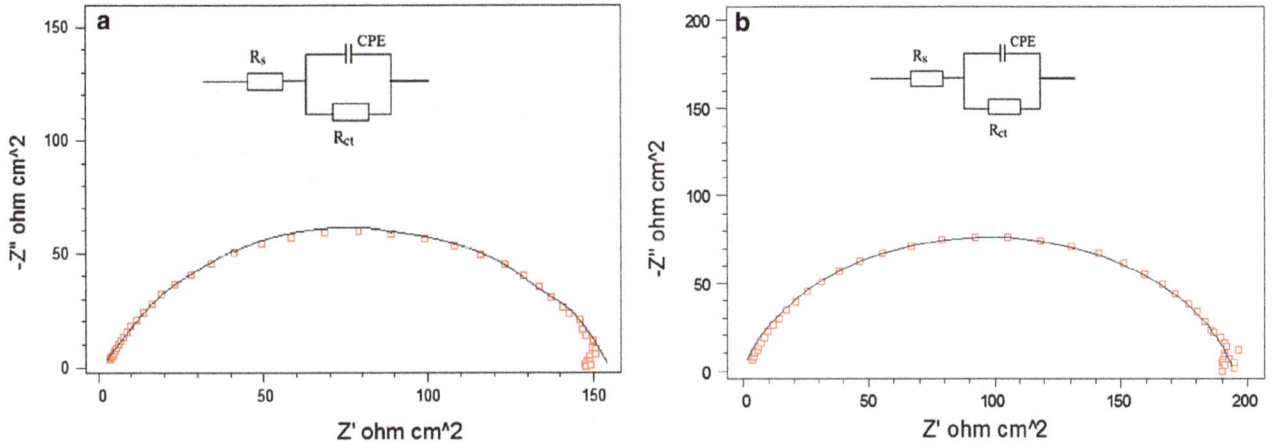

Fig. 8 EIS fitting curves and their corresponding equivalent circuit for CS in the presence of 0.3 mM **a** CPFASC and **b** NPFASC in 1 M HCl at 29 °C

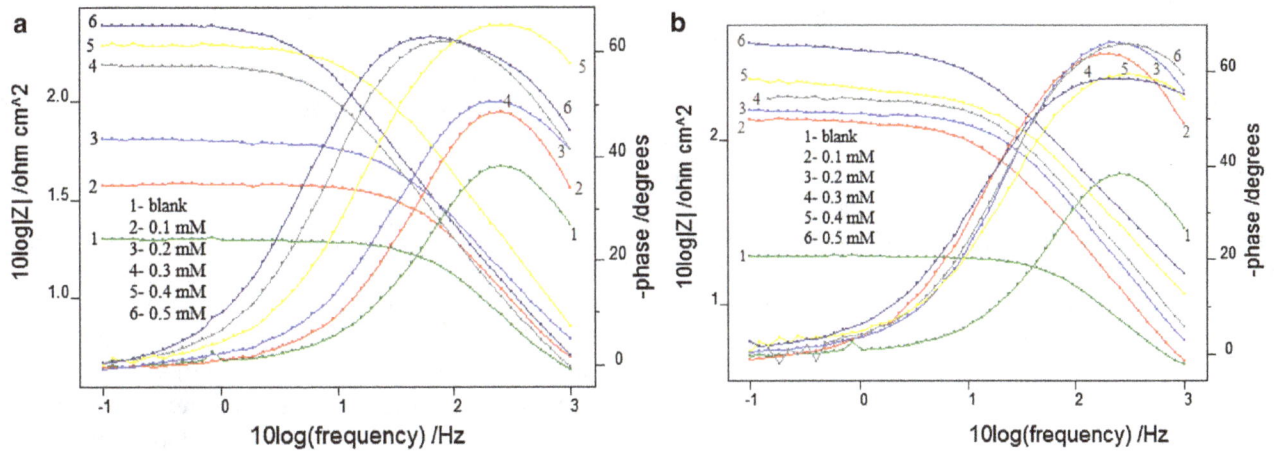

Fig. 9 Impedance and Bode plots of CS in the presence and absence of **a** CPFASC and **b** NPFASC in 1 M HCl at 29 °C

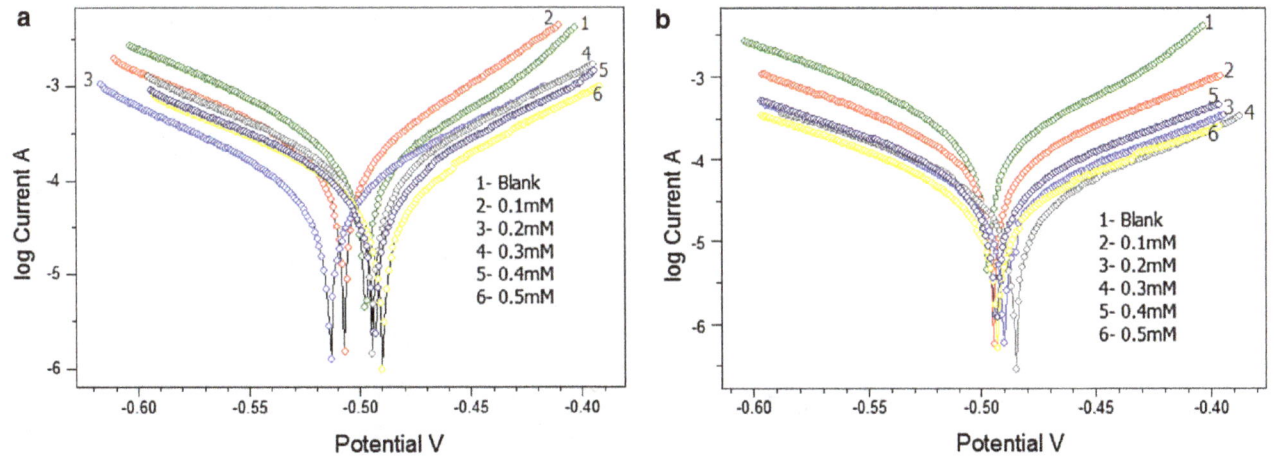

Fig. 10 Tafel plots of CS in the presence and absence of **a** CPFASC and **b** NPFASC in 1 M HCl at 29 °C

Surface analysis

IR spectral studies

The film deposited on the CS surface (treated for 24 h in 1 M HCl in the presence of 0.5 mM semicarbazone) was mechanically removed and subjected to IR spectral analysis. The spectrum was compared with the IR spectrum of the semicarbazone (Fig. 11). The IR spectrum of the surface film was very similar to that of spectrum of CPAFSC. But lowering of frequencies in some regions indicated that the molecules made coordinate type bond with the surface metal atoms. When comparing the spectrum of NPFASC and its surface film, appreciable dissimilarities could be observed. This is an indication of alteration of the basic structure of the adsorbed molecule on the metal surface, i.e., the conversion of NPFASC to APFASC in the presence of Fe and HCl. A sharp peak that appeared at ~ 3500 cm^{-1} in the spectrum of surface film is assignable to the stretching frequency of newly formed amino group in APFASC. It may be concluded that the surface film IR spectrum of NPFASC is the combined spectrum of adsorbed NPFASC and APFASC on Fe surface. Some of the fundamental frequencies of NPFASC were observed in the spectrum of the surface film. A very strong absorption in the lower region of the IR spectrum (~ 800–500 cm^{-1}) of NPFASC film may be attributed to the strong interaction of the molecules with the metal atoms by making coordinate bonds such as Fe–O and Fe–N, etc.

SEM studies

Surface morphological studies of CS specimens were performed using SEM. Figure 12a–d represents the SEM images of metal specimens, i.e., bare, treated with 1 M HCl for 24 h (blank), treated with 0.5 mM CPFASC and 0.5 mM NPFASC in 1 M HCl solution. It was clear that the four surface morphologies were entirely different. The surface of bare specimen was smoother than other specimens. Mild pits and cracks that appeared in the image of bare specimen were due to the effect of polishing. Attack of the aggressive solution on the metal surface made it rough is evident from Fig. 12b. From the surface images, Fig. 12c, d, it can be concluded that heterocyclic hydrazones shield the CS corrosion markedly in acidic medium by forming a protective layer.

AFM studies

Surface interaction of the heterocyclic semicarbazones on CS was further confirmed by AFM studies. Topography of bare specimen, blank, specimens treated with 0.5 mM CPFASC and NPFASC in 1 M HCl for 24 h are given in Fig. 13a–d, respectively. The roughness parameters which characterize the topography of surfaces like average roughness (arithmetic mean of the absolute values of the height of the surface profile, R_a), root mean square roughness (mean squared absolute values of surface roughness profile R_q), ten-point roughness (arithmetic mean of the five highest peaks added to the five deepest valleys over the evaluation length measured R_z) and maximum peak-to-valley height (R_{px}) [32, 33] are provided in Table 4. The roughness parameters for the bare metal specimen were very low due to the smoothness of the surface when compared to the blank specimen, which was in continuous reaction with the aggressive medium. When analyzing the topography of the metal specimens which were in contact with the organic molecules, it is understandable that the roughness parameters lie between the parameters of bare and blank specimens. This is an indication of the adsorption of the molecules on the

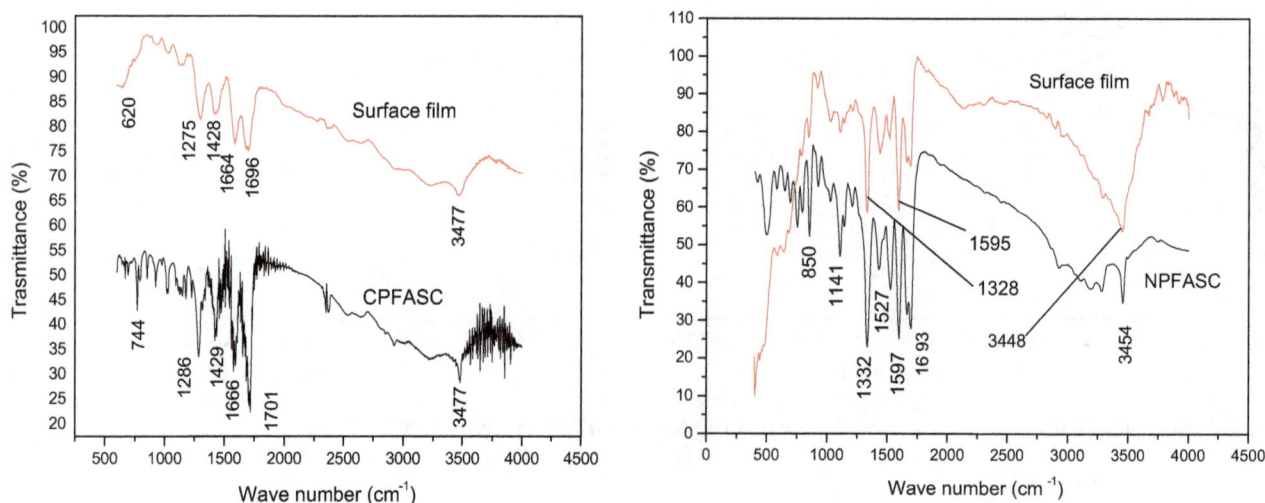

Fig. 11 IR spectra of semicarbazone and the surface film formed on CS in 1.0 M HCl in the presence of 0.5 mM semicarbazone for 24 h

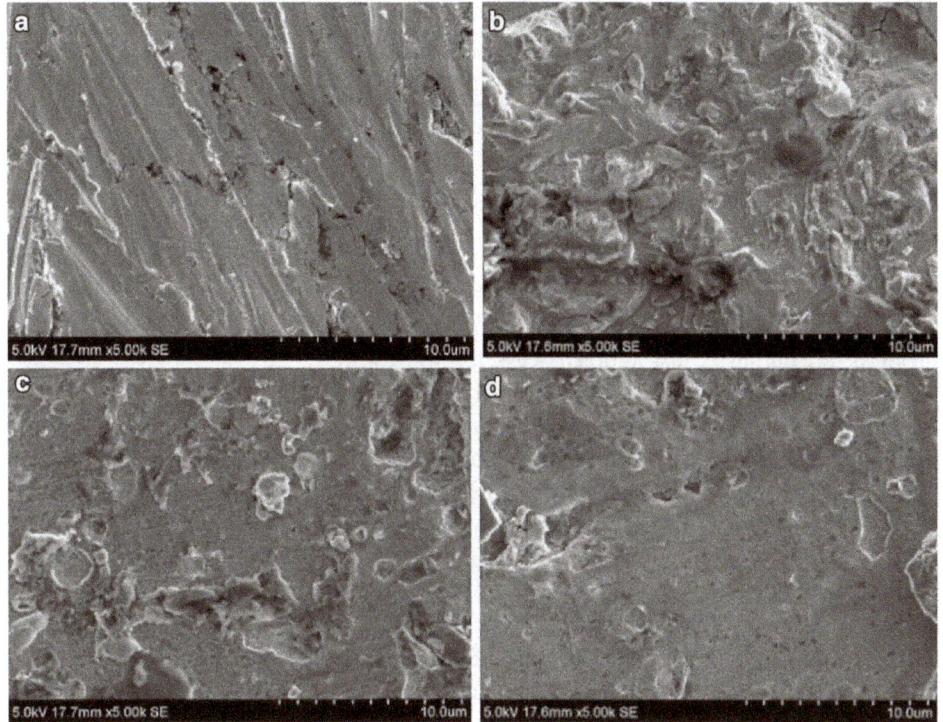

Fig. 12 SEM images of **a** bare CS surface **b** CS treated with 1 M HCl for 24 h **c** CS treated with 1 M HCl in the presence of 0.5 mM CPFASC for 24 h **d** CS treated with 1 M HCl in the presence of 0.5 mM NPFASC for 24 h

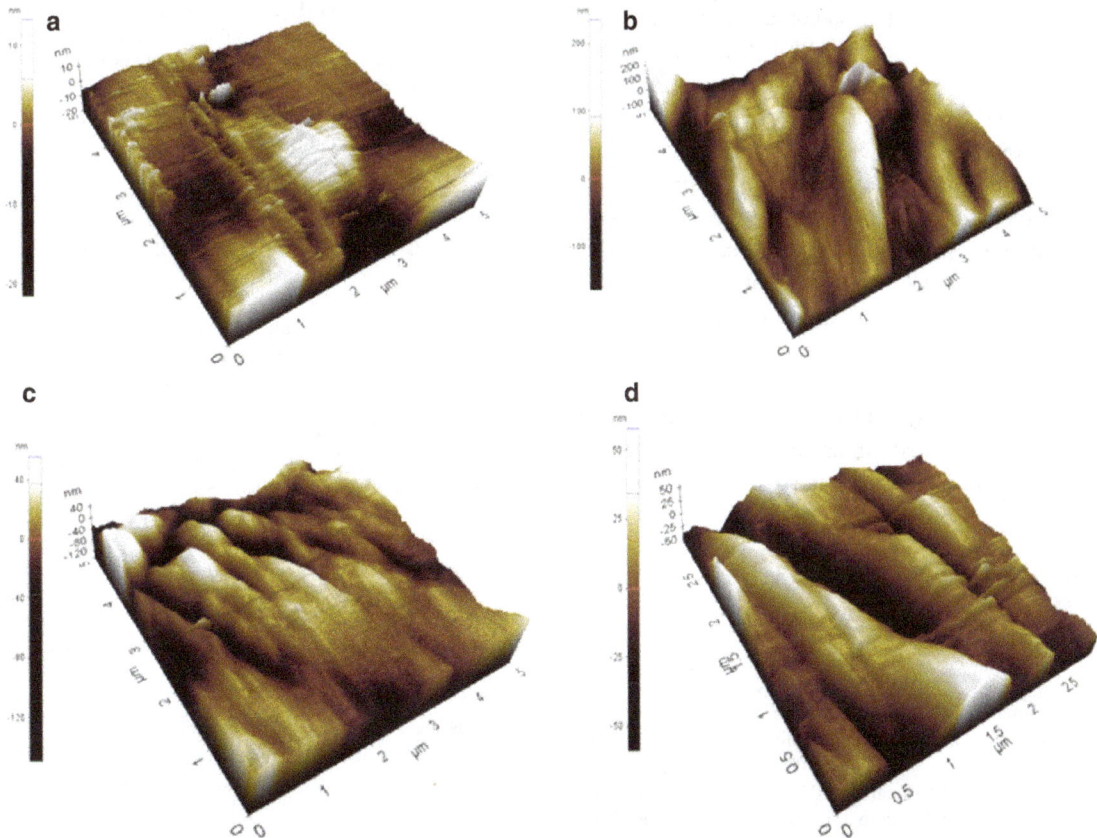

Fig. 13 Topography of **a** bare CS **b** CS treated with 1 M HCl for 24 h **c** CS treated with 1 M HCl in the presence of 0.5 mM CPFASC for 24 h **d** CS treated with 1 M HCl in the presence of 0.5 mM NPFASC for 24 h

Table 4 Roughness parameters of metal specimens by AFM studies

Metal surface	Peak-to-valley height R_{px} (nm)	RMS roughness R_q (nm)	Average roughness R_a (nm)	Ten-point roughness R_z (nm)
Bare metal	34.79	3.02	2.16	30.66
Metal in 1 M HCl (24 h)	399.42	47.11	36.90	380.71
Metal in 1 M HCl with 0.5 mM CPFASC (24 h)	121.77	24.62	16.35	110.24
Metal in 1 M HCl with 0.5 mM NPFASC (24 h)	116.49	17.57	13.80	109.7

metal surface during the dissolution process. It is worthwhile to mention that the AFM parameters of the metal specimen treated with NPFASC are to some extent lower than that of the specimen treated with CPFASC indicating that the former molecule fights well against corrosion than the latter.

Conclusions

The corrosion monitoring studies established that inhibition efficiency of NPFASC is higher than that of CPFASC in 1 M HCl medium, even though the molecule possesses a non-planar geometry and electron withdrawing group. The explanation for the unexpected outcome was the transformation of NPFASC molecule into APFASC in the presence of iron and acid. Mass spectral studies of the surface film confirmed the transformation of NPFASC molecules into APFASC. CPFASC and NPFASC followed Freundlich and El-Awady adsorption isotherms, respectively, on CS surface in HCl medium. Comparatively weak interaction existed between CPFASC and metal surface. Potentiodynamic polarization studies revealed that CPFASC was more active at cathodic sites of corrosion while NPFASC prevented the anodic process of corrosion appreciably. Surface analysis using IR spectroscopy was also performed. The protective layer of inhibiting compounds on CS was visible from the scanning electron micrographs. AFM analysis explored the topography of CS specimens.

References

1. Hosseini M, Mertens SFL, Ghorbani M, Arshadi MR (2003) Asymmetrical Schiff bases as inhibitors of mild steel corrosion in sulphuric acid media. Mater Chem Phy 78(3):800–808
2. Saxena N, Kumar S, Sharma MK, Mathur SP (2013) Corrosion inhibition of mild steel in nitric acid media by some Schiff bases derived from anisalidine. Pol J Chem Tech 15(1):61–67
3. Toliwal SD, Jadav K, Pavagadhi T (2011) Inhibition of corrosion of mild steel in 1 N HCl solutions by Schiff base derived from non-traditional oils. Indian J Chem Tech 18(4):301–308
4. Keleş H, Keleş M (2012) Electrochemical investigation of a Schiff base synthesized by cinnamaldehyde as corrosion inhibitor

on mild steel in acidic medium. Res Chem Intermed. doi:10.1007/s11164-012-0955-5
5. Gopi D, Govindaraju KM, Kavitha L (2010) Investigation of triazole derived Schiff bases as corrosion inhibitors for mild steel in hydrochloric acid medium. J Appl Electrochem 40(7):1349–1356
6. Hosseini SMA, Azimi A (2008) The inhibition effect of the new Schiff base, namely 2,2'-[bis-N(4-choloro benzaldimin)]-1,1'-dithio against mild steel corrosion. Mater Corros 59(1):41–45
7. John S, Joseph A (2012) Electro analytical, surface morphological and theoretical studies on the corrosion inhibition behavior of different 1,2,4-triazole precursors on mild steel in 1 M hydrochloric acid. Mater Chem Phys 133:1083–1091
8. Khaled KF, Abdel Shafi NS (2013) Chemical and electrochemical investigations of L-arginine as corrosion inhibitor for steel in hydrochloric acid solutions. Int J Electrochem Sci 8:1409–1421
9. Martinez S, Štagljar I (2003) Correlation between the molecular structure and the corrosion inhibition efficiency of chestnut tannin in acidic solutions. J mol Struc-Theochem 640(1–3):167–174
10. Silva AB, Elia ED, Gomes JA (2010) Carbon steel corrosion inhibition in hydrochloric acid solution using a reduced Schiff base of ethylenediamine. Corros Sci 52(3):788–793
11. Vinod PR, Joby TK, Shaju KS, Aby P (2013) Study of synergistic effect of iodide on the corrosion antagonistic behaviour of a heterocyclic phenylhydrazone in sulphuric acid medium on carbon steel. ISRN Corrosion. doi:10.1155/2013/390823
12. Racane L, Kulenovic VT, Boykin DW, Karminski-Zamola G (2003) Synthesis of new cyano-substituted bis-benzothiazolyl arylfurans and arylthiophenes. Molecules 8(3):342–348
13. AST G-31-72 (1990) Standard recommended practice for the laboratory immersion corrosion testing of metals. ASTM, Philadelphia
14. Fouda AS, Abdallah M, Medhat M (2012) Some Schiff base compounds as inhibitors for corrosion of carbon steel in acidic media. Prot Metals Phy Chem Surf 48(4):477–486
15. Govindaraju KM, Gopi D, Kavitha L (2009) Inhibiting effects of 4-amino-antipyrine based Schiff base derivatives on the corrosion of mild steel in hydrochloric acid. J Appl Electrochem 39(12):2345–2352
16. Vinod PR, Joby TK, Shaju KS, Aby P (2014) Corrosion inhibition investigations of 3-acetylpyridine semicarbazone on carbon steel in hydrochloric acid medium. Res Chem Intermed 40(8):2689–2701
17. El Azhar M, Mernari B, Traisnel M, Bentiss F, Lagrenée M (2001) Corrosion inhibition of mild steel by the new class of inhibitors [2,5-bis(n-pyridyl)-1,3,4-thiadiazoles] in acidic media. Corros Sci 43(12):2229–2238
18. Ebenso EE (2003) Synergistic effect of halide ions on the corrosion inhibition of aluminium in H_2SO_4 using 2-acetylphenothiazine. Mater Chem Phys 79(1):58–70
19. Li X, Deng S, Fu H (2009) Synergism between red tetrazolium

and uracil on the corrosion of cold rolled steel in H_2SO_4 solution. Corros Sci 51(6):1344–1355

20. Ma Cafferty M, Hackerman N (1972) Double layer capacitance of iron and corrosion inhibition with polymethylene diamines. J Electrochim Soc 119(2):146–154

21. Béchamp MA (1854) De l'action des protosels de fer sur la nitronaphtaline et la nitrobenzine. nouvelle méthode de formation des bases organiques artificielles de Zinin". Annales de chimie et de physique 42:186–196

22. Umoren SA, Eduok UM, Oguzie EE (2008) Corrosion inhibition of mild steel in 1 m H_2SO_4 by polyvinyl pyrrolidone and synergistic iodide additives. Portugaliae Electrochim Acta 26(6):533–546

23. El-Shafei AA, Moussa MNH, El-Far AA (2001) The corrosion inhibition character of thiosemicarbazide and its derivatives for C-steel in hydrochloric acid solution. Mater Chem Phys 70(2):175–180

24. Sudhish KS, Eno EE (2011) Corrosion inhibition, adsorption behavior and thermodynamic properties of streptomycin on mild steel in hydrochloric acid medium. Int J Electrochem Sci 6:3277–3291

25. Cano E, Polo JL, La Iglesia A, Bastidas JMA (2004) Study on the adsorption of benzotriazole on copper in hydrochloric acid using the inflection point of the isotherm. Adsorption 10(3):219–225

26. Bentiss F, Lebrini M, Lagrenée M (2005) Thermodynamic characterization of metal dissolution and inhibitor adsorption processes in mild steel/2,5-bis(n-thienyl)-1,3,4-thiadiazoles/hydrochloric acid system. Corros Sci 47(12):2915–2931

27. Yurt A, Balaban A, Kandemir SU, Bereket G, Erk B (2004) Investigation on some Schiff bases as HCl corrosion inhibitors for carbon steel. Mater Chem Phys 85(2–3):420–426

28. Chaitra TK, Mohana KNS, Tandon HC (2015) Thermodynamic, electrochemical and quantum chemical evaluation of some triazole Schiff bases as mild steel corrosion inhibitors in acid media. J Mol Liq 211:1026–1038

29. Popova A, Christov M (2006) Evaluation of impedance measurements on mild steel corrosion in acid media in the presence of heterocyclic compounds. Corros Sci 48(10):3208–3221

30. Satapathy AK, Gunasekaran G, Sahoo SC, Amit K, Rodrigues PV (2009) Corrosion inhibition by Justicia gendarussa plant extract in hydrochloric acid solution. Corros Sci 51(12):2848–2856

31. Ferreira ES, Giacomelli C, Giacomelli FC, Spinelli A (2004) Evaluation of the inhibitor effect of L-ascorbic acid on the corrosion of mild steel. Mater Chem Phys 83(1):129–134

32. Haugstad G (2012) Atomic force microscopy: understanding basic modes and advanced applications. Wiley, USA

33. Kumar BR, Rao TS (2012) Comparative studies of morphological and microstructural properties of electrodeposited nanocrystalline two-phase Co–Cu thin films prepared at low and high electrolyte temperatures. Dig J Nanomater Bios 7(4):1881–1889

16

Copper activation option for a pentlandite–pyrrhotite–chalcopyrite ore flotation with nickel interest

I. O. Otunniyi[1] · M. Oabile[1] · A. A. Adeleke[1] · P. Mendonidis[1]

Abstract

Introduction Different sulphide ores respond differently in the prospect of copper activation to thiols for froth flotation. With the BCL Selebi-Phikwe ore body, pentlandite–pyrrhotite–chalcopyrite co-occur and there is the constraint of fine pentlandite dissemination in the pyrrhotite, so that liberation is limited. In the presence of pentlandite, pyrrhotite can even be galvanically depressed. In this scenario, response of the ore to copper activation may not be easily predicted. To explore the possibilities, flotation with various dosages of copper addition was investigated for this ore with specific interest in how the nickel recovery responds. The responses of copper, iron, cobalt and quartz were also monitored.

Result Increase in nickel recovery and grade with copper sulphate dosage was obtained, compared with no addition, but the trend showed a maxima after which depression was obtained. The optimum dosage in this instance was between 15 and 30 g/ton for best grade recovery. The response is understandable from the fact that the predominant copper species in the aqueous system can be ion or hydroxide depending on the concentration, among other factors. As ions the copper thus imparts activation, while as hydroxide it will cause depression. Pyrrhotite and chalcopyrite recovery also increased marginally along with the pentlandite.

Conclusion To achieve activation of pentlandite in the ore to thiol collector using copper sulphate, the dosage must be about the optimum; higher dosages will bring about depression.

Keywords Froth flotation · Pentlandite · Pyrrhotite · Chalcopyrite · Copper sulphate activation

Introduction

The pentlandite–pyrrhotite–chalcopyrite ore body at Selebi-Phikwe in Botswana has been mined by the BCL (Bamangwatu Concessions Limited) concentrator for the nickel and copper content since 1973. Various sources cite 0.55–0.71 % Ni and 0.58–0.75 % Cu runoff mine grades, producing 2.5–3.5 % Ni and 3.2–4.5 % Cu concentrates [1–3]. The life of the mine has been projected differently to end by 2010 and 2014, but with deeper exploration and mining efforts, the life of the mine can be extended until 2024 [4, 5]. In this further exploration, cutoff grade for resource reporting was put at 0.3 % Ni equivalent. The life of mine extension therefore comes with the challenge to increase production, grade and improved efficiency to maintain profitability beyond 2014 [5].

Record plant performance data show that nickel recoveries were about 85 %, but this dropped to values below 80 % in the 1990s. Various efforts to improve recoveries include changes in the flotation circuits with the introduction of larger cell to increase residence time of slow-floating minerals; drum magnetic separator treatment of flotation tailings to recover weakly magnetic pyrrhotite containing finely disseminated target pentlandite; replacement of old rougher, scavenger and cleaner tanks at different times; automation of milling and flotation controls systems. The efforts restored recovery to 83–86 % [2].

In these efforts, the chemistry was kept at the butyl xanthate collector with the generic frother, and surfactant

✉ I. O. Otunniyi
iyiolao@vut.ac.za

[1] Metallurgical Engineering, Vaal University of Technology, Vanderbijlpark, South Africa

suite modifications were not explored. The prospect of activation is considered worthy of investigation to make available more options for maximizing grade-recovery outcomes to prolong the mine life. For pentlandite activation to xanthates, the addition of copper, lead or silver can be recommended. Although the xanthate compound of any of these metals, having lower dissociation constants compared to that of nickel, will stabilize the surface xanthate product and enhance recovery, copper has been used mostly for many sulphide mineral beneficiation [6–8]. However, due to vast shades of variations in mineral occurrences from one deposit to another and even within the same ore body, response to well-known flotation schemes or reagents can vary widely [9]; hence, a conclusion pre-empting the response in this instance cannot be made without due investigations. The mineralogy of this ore body and the chemistry of $CuSO_4$ interaction in the presence of co-occurring sulphides will have to be carefully considered.

The Selebi-Phikwe ore body consists of four co-occurring sulphide minerals—pentlandite, chalcopyrite, pyrite and pyrrhotite—along with other minerals, while the major gangues are amphibolites, quartz, mica and feldspar, forming typically about 68 % (Table 1). Cobalt is present in solid solution in the pentlandite to about 0.05 % head grade of the ore. In the presence of xanthate, sulphide minerals float favourably at different pH regimes; pyrrhotite in particular will float at a relatively lower pH compared to pentlandite, such that pentlandite can be selectively floated in pulps with sufficiently alkaline pH [10]. Differential flotation for these sulphides has been demonstrated where the natural floatability of chalcopyrite was enhanced by sulphur (IV) oxide with diethylenetriamine as the complexing agent, while pentlandite and pyrrhotite were depressed [11]. The depression was later reversed for pentlandite by xanthate addition to separate it from pyrrhotite. Using sodium isobutyl xanthate, the floatability of the sulphides obtained from Merensky ore was also shown to differ: highest for chalcopyrite and least for pyrrhotite [12]. This background apparently presents

the possibilities of exploring the rejection of the bulky pyrrhotite (24.1 %) and concentrate pentlandite and chalcopyrite (both only 3.7 %), but the pyrrhotite contains about 14 % of the total nickel content in the ore, as can be inferred from the composition in Table 1. Rejecting the total pyrrhotite content therefore implies that the overall recovery will, by such process design, be capped below 86 %.

The other alternative is to liberate the pentlandite in the pyrrhotite before flotation, but the pentlandite values occur as 8–38 μm exsolution phase in the pyrrhotite, according to the mineral liberation analysis [1]. Grinding down to attain liberation will imply slime and entrainment issues. The option of fine grinding for liberation is therefore not available. From investigation of varying grinds down to 80 % passing 75 μm, about 48 % passing 75 μm has been recommended for the operation [1]. On the other hand, activation is in a fix to activate pyrrhotite, and the outcome of the prospect is tied to how pyrrhotite will respond at the condition optimum for pentlandite flotation.

From various reports, the response of pyrrhotite to $CuSO_4$ activation has been observed to differ among ore deposits, with recovery improvement ranging from 50 % to insignificant [9, 13]. In a study of pyrrhotite from four providences, at pH 7, $CuSO_4$ was found to have no effect on sodium isobutyl xanthate (SIBX) recovery of Nkomati pyrrhotite, while Phoenix pyrrhotite showed a drop in recovery [9]. At pH 10, the electrochemical impedance values at the surface of all the four pyrrhotite samples increased remarkably and SIBX could not float the samples. With $CuSO_4$ addition, the recovery improved glaringly for three of the sulphides; from less than 10 % to above 80 % for Sudbury CNN and Phoenix pyrrhotite, but for Getrude West the effect was negligible [9]. Variation in pyrrhotite behaviour across deposits has been found attributable to its varying magnetisms, crystallography, and non-stoichiometric compositions of its mineralogy [9, 14].

Pentlandite itself, taken as 1.7 wt% in the feed, can be depressed or activated by $CuSO_4$, depending on pulp chemistry [8]. In oxidizing pulp, copper hydroxide precipitates causing depression, while in reducing pulp, copper (II) can be reduced to copper (I), with sulphur oxidation, to produce CuS followed by Cu_2S on the pentlandite surface to achieve activation effect. This has also been observed with other sulphides—pyrite, sphalerite and pyrrhotite [7, 15–17].

From this background, it follows that the response of the ore sulphides to $CuSO_4$ may not be readily predicted, but depends on various circumstances. Also, there is liberation constraint in this instance as the grind is constrained so that the pentlandite in pyrrhotite is not fully liberated. The presence of chalcopyrite can also affect how other sulphides actually respond. To see how the scenario affects

Table 1 Representative mineral content of the BCL nickel–copper ore [1]

Mineral	Composition	Mass %	Contained % Ni
Pentlandite	$(FeNi)_9S_8$	1.7	0.61
Chalcopyrite	$CuFeS_2$	2.0	–
Pyrrhotite	Fe_7S_8, Fe_9S_{10}	24.1	0.1
Pyrite	FeS_2	1.0	0.004
Magnetite	FeO, Fe_2O_3	2.5	–
Gangue	Various	68.7	0.01
Total		100	0.72

the outcome, there can be very many variables to investigate separately, but a careful look at the effect of varying dosages of copper addition at established pH and grinding size should give a picture of the overall effect. The product fractions will be analysed to find how the different constituents respond with time. Recommendations can therefore be made with respect to the nickel interest.

Materials and method

Ore sample collection and size analysis

Milled ore sample used for the investigation was obtained from the BCL concentrator plant. The sample was collected at the cyclone overflow launder feeding the rougher flotation bank, using a sample cutter at 15 min intervals for a period of 2 h. Each time a sample was cut, the density of the pulp was checked for consistency with a density gauge. The total slurry collected, about 33 kg, was stirred and a sample was taken from the bucket, filtered and the residue dried. The dried sample was rolled to break any lumps and then split using a rotary splitter to get representative samples. Sieve analysis of the dry sample was done using $\sqrt{2}$ sieve series from 212 to 38 μm, to establish the feed particle size distribution (PSD) of the ore.

Froth flotation

The flotation investigations were done using a 2 L Denver cell at a pulp relative density of about 1.35 and impeller speed of 1200 rpm. The temperature ranged from 22.7 to 23.8 °C. The pH was maintained at about 9 using lime at the start of each float. In the base case conditioning without $CuSO_4$ addition, treatment 1 (T1) was done using potassium normal butyl xanthate at 75 g/ton, at conditioning time of 3 min. The other three treatments, T2, T3 and T4, were done with copper sulphate dosages of 15, 30 and 45 g/ton, respectively, and the pulp conditioned for 5 min before adding the collector. In each flotation run, five concentrates, C1, C2, C3, C4 and C5, were separately collected over the respective time intervals between 0, 1, 3, 6, 10 and 15 min of flotation. This will enable the kinetics of the response to be assessed for the different target contents. Each reagent condition was run in triplicate, the concentrate fractions and tailing for each run were dewatered, weighed and the mass pulls compared for procedural repeats. The fractions from every run were assayed for nickel, copper, iron and cobalt from hot aqua regia digestion using atomic absorption spectroscopy [18], with the residue considered essentially silica. From the assays, the mass pull and sink values, the feed grades were reconstituted for every run. Cumulative grades and recoveries were computed for the various constituents to analyse the responses to the treatments.

Results and discussion

The PSD analysis of the as-received samples gave 42 %, passing 75 μm. The responses of the different minerals to the flotation conditioning are discussed as follows. For a typical result of the floatation tests, Table 2 shows the data for the duplicate run for T1 (without activation) with the computation. Similar data and computation for the first and triplicate runs for T1, and T2–T4 runs, generate the overall charts discussed following.

Nickel response

Figures 1 and 2 show how nickel cumulative recovery and grade varied with time during the flotation for the varying dosages of the copper sulphate added to the pulp. The overall consideration shows that at 15 g/ton copper sulphate dosage, nickel recovery improved clearly by a maximum of about 6 % without grade compromise. Looking at the details, the kinetics (Fig. 1) shows that in the first minute, nickel response was fastest without activation, giving the highest assay. This slowed down quickly and the cumulative grade dropped to the least for the treatments, while the final recovery was 88 %.

At 15 g/ton activation, the initial response appeared slower than that for no activation, but it was sustained over a longer period, so that the overall recovery clearly increased. The 30 g/ton condition followed a close pattern, but the overall recovery dropped a little while the assay increased. At 45 g/ton the result followed a different trend, with recovery lying lowest for all the duration, while cumulative assays ended highest. These trends are notable for insight into what is happening in the pulp after this conditioning. First, the conditioning reduced the kinetics of nickel response, but gave overall increased recovery. The overall response is obviously an indication of a composite response from different constituents. Certain constituent actually was floating very fast without the activation. The constituent perhaps still reported to the float, though at a slower rate, giving overall increase in recovery.

For the depression at the highest dosage, it implies that copper hydroxide predominates in this pulp at this concentration range. In a sphalerite collectorless flotation at pH 8–10 [19], recovery was observed to drop from above 70 % to below 20 % with increase in Cu^{2+} concentration from 2×10^{-6} to 2×10^{-5} M. Using the instance of predominance diagram of copper in aqueous system shown in Fig. 3, the work showed that, with increasing

Table 2 Floatation data for T1 (second run) and analyses

T1–2

Fraction	Time, min	Mass, g	Mass, %	Cum. mass, g	Cum. mass, %	Grade, %						Mass distribution (g) (mass × grade)					
						Ni	Cu	Co	SiO2	S	Fe	Ni	Cu	Co	SiO2	S	Fe
Conc 1	1	112.20	7.36	112.20	7.36	3.810	5.570	0.219	6.800	35.900	46.000	4.275	6.250	0.246	7.630	40.280	51.612
Conc 2	3	98.40	6.45	210.60	13.81	1.910	1.260	0.112	13.500	28.700	47.400	1.879	1.240	0.110	13.284	28.241	46.642
Conc 3	6	65.10	4.27	275.70	18.07	0.940	0.320	0.068	22.800	19.000	35.700	0.612	0.208	0.044	14.843	12.369	23.241
Conc 4	10	39.30	2.58	315.00	20.65	0.790	0.240	0.060	22.400	19.300	36.800	0.310	0.094	0.024	8.803	7.585	14.462
Conc 5	15	23.70	1.55	338.70	22.20	0.560	0.170	0.052	28.900	12.900	27.200	0.133	0.040	0.012	6.849	3.057	6.446
Tailing		1186.70	77.8	1525.40	100.00	0.08	0.02	0.04	47.58	1.40	8.91	0.948	0.293	0.448	564.674	16.614	105.703
Feed		1525.40	100.00			0.535	0.533	0.058	40.388	7.090	16.265	8.157	8.126	0.884	616.082	108.146	248.106
Concentrate grade, %						2.129	2.312	0.129	15.178	27.024	42.044						
Recovery, %						88.383	96.391	49.307	8.344	84.638	57.396						

Fraction	Recovery, %						Cumulative recovery, %						Cumulative grade, %					
	Ni	Cu	Co	SiO2	S	Fe	Ni	Cu	Co	SiO2	S	Fe	Ni	Cu	Co	SiO2	S	Fe
Conc 1	52.41	76.91	27.78	1.24	37.25	20.80	52.41	76.91	27.78	1.24	37.25	20.80	3.810	5.570	0.219	6.800	35.900	46.000
Conc 2	23.04	15.26	12.46	2.16	26.11	18.80	75.45	92.17	40.24	3.39	63.36	39.60	2.922	3.556	0.169	9.930	32.536	46.654
Conc 3	7.50	2.56	5.01	2.41	11.44	9.37	82.95	94.73	45.25	5.80	74.80	48.97	2.454	2.792	0.145	12.969	29.340	44.068
Conc 4	3.81	1.16	2.67	1.43	7.01	5.83	86.76	95.90	47.91	7.23	81.81	54.80	2.247	2.474	0.135	14.146	28.087	43.161
Conc 5	1.63	0.50	1.39	1.11	2.83	2.60	88.38	96.39	49.31	8.34	84.64	57.40	2.129	2.312	0.129	15.178	27.024	42.044

Fig. 1 Nickel cumulative
a recovery and **b** grade over
time at varying dosages of
$CuSO_4$

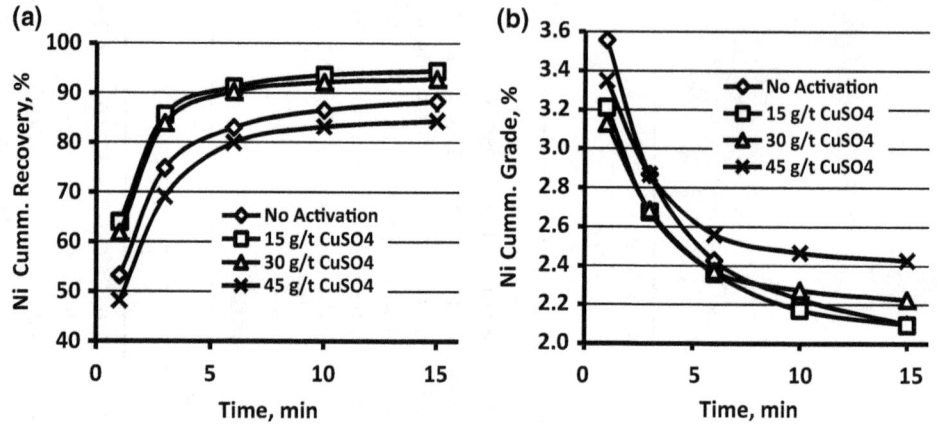

occurs [19]. Hence, copper hydroxide prevailed and sphalerite was decreased. In Figs. 1 and 2, depression at the highest Cu^{2+} dosage can therefore be attributed to increasing copper hydroxide predominance in the pulp at such a concentration range. However, increased recovery at lower dosage is clearly evident from the result, with best recovery at 94.5 % at enrichment ratio of about 3.0. For an insight into how other components contributed to the recovery, the response of the other minerals, particularly pyrrhotite that contains substantial amount of the total nickel content, will be informative.

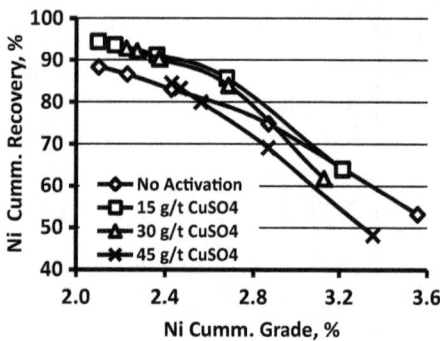

Fig. 2 Nickel cumulative recovery versus cumulative grade over time at varying dosages of $CuSO_4$

Pyrrhotite response

The Fe analysis shows that cumulative recovery increases by more than 10 % with the addition of 15 and 30 g/ton copper sulphate to the pulp (Fig. 4a). By 45 g/ton dosage, the cumulative recovery started dropping clearly, although Fe grade still increased with dosage up to this level (Fig. 4b). An optimum dosage therefore exists after which recovery will be traded off for grade. The observed depression is again attributable to copper hydroxide increasing predominance, such that only the most floatable grains were recovered at the higher dosages. The combined grade–recovery plot (Fig. 5) shows that the best dosage should not exceed 30 g/ton. Since the other iron-bearing minerals contribute a relatively small amount of iron compared with pyrrhotite iron content in the ore (Table 1), it is obvious that that the conditioning activated pyrrhotite. The low recovery of pyrrhotite relative to pentlandite is understandable, because the inactivated pulp was optimized for pentlandite ab initio. Moreover, it is also known that pentlandite presence with pyrrhotite in a pulp can galvanically depress pyrrhotite. With Fourier transform infrared (ATR) spectroscopy, it was concluded that galvanic interactions in a pentlandite–pyrrhotite mixed mineral pulp enhance preferential formation of hydroxide on pyrrhotite and dixanthogen on pentlandite, which implies

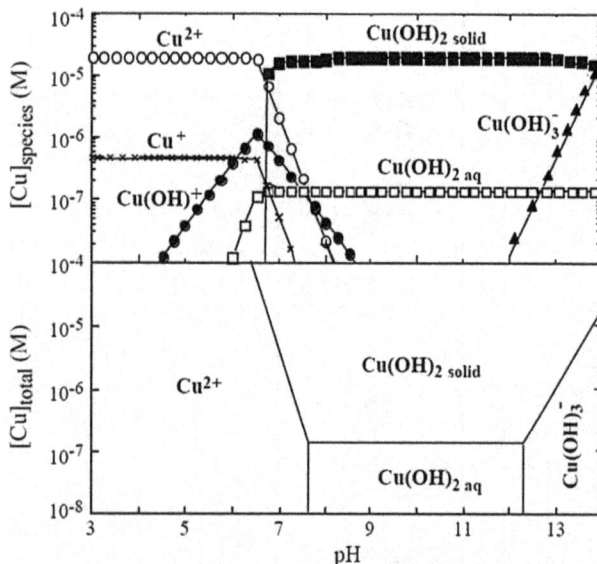

Fig. 3 Copper predominance area diagram at Eh 250 V, SHE (*below*), with concentration of each copper species as a function of pH at a fixed total copper concentration of 2×10^{-5} M (*on top*) [16]

concentration, Cu^{2+} field reduces, while $Cu(OH)_{2solid}$ field expands over a wider pH range. Depending on pH, $Cu^{2+} \rightarrow Cu(OH)_{2solid}$ or $Cu(OH)_{2aq} \rightarrow Cu(OH)_{2solid}$ precipitation

Fig. 4 Iron cumulative
a recovery and **b** grade over
time at varying dosages of
$CuSO_4$

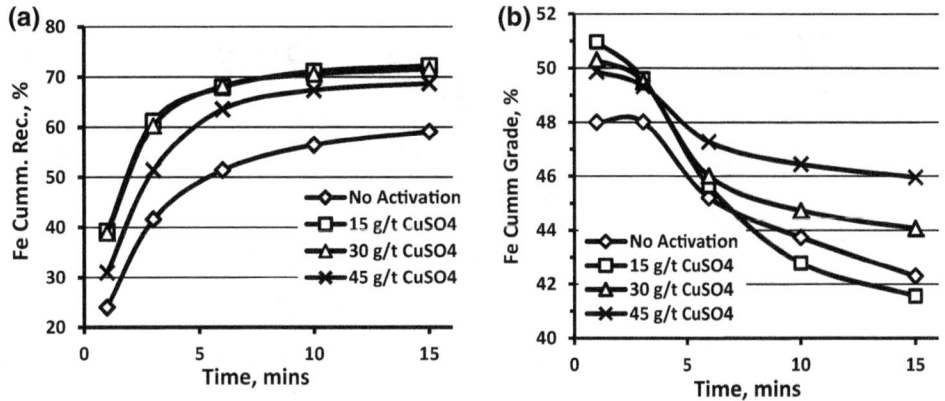

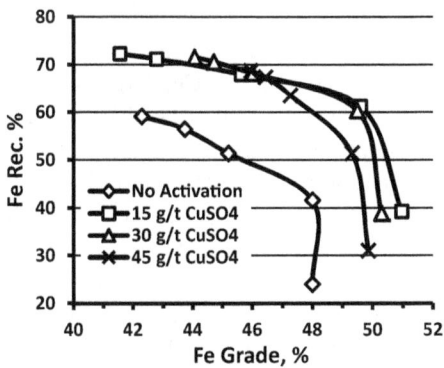

Fig. 5 Iron cumulative recovery versus cumulative grade over time at varying dosages of $CuSO_4$

enhanced floatability of pentlandite at the expense of pyrrhotite in the same pulp [20]. With other flotation implications of the mineralogical differences, the recovery of the two minerals cannot be maximized in the same pulp and the extent of increase in recovery obtained here can be considered satisfactory.

Linking the pyrrhotite recovery to nickel recovery, with the 10 % increase in pyrrhotite recovery after the conditioning, about a tenth more of the pentlandite in pyrrhotite was recovered. This also implies that there was about 1.4 % increased recovery of the total nickel content, given that 14 % of total nickel is in pyrrhotite (Table 1). At overall 72 % highest recovery of pyrrhotite at enrichment ratio of 3.1, about 4.2 % of nickel will still be lost with pyrrhotite, except that flotation was selective of grains containing pentlandite, where partial exposure is possible. Further effort to recover more nickel will have to consider rougher tailing reconditioning to recover the pyrrhotite.

Chalcopyrite response

Although chalcopyrite recovery was good without activation (at 96 %), copper analysis shows that some increase in

recovery (to 97.6 %) was still obtained at the 15 g/ton dosage with clearly better grade (Fig. 6a, b). At the higher dosages, recoveries dropped. By the 45 g/ton dosage level, recovery was well below the value obtained without activation. The copper grade was highest at the low recovery, indicating drop in mass pull, with only the most floatable grains going into the concentrate. At the highest recovery of 97.7 %, the grade was 3.5 %, indicating an enrichment ratio of 5.0.

Cobalt recovery

The cumulative recovery of cobalt increased from about 47 % without activation to above 60 % at the higher activation dosage, with assays about 0.11 % (Fig. 7). Since cobalt is in solid solution in the pentlandite and improved recovery of the target sulphides was obtained, the increased recovery of cobalt is therefore normal and it is also proportionate.

Silica entrainment

Recovery and grades of silica under the treatments were assessed for an indication of the gangues response. At 15 g/ton copper sulphate dosage, final silica recovery in the flotation concentrate is above the recovery under the condition without copper addition (Fig. 8a). Since the cumulative grade of silica in the concentrate increased with flotation time all through (Fig. 8b), this recovery pattern characterizes entrainment [21, 22]. Since water recovery suffices to sustain entrainment, entrained matter forms the larger fraction of the mass pull after hydrophobic flotation response has subsided, explaining increasing silica grade with increasing recovery. At the higher copper sulphate dosages, the entrainment recovery reduced to lower values. This easily reconciles with the general depression observed for the 45 g/ton dosage.

Fig. 6 Copper cumulative recovery versus **a** time and **b** cumulative grade at varying $CuSO_4$ dosages

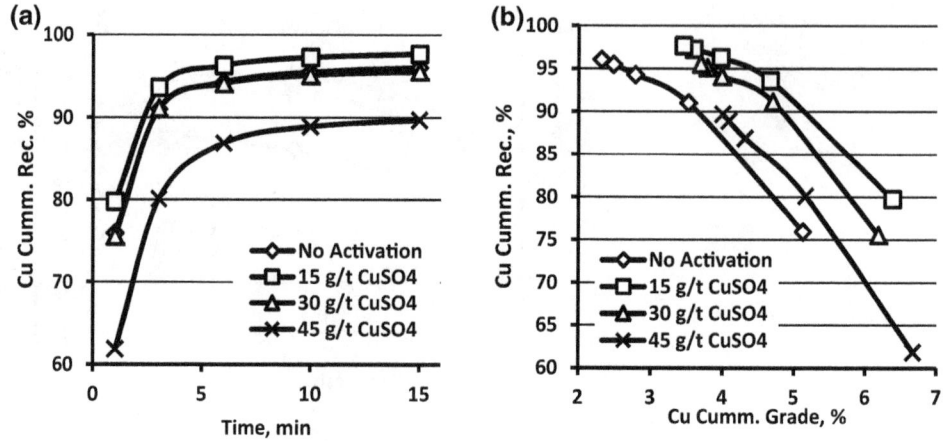

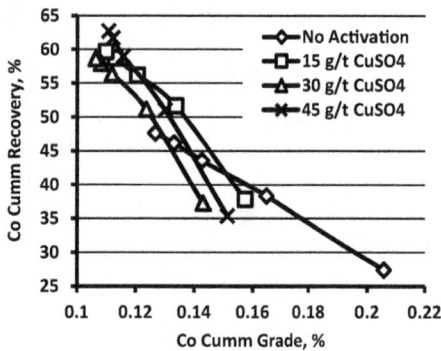

based on the change in the predominance of copper, from Cu^{2+} to $Cu(OH)_2$. Hydrophilic $Cu(OH)_2$ precipitates at the surface of sulphide grains and bring about depression with increase in the copper hydroxide species in the pulp. This frustrates formation of the metal polysulfide on the grains that needs activation, as well as xanthate interaction for all the sulphides, such that even chalcopyrite is depressed. About this optimum dosage, the Fe content in the concentrate will increase with pyrrhotite recovery. This is a necessary compromise to recover the nickel content in the pyrrhotite. However, the pyrrhotite recovery is still relatively low, as the recovery of pentlandite and pyrrhotite cannot be maximized in the same pulp.

Fig. 7 Cobalt cumulative recovery versus cumulative grade at varying $CuSO_4$ dosages

Overall implication

Combining the observations, it is obvious that copper sulphate addition to this pulp can achieve improved grade and recovery of the target values of nickel and copper. The dosage, however, has to be between 15 and 30 g/ton to optimize grade recovery. The depression after the optimum dosage of copper sulphate is obtained can be explained

Conclusion

Increased recoveries of nickel was obtained in the BCL concentrator Selebi-Phikwe pentlandite–pyrrhotite–chalcopyrite ore flotation up to a maximum copper sulphate dosage, above which depression was observed and recoveries fell below that for the condition without any copper addition. Nickel recovery with increasing copper sulphate

Fig. 8 Silica cumulative recovery versus **a** cumulative grade and **b** time at varying $CuSO_4$ dosages

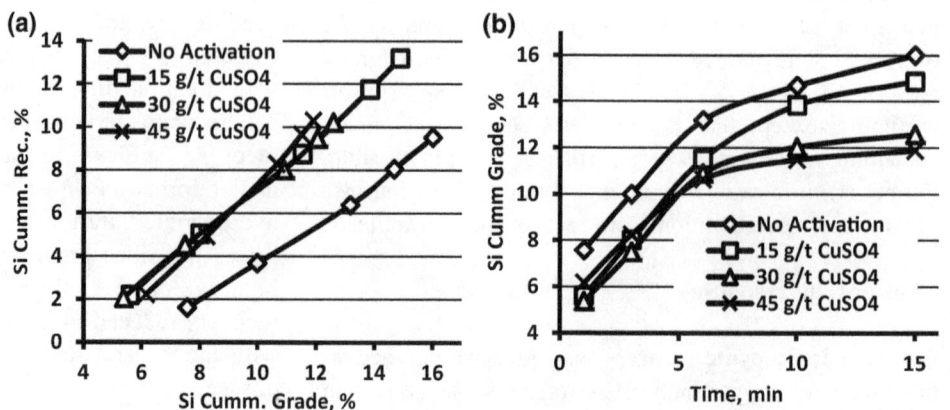

dosage therefore shows a maxima. A dosage range about this maxima dosage is the critical process parameter for successful activation of the ore with copper sulphate. The depression at higher dosage is explicable based on predominance change of copper with increasing concentration in the pulp. However, this increased recovery is with a compromise that pyrrhotite must be recovered along with pentlandite, due to dissemination constraints. Despite chalcopyrite presence, pyrrhotite recovery still improved, as chalcopyrite recovery also increased marginally. Because response to copper sulphate addition in flotation can vary from possible depression to activation in sulphide ores, it is notable in this ore that despite flotation constraints imposed by the mineralogy, activation with copper sulphate to increase nickel recovery can still be achieved, but within the careful dosage range.

References

1. Mulaba-Bafubiandi AF, Medupe O (2007) An assessment of pentlandite occurrence in the run of mine ore from BCL mine (Botswana) and its Impact on the flotation yield. In: Proc. of the fourth Southern African conference on base metals, Swakopmund, Namibia, SAIMM, pp 57–76
2. Simwaka M, Gumbie M, Moswate P, Moroka M, Keitshokile DC, Dzinomwa G (2009) Milestones in the improvement of concentrator nickel. In: Proc. fifth base metal conference, Kasane, Botswana, SAIMM, pp 359–368
3. Tripathi N, Peek E, Stroud M (2011) Advanced process modelling at the BCL smelter: improving economic and environmental performance. JOM 63(1):63–67
4. NDP9 (2002) Mineral development. In: Botswana National Development Plant 9, Publication of Government of Botswana (online)
5. BCL (2011) Strategy, monitoring and evaluation. www.bcl.bw/index.php?id=34. Assessed 10 Nov 2014
6. Finkelstein NP (1997) The activation of sulphide minerals for flotation: a review. Int J Miner Process 52:81–120
7. Chandra AP, Gerson AR (2009) A review of the fundamental studies of the copper activation mechanisms for selective flotation of the sulphide minerals, sphalerite and pyrite. Adv Colloid Interface Sci 145:97–110
8. Peng YJ, Seaman D (2012) Effect of feed preparation on copper activation in flotation of Mt Keith pentlandite. Miner Process Extr Metall Trans Inst Min Metall C 121(3):131–139
9. Ekmekçi Z, Becker M, BagciTekes E, Bradshaw E (2010) An impedance study of the adsorption of CuSO4 and SIBX on pyrrhotite samples of different provenances. Miner Eng 23:903–907
10. Senior GD, Shannon LK, Trahar WJ (1994) The flotation of pentlandite from pyrrhotite with particular reference to the effects of particle size. Int J Miner Process 42(3–4):169–190
11. Kelebek S, Wells PF, Fekete SO (1996) Differential flotation of chalcopyrite, pentlandite and pyrrhotite in ni–cu sulphide ores. Can Metall Q 35(4):329–336
12. Wiese JG, Harris PJ, Bradshaw DJ (2005) The influence of the reagent suite on the flotation of ores from the Merensky Reef. Miner Eng 18:189–198
13. Bradshaw DJ, Buswell AM, Harris PJ, Ekmekci Z (2006) Interactive effects of the type of milling media and copper sulphate addition on the flotation performance of sulphide minerals from Merensky ore Part I: pulp chemistry. Int J Miner Process 78:153–163
14. Becker M, de Villiers JPR, Bradshaw DJ (2008) Evaluation of pyrrhotite from selected Ni and PGE ore deposits and the influence of its mineralogy on flotation performance. In: 9th international congress for applied mineralogy, Brisbane, AusIMM, pp 401–409
15. Trahar WJ, Dunkin HH (1954) The effect of an excessive addition of copper sulphate on the flotation of sphalerite. In: Proceedings of the AusIMM 173:13–29
16. Clarke P, Fornasiero D, Ralston J, Smart RSC (1995) A study of the removal of oxidation products from sulphide mineral surfaces. Miner Eng 8:1347–1357
17. Dávila-Pulido GI, Uribe-Salas A, Nava-Alonso F (2012) Revisiting the chemistry and kinetics of sphalerite activation with Cu(II): a contact angle study. Open Miner Process J 5:1–5
18. Willard Horbart H, Merritt LL, Dean JA, Settle Frank A (1988) Instrumental methods of analysis, 7th edn. Wardsworth Inc, California
19. Fornasiero D, Ralston J (2006) Effect of surface oxide/hydroxide products on the collectorless flotation of copper-activated sphalerite. Int J Miner Process 78:231–237
20. Bozkurt V, Xu Z, Finch JA (1998) Pentlandite/pyrrhotite interaction and xanthate adsorption. Int. J. Miner Process 52:203–214
21. Otunniyi IO, Groot DR, Vermaak MKG (2013) Particle size distribution and water recovery study under the natural hydrophobic response flotation of printed circuit board comminution fines. Miner Metall Process J 30(2):85–90
22. Zheng X, Franzidis JP, Johnson NW (2006) An evaluation of different models of water recovery in flotation. Miner Eng 19:871–882

17

Dodecyl methacrylate and vinyl acetate copolymers as viscosity modifier and pour point depressant for lubricating oil

Pranab Ghosh[1] · Mainul Hoque[1] · Gobinda Karmakar[1] · Malay Kr. Das[2]

Abstract The article presents application of homo polymer of dodecyl methacrylate (DDMA) and its copolymers with vinyl acetate (VA) as multifunctional additives for lubricant formulation. Homo polymer of DDMA and five copolymers of DDMA with VA at different molar ratios were synthesized by free radical polymerization method using azobisisobutyronitrile (AIBN) as initiator. The characterization of the polymers was carried out through FTIR, NMR and GPC (gel permeable chromatography) analysis. The performance of all the polymers as viscosity index improver (VII) or viscosity modifier and pour point depressant (PPD) additive in two different base oils (mineral) were evaluated. The mechanism of action of the polymers as pour point depressant was studied by photo micrographic analysis. Rheological study of the formulated lubricant was also carried out and reported. The thermal stability of the polymers was determined by thermogravimetric analysis (TGA). It was found that thermal stability, VI and molecular weights of copolymers are higher than the homopolymer which showed better PPD property.

Keywords Copolymer · Molecular weight · Viscosity modifier · Pour point depressant · Rheology

✉ Pranab Ghosh
pizy12@yahoo.com

1 Natural Product and Polymer Chemistry Laboratory, Department of Chemistry, University of North Bengal, Darjeeling 734013, India

2 Department of Physics, University of North Bengal, Darjeeling 734013, India

Introduction

Lubricating oils play a very important role in automobile industry. It keeps the moving parts lubricated and protects them against rust and corrosion. Lubricating oils alone cannot satisfy all the requirements of modern engines. Some additives are to be blended with lube oil to improve the overall performance of the lubricant [1]. The role of additives in lubricant is very significant. They optimize the performance of lubricant and generally fall into two major categories viz. surface active additive and performance enhancing additive [2]. The first one protects the metal surfaces of the engine from corrosion, such as antiwear, anti-rust and extreme pressure additives [3]. The second type reinforces the base stock performance, such as viscosity index improver (VII) [4, 5], pour point depressant (PPD) [6, 7], antioxidant [8], and detergent–dispersant [9]. Generally, in the formulation of a high performance lubricant, different types of additives at different percentages are blended with the base stocks. This increases the overall cost of the furnished lubricant. The addition of additives having multifunctional character to the base fluid may lead to formulate a cost effective as well as better performing lubricant. Therefore, research in this area has attracted much attention. In this work we have synthesized multifunctional additives which showed excellent VII and PPD performances. Lot of works in this direction has carried out so far. Kamal and his group [10] in their work have shown the application of copolymers of vinyl acetate (VA) and esters of acrylic acid as viscosity index improver for lubricating oil. They also studied the rheological properties of lube oil with and without polymeric additives. Al-Sabagh and his group [11] have mentioned the application of copolymers of VA with n-alkyl itaconate as pour point depressants in lubricating oil. The rheological properties were also studied by the same research team. Abdel-Azim et al. [12] has synthesized the copolymers of dialkyl fumarate with vinyl acetate and recognized that the copolymer of didodecyl fumarate with VA is the most effective as PPD for lubricating oil. In 2008, Nassar [13] had prepared six polymers at different molar ratios of 2-ethylhexyl methacrylate and vinyl acetate and studied the VI property of the lubricants. He reported that most efficient VI property is obtained when the ratio of acrylate and vinyl acetate is 1:0.2. The potential application of copolymer of vinyl acetate (VA) as pour point depressant for crude oil is also well documented. VA—α olefin copolymers [14] and VA—methacrylate copolymers [15] were used as pour point depressant for crude oil. The copolymers of vinyl acetate, styrene and n-butyl acrylate having different monomer ratios were used to study the rheological behaviours of Mexican crude oil [16]. Borthakur and his group [17] have shown the application of alkyl fumarate and vinyl acetate copolymer in combination with alkyl acrylate as a flow improver for high waxy Borholla crude oil. Machado et al. [18] studied the influence of ethylene vinyl acetate copolymers on viscosity and pour point of a Brazilian crude oil. Al-Shafy and Ismail prepared an ester of polyethylene acrylic acid with 1-docosanol and finally it was grafted with vinyl acetate to produce graft ester. The grafted product was used as flow improver for Egyptian waxy crude oils [19].

Although there exists lot of works on the additive performance of copolymers of VA and different acrylates, reports regarding their application as multifunctional additive like PPD and VII for lube oil are scanty. Therefore, in this work copolymers of DDMA with VA at different molar ratio were synthesized and their performances are evaluated. Here efficiencies of the polymers as viscosity index improver and pour point depressant in two different types of mineral base oils (SN150 and SN500) were carried out according to ASTM standards. Photo micrographic images were taken to study their mechanism of action as pour point depressant. Rheological properties of the lubricants were also studied during the work by a rheometer. Homo polymer of dodecylmethacrylate (DDMA) was also synthesized along with the copolymers for comparison of the results.

Experimental

Chemicals used

Methacrylic acid (MA, 99%, LOBA Cheme Pvt. Ltd.), vinyl acetate (VA, 99%, s.d fine chem Ltd.), dodecyl alcohol (DA, 98%, SRL Pvt. Ltd.) were used without purification. Toluene (99.5%, Merck Specialties Pvt. Ltd.),

hexane (99.5%, s.d fine cheme Ltd.) and methanol (98%, Thomas Baker Pvt. Ltd.) were used after distillation. Hydroquinone (HQ, 99%, Merck Specialties Pvt. Ltd.) and azobisisobutyronitrile (AIBN, 98%, Spectrochem Pvt. Ltd.) were purified by recrystallization before use. Conc. H_2SO_4 (98%, Merck Specialties Pvt. Ltd.) was used as received. Base oils were collected from IOCL and BPCL, India and physical properties of them are given in Table 1.

Preparation of ester and its purification

Dodecylmethacrylate (DDMA) was prepared by reacting methacrylic acid with dodecyl alcohol in 1.1:1 molar ratio in presence of conc. H_2SO_4 as catalyst, 0.25% (w/w) hydroquinone (with respect to the total amount of the reactants) as polymerization inhibitor and toluene as solvent in a Dean Stark apparatus. The process of esterification was carried out by the procedure as reported in the earlier publication [20]. The purification of the ester was carried out by adding a suitable amount of charcoal to the prepared ester and refluxed for 3 h and then filtered off. The filtrate was treated with dilute (0.5 N) NaOH solution in a separating funnel to remove the unreacted acid and hydroquinone. The process was repeated several times and finally washed with distilled water. The purified ester was passed through sodium sulphate and left overnight over calcium chloride to dry completely and used in the polymerisation process.

Preparation of copolymers and homopolymer

The monomers, VA and DDMA, at different molar ratios (Table 2) were subjected to free radical polymerization using AIBN radical initiator. The polymerization was carried out in a three-necked round-bottom flask fitted with a magnetic stirrer, condenser, thermometer and an inlet for the introduction of nitrogen. In the flask, the mixture of DDMA and VA at a definite molar ratio in toluene was heated to 363 K for half an hour. AIBN (0.5% w/w with respect to the total monomer) was then added and heated for 6 h keeping the temperature constant at 363 K. At the end of the reaction time, the mixture was poured into cold

Table 1 Physical properties of the mineral base oil

Properties	SN150	SN500
Density (g cm^{-3}) at 40 °C	0.84	0.87
Viscosity at 40 °C in cSt	23.502	107.120
Viscosity at 100 °C in cSt	3.980	10.322
Viscosity index	85.15	81.5
Cloud point, °C	−4	+1
Pour point, °C	−6	−0.5

Table 2 Molar ratio and molecular weights of the prepared polymers

Polymer code	Molar ratio of monomers		Initiator	Average molecular weights		
	DDMA	VA		M_n	M_w	PDI
P-1	1	0	AIBN	17,824	24,588	1.379
P-2	1	0.075	AIBN	21,575	33,018	1.530
P-3	1	0.150	AIBN	24,013	35,842	1.492
P-4	1	0.225	AIBN	56,866	66,210	1.164
P-5	1	0.300	AIBN	61,990	81,219	1.310
P-6	1	0.375	AIBN	81,070	88,008	1.085

M_n is number average molecular weight, M_w is weight average molecular weight, *PDI* polydispersity index, *DDMA* dodecylmethacrylate, *VA* vinyl acetate, *AIBN* azobisisobutyronitrile

methanol with stirring to terminate the polymerization and precipitate the polymer. The polymer was further purified by repeated precipitation of its hexane solution by methanol followed by drying under vacuum at 313 K. The homopolymer of DDMA was also prepared and purified in the same procedure. The synthesis of all the polymers is mentioned in Scheme 1.

Measurements

Spectroscopic measurements

IR spectra were recorded by Shimadzu FTIR 8300 spectrometer using 0.1 mm KBr cell at room temperature within the wave number range of 400–4000 cm^{-1}. NMR spectra were recorded in Bruker Avance 300 MHz FT-NMR spectrometer using a 5 mm BBO probe. CDCl$_3$ was used as solvent and tetramethylsilane (TMS) as reference material.

Determination of the molecular weight

The number average molecular weight (M_n) and weight average molecular weight (M_w) were measured by GPC instrument (polystyrene calibration) equipped with a 2414 detector, waters 515 HPLC pump and 717 plus auto sampler. Sample solutions (0.4% w/v in HPLC grade THF) are prepared by dissolving ~4 mg of polymer per ml THF and filtering (0.45-μm Millipore PTFE) to remove suspended particulates. The pump flow rate is 1.0 mL/min with THF as the carrier solvent, and injection volumes are set to 20 μL. The polydispersity index [21] which indicates the nature of the distribution of the molecular weights in the polymers was also calculated.

Determination of thermo gravimetric analysis (TGA) data

The thermo-oxidative stability of all the polymers was determined by a thermo gravimetric analyzer (Shimadzu

Scheme 1 Esterification, homopolymerization and copolymerization reaction

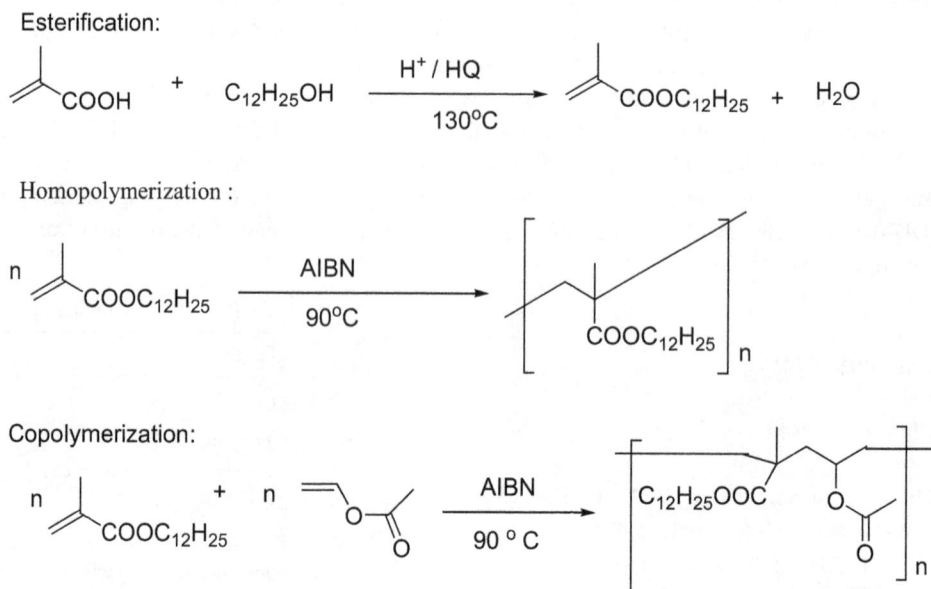

Esterification:

Homopolymerization :

Copolymerization:

TGA-50) in air using an alumina crucible at a heating of rate of 10 °C/min. A comparison of thermal stabilities of homo polymer with the copolymers was explained by this study.

Performance evaluation as viscosity index improvers

Viscosity indices of the lubricating oils (SN150 and SN500) blended with polymers at different concentration levels [ranging from 1% to 5% (w/w)] were calculated by measuring kinematic viscosity (KV) values at 313 and 373 K. The ASTM D445 method was applied to determine the KV values and ASTM D 2270-10 method was applied to determine the VI values.

Performance evaluation as pour point depressants

Pour points were determined using the cloud and pour point tester (model WIL-471, India) according to ASTM D 97-09 method. The performance of additives as PPD was investigated through variation of their concentration [from 1 to 5% (w/w)] in the formulated lubricants.

Photographic analysis

The photomicrograph images showing wax behaviour of the lube oil (SN150, pour point = −6 °C) without and with polymers (4%, w/w) have been recorded at 0 °C. A Banbros polarizing microscope (model BPL-400B) was used for this purpose and the adopted magnification was 200X.

Rheological study

Rheological study of the homo polymer and copolymers at 5% (w/w) concentration in SN150 oil was performed using Brookfield rheometer (Model DV-III ultra). Dynamic viscosity (cp) and shear rate (s^{-1}) were measured at two temperatures, 40°C and 100 °C.

Results and discussion

Spectroscopic analysis

The homopolymer of DDMA exhibited IR absorption band at 1722.3 cm^{-1} for the ester carbonyl group. Peaks at 2853.5 and 2924 cm^{-1} were for the alkyl (CH$_3$CH$_2$–) groups. Peaks at 1456.2, 1435.9, 1377.3, 1330.2, 1296.1 and 1164.0 cm^{-1} were due to CO stretching vibration and absorption bands at 1013.5, 937.3, 812.9, 715.5, and 649.0 cm^{-1} were due to bending of C–H bonds.

In the ^{1}H NMR of homopolymer, methyl protons appeared in the range of δ_H 0.881–1.027 ppm, methylene protons in the range of 1.283–1.812 ppm for all alkyl groups. A broad peak at 3.930 ppm indicated the protons of –OCH$_2$ group. Absence of any significant peaks in the range of 5–6 ppm in the spectrum confirmed completion of the polymerisation process. In the ^{13}C NMR of homopolymer, peaks at δ_C 176.73–177.84 ppm indicated the presence of ester carbons. The peaks at 63.06–65.40 ppm confirmed the presence of –OCH$_2$ carbon. Peaks in the range of 14.13–45.22 ppm represent all sp^3 carbon atoms of alkyl groups. No significant peaks in the range of 120–150 ppm indicated the absence of sp^2 carbon atoms and therefore supported formation of the polymers.

The spectral data (IR and NMR) of all the five copolymers (P-2 to P-6) are similar. In the IR spectra, peaks at 1732.9–1735.2 and 1716.5–1720 cm^{-1} indicated the presence of ester carbonyl groups in the copolymers due to vinyl acetate and DDMA moiety, respectively. Peaks at 2853.6–2854 and 2922.9–2925.4 cm^{-1} were for the CH$_3$-CH$_2$– groups. The peaks at 1456.2–1456.8 cm^{-1}, 1377.1–1377.8 cm^{-1}, 1368.4–1369 cm^{-1}, 1321.1–1321.9 cm^{-1}, 1296.1–1296.8 cm^{-1}, 1238.2–1238.8 cm^{-1}, 1163.0–1164.2 cm^{-1}, 1065.5–1066 cm^{-1}, and 1011.6–1012 cm^{-1} were due to CO stretching vibration and absorption bands at 814.9–816 and 721.3–721.9 cm^{-1} were due to bending of C–H bond. It is observed from the IR data of five copolymers that with increasing the percentage of vinyl acetate moiety in the copolymers the peak intensity olefinic groups gradually decreases.

In ^{1}H NMR of copolymers, a broad peak δ 3.926–4.158 ppm indicated the protons of –OCH$_2$ and –OCH$_3$ groups. The hydrogen attached to sp3 carbons appeared in the range of 0.858–2.637 ppm. Absence of any significant peaks in the range of 5–6 ppm indicated the disappearance of C=C bonds and confirmed formation of the copolymers.

In ^{13}C NMR, peaks at δ_C 176.60–176.70 ppm indicated the presence of ester carbonyl groups. The peaks appeared from 64.66 to 65.06 ppm indicated the presence of –COCH$_3$ methyl carbons and –OCH$_2$ carbons. The peaks ranging from 14.08 to 45.09 ppm represented all other sp^3 carbons. No significant peaks in the range of 120–150 ppm indicated the absence of sp^2 carbons and confirmed the formation of the copolymers. (All IR and NMR spectra are given in supporting information).

Analysis of molecular weight data

The experimental values of M_n and M_w of the homo and copolymers are given in Table 2. From the values, it is found that molecular weight of homopolymer is less than the copolymers. The molecular weights of copolymers increase with increasing the percentage of vinyl acetate moiety in the prepared copolymers. The comparatively higher PDI values of P-2 and P-3 indicated that their molecular weight distributions are wider and the copolymers are more branching compared to others. On the other hand, the lowest PDI value of P-6 indicated that the molecular weight distribution of the copolymer is narrow compared to others and the polymer is expected to be more linear which is also reflected of its highest thermal stability.

Analysis of thermo gravimetric data

From the TGA values (Fig. 1), it is found that homo polymer of DDMA (P-1) is thermally less stable than the copolymers. Generally, copolymers having vinyl acetate (VA) as a monomer has a head to tail structure. It may be due to fact that VA molecules add to the growing chain of the polymer through its –CH$_2$– group and such type of addition involves less steric effect [22]. Polymers formed in this process always have a higher molecular weight and become thermally more stable [23]. In case of P-1, decomposition started at 150 °C with 12.5% weight loss and 75% weight was lost at 400 °C. Among the copolymers, decomposition of thermally most stable P-6 started at 220 °C with 10% weight loss and found 50% weight loss at 400 °C. It may be due to higher molecular weight and lower PDI value [24] of P-6. The order of thermal stability of the prepared polymers is P-1 < P-2 < P-3 < P-5 < P-4 < P-6.

Analysis of viscosity index data

Viscosity index (VI) values of the lubricant compositions prepared by blending polymers at different concentration levels [1–5% (w/w)] with two different types of base oils were determined from the kinematic viscosity values of the blends measured at 40 °C and 100 °C. The experimental

results are given in Fig. 2 (for SN150) and Fig. 3 (for SN500). It is found that VI values of all the polymers increase with increasing concentration of the additives in base oils. With increase in temperature, the lube oil viscosity decreases. But at higher temperature, due to expansion of the additive molecules, the size of the solvated additives increases. This increase in size of the solvated additives counterbalances the reduction of viscosity of the lubricants at higher temperature. Increase in concentration of polymers in lube oil also leads to increase of the total volume of solvated polymer molecules, thus exerting higher thickening effect and improves the VI property [25–27]. From the experimental values, it is found that the VI values of homo polymer (P-1) are less than copolymers. It may be due to low molecular weight of P-1 compared to others. The VI of copolymers increases with increasing the percentage of vinyl acetate which may be due to increase of molecular weight of the copolymers P-2 to P-6 [28]. The relation between molecular weights and VI values of the synthesized polymers are also compared with the copolymers of 2-ethylhexyl methacrylate and VA published previously [13] and mentioned in Table 3. It is found that VI values increase with increasing the molecular weight in the polymers. For higher molecular weight polymers, due to higher degree of polymerization, the polymer units will be larger, i.e., their hydrodynamic volumes will definitely be greater that may be the reason of their higher VI values.

Analysis of pour point data

Pour points of the lubricant compositions prepared blending the polymers at different concentrations ranging from 1% to 5% (w/w) with the base stocks were tested and experimental values are given in Figs. 4 and 5. From the values, it is observed that all the polymers can be used effectively as PPD and the efficiency increases with

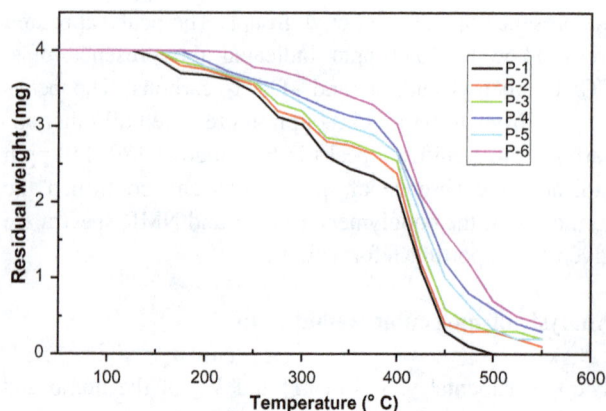

Fig. 1 Thermal degradation of polymers

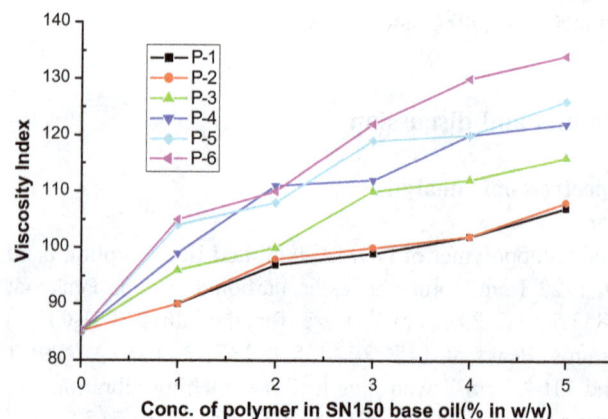

Fig. 2 Variation of viscosity index of the base oil (SN150) blended with additives at different concentrations

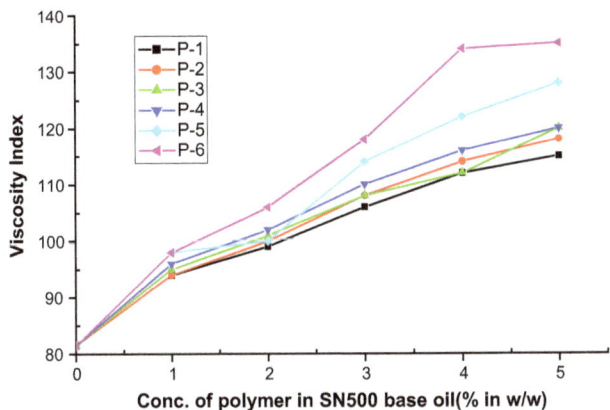

Fig. 3 Variation of viscosity index of the base oil (SN500) blended with additives at different concentrations

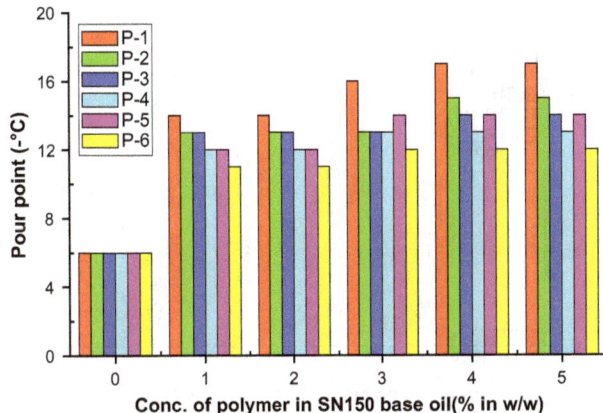

Fig. 4 Variation of pour point of the base oil (SN150) blended with additives at different concentrations

increasing concentration of the polymers in base oil up to 4%. This means that within this concentration range, at lower temperature, the polymers may interact with the paraffinic wax of the lube oil and change their crystal sizes [29]. The homo polymer P-1 has greater efficiency as PPD compared to the copolymers. It may be due to higher polarity of P-1 compared to others. The highly polar polymers are greatly adsorbed by the wax crystals present in lube oil, and therefore efficiency as PPD of the polymers improves [30]. Among the five copolymers, the polymer P-2, P-3 and P-5 have greater efficiency as PPD. The higher polarity of may be due to higher PDI values of these three copolymers [31]. A polymer having higher polydispersity index (PDI) is generally branched chain polymer [32]. The branched chain polymers, due to higher polarity, can easily prevent formation of three-dimensional crystal

network structures of the wax particles at low temperature than the linear chain polymers of lower PDI values.

Analysis of rheological study

Rheological study was performed with lube oil (SN150) without and with 5% (w/w) polymers. The values of dynamic viscosity against shear rate at 40 °C and 100 °C are given in Figs. 6 and 7 respectively. The viscosity of the lube oil without any additives is approximately constant with increasing the shear rate. This indicates that pure lube oil behaves like a Newtonian fluid at any shear rate [33]. For lube oil with additives, it is found that at low shear rate ($5-30 \ s^{-1}$), the viscosity gradually decreases and the lubricant behaves like a non- Newtonian fluid [34]. However, at high shear rate, viscosity of the lubricants

Table 3 Comparison of viscosity index values of the synthesized polymers with other polymers published earlier [13]

Polymer code	Molar ratio of monomers		av. mol. wt.	VI			
	Acrylate	VA		0%	1%	2%	3%
P-1	1	0	24,588	85.2	90	97	99
S-1	1	0	13,000	93	120	125	127
P-2	1	0.075	33,018	85.2	90	98	100
S-2	1	0.2	106,563	93	150	156	160
P-3	1	0.15	35,842	85.2	96	100	110
S-3	1	0.6	67,544	93	142	145	152
P-4	1	0.225	66,210	85.2	99	111	112
S-4	0.2	1	14,471	93	124	126	129
P-5	1	0.3	81,219	85.2	104	108	119
S-5	0.6	1	19,539	93	125	130	130
P-6	1	0.375	88,008	85.2	105	110	122
S-6	1	1	20,012	93	129	137	142

S-1 is homopolymer of 2-ethylhexyl methacrylate and S-2 to S-6 are the copolymers of 2-ethylhexyl methacrylate with vinyl acetate (VA) at different molar ratio [13]. The VI of polymers in our present investigation was calculated according to ASTM D 2270-10 method and for the polymers published in the article was determined according to ASTM D 2270-87 method

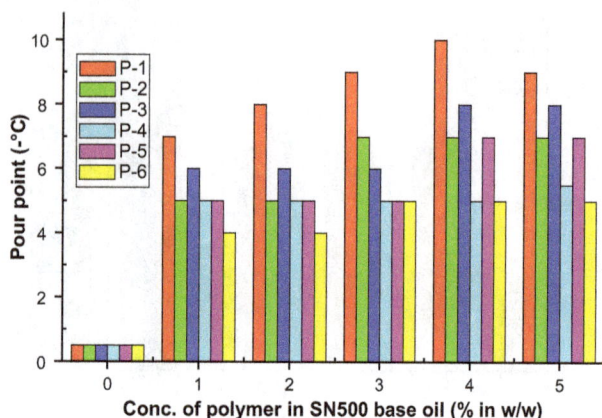

Fig. 5 Variation of pour point of the base oil (SN500) blended with additives at different concentrations

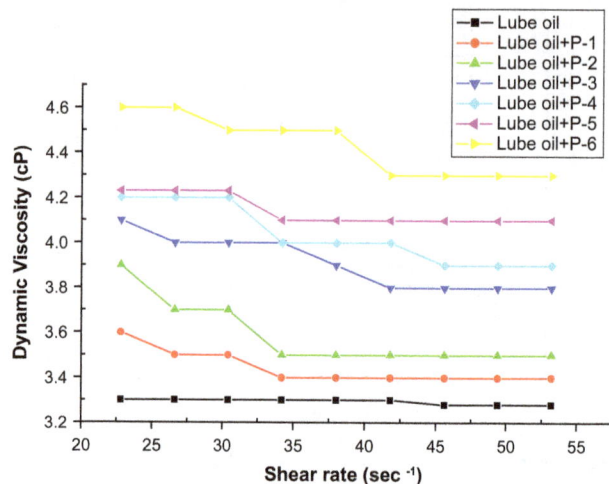

Fig. 6 Variation of dynamic viscosity with shear rate at 40 °C

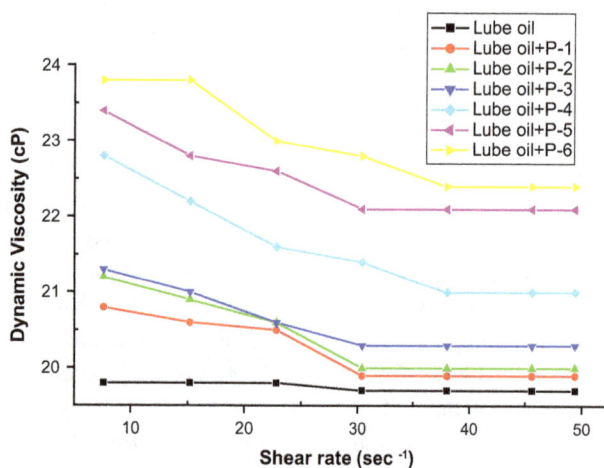

Fig. 7 Variation of dynamic viscosity with shear rate at 100 °C

DDMA which results in their higher average molecular weights and therefore greater volume.

Analysis of photo micrographic image

Photo micrographic analysis of the lube oil (SN150) without and with 4% (w/w) additives was carried out to study the effect of additives as PPD and its mechanism. The photograph images are shown in the Fig. 9a–g. In Fig. 9a, the photograph of lube oil (pour point = −6 °C) without any additives, there is a large number of cyclic crystalline and some needle shaped waxes. The Fig. 9b–g represented the base oil blended with 4% of P-1, 4% of P-2, 4% of P-3, 4% of P-4, 4% of P-5 and 4% of P-6, respectively. Greater wax modification is found when the lube oil is blended with P-1(homo polymer) and least wax modification is observed in case of P-6. This indicates that homo polymer of DDMA is better as PPD compared to its copolymers with VA. This is in agreement with the pour point values determined by ASTM D97-09 method. Length of the largest crystalline waxes in Fig. 9a–g are 447.3, 120.6, 192.2, 223, 242.5, 226.8 and 268.7 µm, respectively.

Conclusion

From the above study it is found that polymers are effective as viscosity index improver and pour point depressant for lube oil. Viscosity index property of homo polymer is lower than copolymers. VI property increases with increase in the percentage of vinyl acetate in the copolymers. The efficiency as pour point depressant was found higher in case of homo polymer compared to copolymers. Among the copolymers, the polymers which have higher PDI values, results better PPD property. From rheological study, it

approximately remains constant and behaves like Newtonian fluids [35]. At low temperature, viscosity modifiers in lube oil exist as spherical coil having random orientation and exert high viscosity in the absence of shear. When shear is applied, the additives start rearranging themselves in the direction of flow and viscosity of the lubricants decrease [36]. At high shear rate, all the polymers are arranged in the direction of flow and there is a negligible change in viscosity. At higher temperature, viscosity modifiers exist in expanded form (Fig. 8) and hence polymers are easily arranged in the direction of flow under shear and viscosity approximately remains constant under high shear rate for all polymer blended lube oil. In this way, the copolymers counterbalance decrement of lube oil viscosity under high temperature. The decrease of viscosity was caused due to applying of shear and changing temperature from 40 °C to 100 °C. The higher of dynamic viscosities of the polymers from P-6 to P-1 at any shear rate is may be due to incorporation of VA in the backbone of

Fig. 8 Effect of temperature on polymer in lube oil

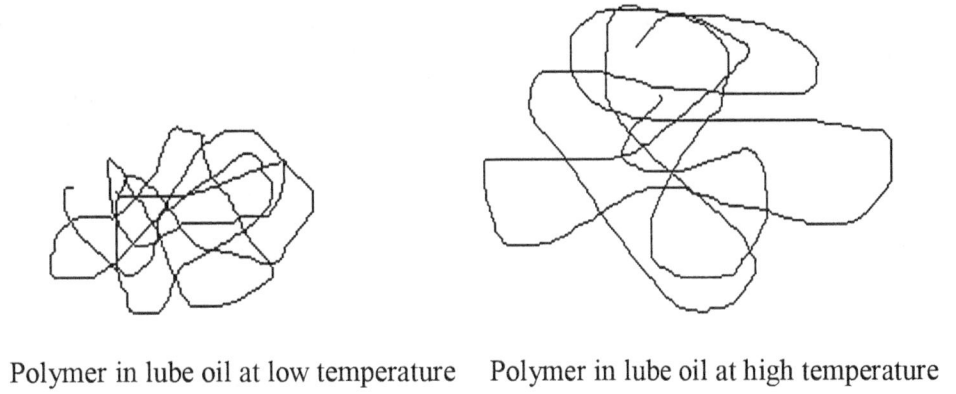

Polymer in lube oil at low temperature   Polymer in lube oil at high temperature

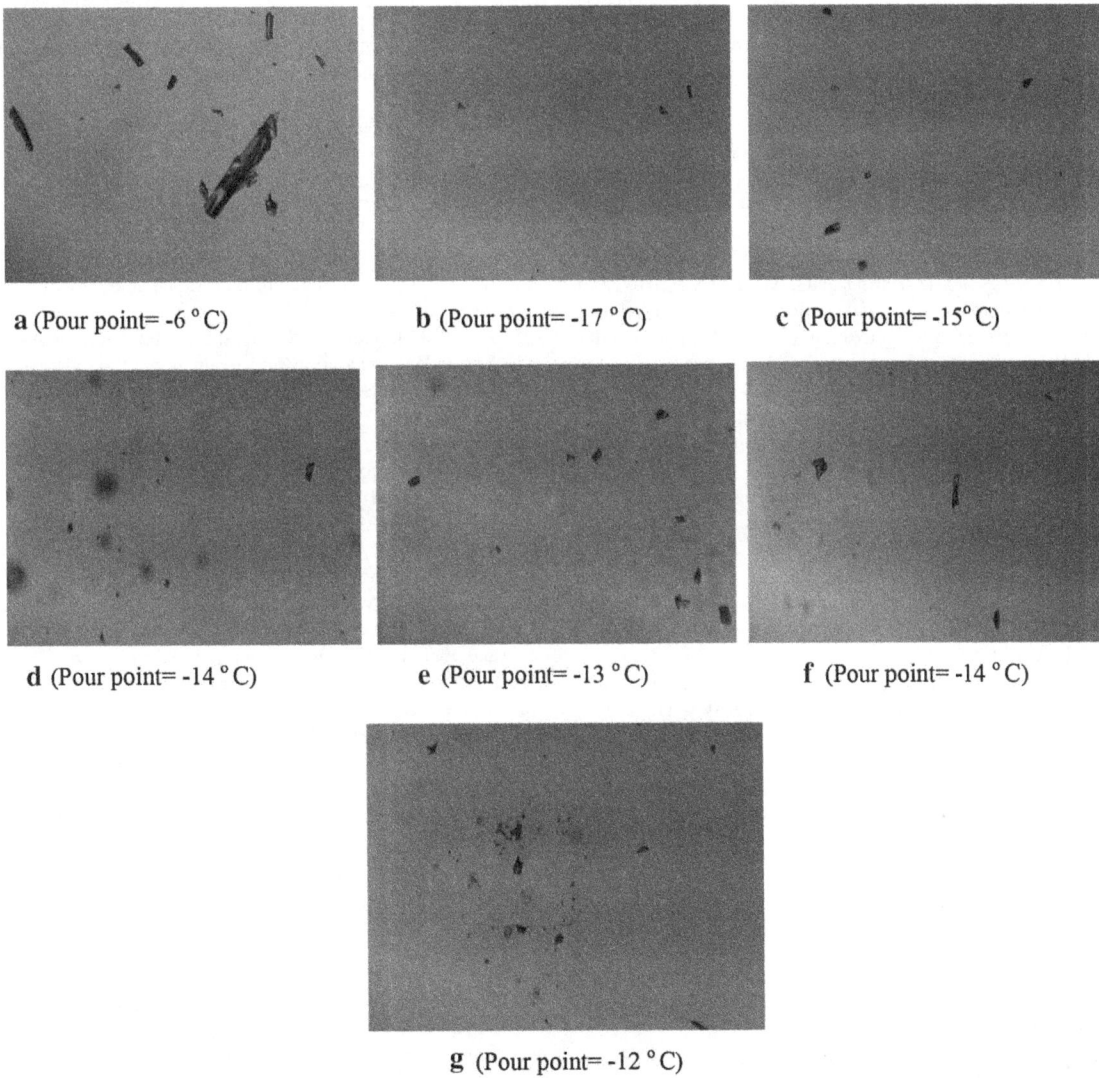

a (Pour point= -6 °C)   b (Pour point= -17 °C)   c (Pour point= -15°C)

d (Pour point= -14 °C)   e (Pour point= -13 °C)   f (Pour point= -14 °C)

g (Pour point= -12 °C)

Fig. 9 Photomicrograph images of **a** pure lube oil (SN150), **b** lube oil + 4% (w/w) of P-1, **c** lube oil + 4% (w/w) of P-2, **d** lube oil + 4% (w/w) of P-3, **e** lube oil + 4% (w/w) of P-4, **f** lube oil + 4% (w/w) of P-5, **g** lube oil + 4%(w/w) of P-6

was found that pure lube oil is a Newtonian fluid at any shear rate but polymer doped lube oil is non-Newtonian fluid at low shear and Newtonian fluid at high shear rate.

Acknowledgements Authors thank UGC, New Delhi for financial support. Thanks also to IOCL, India for supplying base oil.

References

1. Ghosh P, Das T (2011) Copolymer of decyl acrylate and styrene: synthesis, characterization and viscometric studies in different base stock. Adv Appl Sci Res 2:272–283
2. Mortier RM, Orszulik SY (1997) Chemistry and technology of lubricants, 2nd edn. Chapman & Hill, London
3. Miwa TK, Rothfus JA, Dimitroff E (1979) Extreme-pressure lubricant tests on jojoba and sperm whale oils. J Am Oil Chem Soc 56:765–770
4. Akhmedov AI (1994) Copolymer of alkyl methacrylates with styrene as V. I. improvers for lubricating oils. Chem Tech Fuels Oil 30:34–37
5. Jerbic IS, Vukovic JP, Jukic A (2012) Production and application properties of dispersive viscosity index improvers. Ind Eng Chem Res 5:11914–11923
6. Jung KM, Chun BH, Park SH, Lee CH, Kim SH (2011) Synthesis of methacrylate copolymers and their effects as pour point depressants for lubricating oil. J Appl Polym Sci 120:2579–2586
7. Ghosh P, Das T, Nandi D, Karmakar G, Mandal A (2010) Synthesis and characterization of biodegradable polymer—used as a pour point depressant for lubricating oil. Int J Polym Mater 59:1008–1017
8. Nassar AM, Ahmed NS, Abd El-Aziz KI, Abdel-Azim AAA, El-Kafrawy AF (2006) Synthesis and evaluation of detergent/dispersant additives from polyisobutylene succinimides. Int J Polym Mater 55:703–713
9. Selezneva IE, Levin AY, Monin SV (1999) Detergents–Dispersants additives for motor oils, alkylphenolates. Chem Technol Fuels Oils 35:389–395
10. Abdir MF, Ashour FEl-Z, Ahmed NS, Kamal RS, El-Zahed SM (2014) The effect of some additives on the rheological properties of engine lubricating oil. Int J Eng Res Appl 4:169–183
11. Al-Sabagh AM, Khalil TM, Sabaa MW, Khidr TT, Saad GR (2012) Poly (n-alkyl itaconate-co-vinyl acetate) as a pour point depressant for lube oil in relation to rheological flow properties. J Dispers Sci Technol 33:1649–1660
12. Abdel-Azim AAA, Abdel-Aziem RM (2001) Polymeric additives for improving the flow properties and viscosity index of lubricating oils. J Polym Res 8:111–118
13. Nassar AM (2008) The behaviour of polymers as viscosity index improvers. Pet Sci Technol 26:514–522
14. Abou El-Naga HH, Abd El-Azim WM, Ahmed MM (1985) Polymeric additives for pour point depression of residual fuel oils. J Chem Technol Biotechnol 35:241–247
15. Pederson KS, Ronningsen HP (2003) Influence of wax inhibitors on wax appearance temperature, pour point and viscosity of waxy crude oil. Energy Fuels 17:321–328
16. Castro LV, Vazquez F (2008) Copolymers as flow improvers for Mexican crude oil. Energy Fuels 22:4006–4011
17. Borthakur A, Laskar NC, Mazumdar RK, Rao KV, Subramanyam B (1995) Synthesis and evaluation of alkyl fumarate—vinyl acetate copolymers in combination with alkyl acrylate as flow improvers for Borholla crude oil. J Chem Technol Biotechnol 62:75–80
18. Machado Andre LC, Lucas Elizabete F (2007) The influence of vinyl acetate content of the poly (ethylene-co-vinyl acetate) (EVA) additive on the viscosity and the pour point of a Brazilian crude oil. Pet Sci Technol 19:197–204
19. Al-Shafy HI, Ismail EA (2014) Studies on the influence of polymeric additives as flow improvers for waxy crude oil. IOSR J Eng 4:54–61
20. Ghosh P, Das M, Upadhyay M, Das T, Mandal A (2011) Synthesis and evaluation of acrylate polymers in lubricating oil. J Chem Eng Data 56:3752–3758
21. Rogosic M, Mencer HJ, Gomzi Z (1996) Polydispersity index and molecular weight distributions of polymers. Eur Polym J 32:1337–1344
22. Marvel CS, Riddle EH (1940) The structure of vinyl polymers. J Am Chem Soc 62:2666–2670
23. Grassie N, Melville HW (1949) The thermal degradation of polyvinyl compounds. Proc R Soc Lon A 199:1–13
24. Ghosh P, Karmakar G (2012) Synthesis and characterisation of Polymyristyl acrylate as a potential additive for lubricating oil. Am J Polym Sci 2:1–6
25. Tanveer S, Prasad R (2006) Enhancement of viscosity index of mineral base oils. Ind J Chem Technol 13:398–403
26. Abdel-Azim AAA, Malcolm BH (1984) Interaction parameters in ternary polystyrene solution at high temperature. Polymer 25:803–807
27. Abdel-Azim AAA, Malcolm BH (1983) Viscometric behaviour of polystyrene in tetralin/cyclohexane mixture. Polymer 24:1429–1433
28. Nassar AM, Ahmed NS, Kamal RS, Abdel-Azim AAA, El-Nagdy EI (2005) Preparation and evaluation of acrylate polymers as viscosity index improvers for lube oil. Pet Sci Technol 23:537–546
29. El-Gamal IM, Atta AM, Al- Sabbagh AM (1997) Polymeric structures as cold flow improvers for waxy residual fuel oil. Fuel 76:1471–1478
30. Nassar AM, Ahmed NS (2006) The behavior of α-olefins butyl acrylate copolymers as viscosity index improvers and pour point depressants for lube oil. Int J Polym Mater 55:947–955
31. Al-Sabagh AM, Sabaa MW, Saad GR, Khidr TT, Khalil TM (2012) Synthesis of polymeric additives based on itaconic acid and their evaluation as pour point depressants for lube oil in relation to rheological flow properties. Egypt J Pet 21:19–30
32. Chunxia H, Stephane C, Paula W-A, John MD (2012) Molecular structure of high melt strength polypropylene and its application to polymer design. Polymer 44:7181–7188
33. Hassaneau MHM, Bartz WJ, Abou El-Naga HH (1997) A study of the rheological behaviour of multigrade oil. Lubr Sci 10:43–58
34. Chou CC, Lee SH (2008) Rheological behaviour and tribological performance of a nanodiamond-dispersed lubricant. J Mater Proc Technol 201:542–547
35. Punit K, Khonsari MM (2009) On the role of lubricant rheology and piezo-viscus properties in line and point contact EHL. Tribol Int 42:1522–1530
36. Ahmed NS, Nassar AM, Nassar RM, Abdel Raouf ME, El-Kafrawy AF (2014) The rheological properties of lube oil with terpolymeric additives. Pet Sci Technol 32:2115–2122

18

DC electrical conductivity and rate of ammonia vapour-sensing performance of synthetic polypyrrole–zirconium(IV) phosphate cation exchange nanocomposite

Asif Ali Khan[1] · Rizwan Hussain[1] · Umair Baig[1]

Abstract Electrically conductive polypyrrole–zirconium(IV) phosphate (PPy–ZrP) cation exchange nanocomposites have been synthesized for the first time by in situ chemical oxidative polymerization of pyrrole in the presence of zirconium(IV) phosphate (ZrP). Fourier Transform Infra-red spectroscopy (FTIR), field emission scanning electron microscopy, transmission electron microscopy, X–ray diffraction, thermogravimetric analysis, differential thermal analysis, derivative thermogravimetry and elemental analysis were used to characterize PPy–ZrP cation exchange nanocomposite. The composite showed good ion-exchange capacity (1.60 meq g^{-1}), DC electrical conductivity (0.33 S cm^{-1}) and isothermal stability in terms of DC electrical conductivity retention under ambient condition up to 100 °C. PPy–ZrP cation exchange-nanocomposite-based sensor was fabricated for the detection of ammonia vapours of aqueous ammonia. The resistivity of the nanocomposites increases on exposure to high-concentration ammonia vapours at room temperature (25 °C). The rate of reaction for ammonia vapour-sensing on PPy–ZrP was observed as second order.

Keywords Composite · Thermogravimetric analysis (TGA) · X-ray diffraction · Transmission electron microscopy (TEM) · Electrical properties

✉ Asif Ali Khan
asifalikhan2008@gmail.com

[1] Analytical and Polymer Research Laboratory, Department of Applied Chemistry, Faculty of Engineering and Technology, Aligarh Muslim University, Aligarh, UP 202002, India

Introduction

In recent years electrically conducting polymers have received much attention for use as advanced materials due to their good physical attributes [1–3]. Among the various conducting polymers such as polypyrrole, polythiophene, polyaniline, etc., polypyrrole is an especially promising electrically conducting polymer for commercial applications due to its high conductivity, good environmental stability and ease in synthesis. In addition, polypyrrole is one of the most familiar conducting polymers that show many advantages in recombining millimicron particles to give nanocomposites [4]. Nanocomposite show new properties due to synergism between the constituents [5–10] and more available surface. Because of the new properties, nanocomposites may find applications in various fields such as device fabrication [11], photo catalysis [12], solar cells [13], fuel cell [14], biomedical and sensing application [15].

Various polypyrrole-based composites such as palladium–polypyrrole nanocomposite, polypyrrole-Au nanocomposite, polypyrrole-TiO$_2$ nanocomposite, polypyrrole-manganese oxide composite, etc. are synthesized and used for ammonia sensing [16–18], DNA bio sensing, in super capacitor electrode, etc. [17]. However, conducting polymer-based ion exchangers with polyvalent sites have been poorly reported in the field of gas sensing [6–8]. Conducting ion-exchange materials having millimicron particles are considered an advanced class of materials because of its excellent ion-exchange behaviour and their analytical as well as electro-analytical applications [19–21].

Thus, in this work, we have synthesized a new electrically conductive ion-exchange nanocomposite PPy–ZrP by in situ oxidative chemical polymerization technique and used it as new sensing material for ammonia vapour at room temperature. To the best of our knowledge, this is the first attempt to synthesize a new electrically conductive PPy–ZrP cation exchange nanocomposite by using in situ oxidative chemical polymerization technique and applied for ammonia vapour sensing characteristics.

Experimental

Chemical, reagents and instruments

The following reagents and instruments were used:

The Pyrrole monomer (98%) from Spectrochem (India Ltd.), anhydrous Iron(III)-chloride (FeCl$_3$), methanol HPLC grade Ortho phosphoric acid (H$_3$PO$_4$) and Zirconium oxychloride were used as received from Qualigens (India Ltd.).All other reagents and chemicals were of analytical grade.

Ultrasonic vibrations (SC-I, Chengdu Jiuzhou Ultrasonic Technology Co.), Fourier Transform Infra-red spectroscopy (FTIR) (Perkin Elmer 1725 instrument), Transmission electron microscopy (TEM) (JEOL TEM, JEM 2100F), field emission scanning electron microscopy (FE-SEM) and energy-dispersive analyzer unit (EDAX) (LEO 435-VF), X-ray diffraction (XRD) (PHILIPS PW1710 diffractometer), Thermal analysis (TGA, DTA and DTG) (thermal analyzer-V2.2A DuPont 9900).

Synthesis

Zirconium(IV) phosphate

Preparation of ZrP was carried out by taking different ratios of zirconium oxychloride solution and aqueous solution of orthophosphoric acid (prepared in demineralized water) under varying conditions given in Table 1. The reaction mixture was thoroughly stirred with a magnetic stirrer at room temperature (25 °C), the solution containing precipitate was stirred for 1 h and was refluxed at 75–80 °C for 24 h. The resulting precipitate was decanted and washed with demineralized water (DMW), filtered by suction and dried at 50 ± 2 °C for 24 h. The excess of acid was removed by repeated washing with DMW. Finally, the

material was dried in an oven at 50 ± 2 °C for 4 h and ground by pastel mortar to obtain a fine powder of ZrP.

Polypyrrole–zirconium(IV) phosphate nanocomposite

PPy–ZrP nanocomposites were prepared by in situ chemical oxidative polymerization [5–8] of pyrrole in the presence of ZrP particles. A schematic representation of the formation of PPy–ZrP nanocomposite is shown in Scheme 1. A certain amount of ZrP (dried at 50 °C for 2 h before use) was dispersed in 100 ml of double-distilled water (DDW) under ultrasonic vibrations (SC-I, Chengdu Jiuzhou Ultrasonic Technology Co.) at room temperature for 1 h. This ZrP dispersed solution was then diverted into a 500-mL single-necked, round-bottom flask equipped with a magnetic Teflon-coated stirrer, and a certain amount of pyrrole monomer was added. The mixture was stirred for 30 min for the adsorption of pyrrole on the surface of ZrP particles. 2 g Ferric chloride in 100 ml of DDW was added to the dispersion. This reaction mixture was stirred for an additional 24 h under the same condition. The resultant PPy–ZrP nanocomposite powder was filtered using a Buchner funnel and then washed with DMW to remove unreacted oxidant. It was further washed thoroughly with methanol to remove any unreacted polymer. The obtained powders were dried completely at 50 °C for further analysis. Pure PPy was synthesized by a similar method as the PPy–ZrP composites prepared without the ZrP particles. The condition of preparation and their ion-exchange capacity (IEC) of the cation exchange nanocomposite samples are given in Table 2.

Ion-exchange capacity (IEC)

The column method was used for the determination of the IEC of each sample; IEC generally expresses the measure of the H$^+$-ion liberated by the nanocomposite cation exchanger to flow through the neutral salt. To calculate IEC 1 g of dry PPy–ZrP (in H$^+$-form) was loaded into a glass column having an internal diameter ~1 cm with a glass wool supported at the bottom. The bed length was approximately 1.5 cm long. 1 Mol L^{-1} sodium nitrate (NaNO$_3$) as eluents was used to elute the H$^+$ ions completely from the cation exchange column, keeping a very

Table 1 Conditions of preparation and the ion-exchange capacity ZrP cation exchanger

Sample code	Mixing volume ratio (v/v)		Appearance of the sample	Na$^+$ ion exchange capacity in (meq g^{-1})
	Zirconium oxychloride (0.1 mol L^{-1})	Ortho-phosphoric acid		
ZrP-1	1	1 (1 mol L^{-1})	White	1.25
ZrP-2	2	1 (2 mol L^{-1})	White	1.05
ZrP-3	3	1 (3 mol L^{-1})	White	0.85

(a)

(b)

Zirconium(IV)phosphate nanorods

Pyrrole

in-situ oxidative chemical polymerization

FeCl₃, H₂O

Polyprrole-Zirconium(IV)phosphate nanocomposite

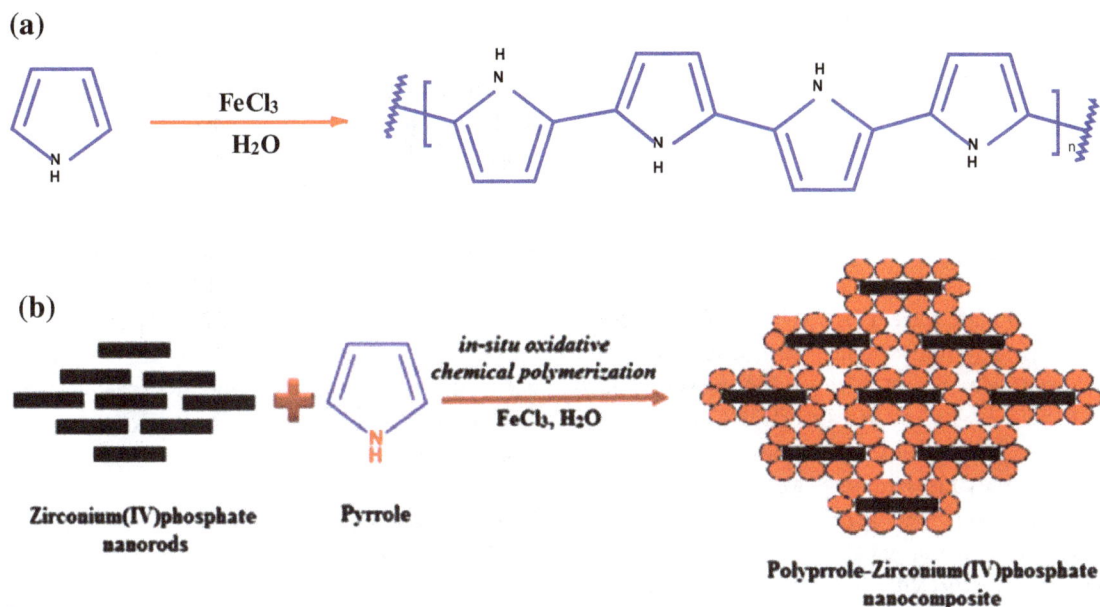

Scheme 1 Schematic diagram of the formation mechanism of **a** PPy and **b** PPy–ZrP nanocomposite

Table 2 Conditions of preparation and the ion-exchange capacity PPy–ZrP nanocomposite cation exchange

Sample code	Zirconium(IV) phosphate (g) (sonicated in 100 mL DDW)	Iron(III)-chloride (g) (in 100 mL DDW)	Pyrrole monomer (mL)	Na^+ ion exchange capacity in (meq g^{-1})	DC electrical conductivity (S cm^{-1})
PZrP-1	2.0	2.0	1.0	0.70	2.01×10^{-1}
PZrP-2	2.0	2.0	2.0	0.90	2.51×10^{-1}
PZrP-3	2.0	2.0	3.0	0.70	3.33×10^{-1}
PZrP-4	2.0	2.0	4.0	0.70	4.21×10^{-1}
PZrP-5	2.0	2.0	5.0	0.70	4.88×10^{-1}
PZrP-6	2.0	2.0	7.0	1.25	4.21×10^{-1}
PZrP-7	4.0	2.0	7.0	1.60	3.12×10^{-1}

slow flow rate (~ 0.5 ml min^{-1}). The effluent was titrated against a standard 0.1 M L^{-1} NaOH solution using phenolphthalein indicator. Table 2 shows the ion-exchange capacity values of the different samples.

Characterization

The Fourier transform infra-red spectroscopy (FTIR) spectra were recorded using Perkin Elmer 1725 instrument. Field emission scanning electron microscopy (FE-SEM) was used to study the surface morphology of the material using LEO 435-VF model electron microscope. Transmission electron microscopy (TEM) was performed by JEOL TEM (JEM 2100F) instrument. X-ray diffraction (XRD) data were recorded by PHILIPS PW1710 diffractometer with Cu Kα radiation at 1.540 Å in the range of $5° \leq 2\theta \leq 70°$ at 40 kV. The thermal stability was investigated by thermal analysis (TGA, DTA and DTG) using

thermal analyzer-V2.2A DuPont 9900. The samples were heated in alumina crucible from 30 to 1000 °C at the rate of 10 °C min^{-1} in the nitrogen atmosphere at the flow rate of 200 mL min^{-1}. The elemental analysis of PPy, ZrP and PPy–ZrP cation exchange nanocomposite (PPy–ZrP) was performed using energy-dispersive analyzer unit (EDAX) attached with FE-SEM.

Electrical conductivity and ammonia-sensing measurements

For electrical conductivity measurements and sensing experiments, 0.2 g material from each sample was palletized at room temperature with the help of a hydraulic pressure instrument at 25 KN pressure for 10 min. DC electrical conductivity of the nanocomposite was measured using a four-in-line probe. The conductivity (σ) was calculated using the following equations [5–8]:

$$\rho = \rho^\circ / G_7(W/S) \tag{1}$$

$$G_7(W/S) = (2S/W)\ln 2 \tag{2}$$

$$\rho^\circ = \left(\frac{V}{I}\right)2\pi S \tag{3}$$

$$\sigma = 1/\rho, \tag{4}$$

where G_7 (W/S) is a correction divisor which is a function of the thickness of the sample as well as probe-spacing where I, V, W and S are current (A), voltage (V), thickness of the film (cm) and probe spacing (cm), respectively. In isothermal ageing experiments, the nanocomposite pellets were heated at 50, 70, 90, 110 and 130 °C in a proportional integral directive (PID) controlled temperature oven. The electrical conductivity measurements were performed at an interval of 10 min. In cyclic ageing experiments, the DC electrical conductivity was measured in the temperature range of 40–150 °C repeatedly for five times at an interval of 1 h. Ammonia-sensing measurements were done by monitoring the resistivity of the nanocomposite using the Laboratory made set-up for ammonia sensing based on four-in-line probe electrical conductivity measuring instrument [6, 9].

Ammonia sensing kinetics

For sensing kinetics 0.2 g selected pelletized material was taken at 20 and 25 °C temperatures and resistivity response was recorded using four in line probe in atmosphere of ammonia vapours with respect to time.

Results and discussion

In this study various samples of PPy–ZrP cation exchange nanocomposite were prepared by in situ chemical oxidative polymerization of pyrrole in the presence of ZrP nanoparticles under different conditions (see Table 2). PZrP-7 sample shows better Na^+ ion exchange capacity (1.60 meq g^{-1}) as compared to the inorganic ZrP (sample ZrP-1, Table 1) (1.25 meq g^{-1}). The IEC of PPyZrP composite was increased due to the addition of conductive polymer into the inorganic material (ZrP) which increases the surface area of the material; thus, exchangeable ionic sites were increased. Due to the better ion exchange capacity and electrical conductivity, sample PZrP-6 (Table 2) was selected for ammonia sensing.

The variation in conductivity with the loading of different amount pyrrole monomer is shown in Table 2. At 7% loading of pyrrole monomer high improvement in electrical conductivity and IEC was observed. Electrical conductivity increases significantly up to 5% loading of pyrrole monomer and decreases slightly at 7% loading. It

means that by adding 5% PPy the percolation threshold might be achieved after further addition of PPy: no significant change in conductivity was observed. Since conductivity increases due to increase in concentration of conducting particles, it is to be well understood that conductivity depends significantly on the carrier transport through the conducting fillers. However, the formation of percolation network within the matrix of the composite also affects the conductivity. Thus after getting percolation threshold, further addition of PPy may change the network of the matrix and further addition of inorganic part (ion exchange material) to get better ion exchange capacity may also decrease the electrical conductivity due to its insulating property.

Temperature dependence of DC electrical conductivity of the PPy and PPy–ZrP

The electrical conductivity of PPy and PPy–ZrP cation exchange nanocomposite was measured with increasing temperatures from 30 to 150 °C. Arrhenius Plot (ln σ_{dc} verses 1000/T) of electrical conductivity of PPy and PPy–ZrP cation exchange nanocomposite were obtained as shown in Fig. 1. Significant change in electrical conductivity of the nanocomposite was observed with the rise in temperature; electrical conductivity of PPyZrP increasing with increase in temperature can be explained by "thermal activated behaviour" [22]. The conduction mechanism in the conducting polymers is explained in terms of polaron and bipolaron formation. Polymer at low level of oxidation of the gives polaron and at high level of oxidation gives bipolaron. Both polarons and bipolarons are mobile and move along the polymer chain by the rearrangement of

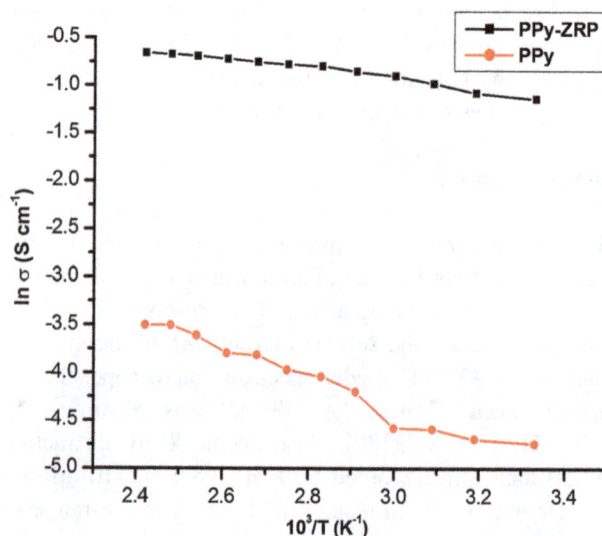

Fig. 1 Plot of ln σ_{dc} versus 1000/T for PPy and PPy–ZRP nanocomoposite

double and single bonds in the conjugated system. The mechanism of charge transport in polymer with non-degenerate ground state is mainly explained by conduction of polarons and bipolarons. The magnitude of the conductivity is dependent on the number of charge carriers available and their mobility. It has been observed that mobility of charge carriers increases with the increase in temperature leading to the increase in conductivity similar to that reported for PANI/WO$_3$ and PANI/CeO$_2$ nanocomposites [22, 23]. Another factor which also affects the electrical conductivity is the molecular alignment of the chains within the entire system.

FTIR studies

The FT-IR spectra of PPy, ZrP and PPy–ZrP nanocomposite are shown in Fig. 2. The FT-IR spectrum of PPy, in the fingerprint region of PPy, shows an absorption peak at 902 cm^{-1} which is characteristic of C–H out-of-plane deformation vibration, confirming the formation of PPy by the monomer. The bands at 1300, 3000 and 1500 cm^{-1} is attributed to the C–N in-plane, N–H starching and the bands at 1167 and 1041 cm^{-1} are related to the C–H bending modes while the strong absorption band obtained at 1449 and 1539 cm^{-1} corresponds to the C–C stretching and C=C bending vibration in the pyrrole ring. Some other peaks in the fingerprint region (600–1500 cm^{-1}) can be attributed to the ring stretching and C–H in plane deformation mode. The PPy–ZrP nanocomposite shows nearly identical values and positions of the main IR bands in the range of 450–4000 cm^{-1}. Compared with FTIR spectra of PPy, a strong band at 3400 cm^{-1} may be attributed to the –OH stretching frequency and a broad band between 1250 and 900 cm^{-1} with a peak of intensity at 1042 cm^{-1} is due to presence of ionic phosphate group and

peak at 795 cm^{-1} is attributed to M–O bonding. The band at 3000 cm^{-1} is attributed to the N–H starching and the stretching vibration of C–N observed at 1325 cm^{-1} indicates that the polymerization of PPy has been successfully achieved on the surface of the ZrP milimicron particles.

X-ray diffraction studies

Figure 3 shows the XRD patterns of PPy, ZrP and PPy-ZrP nanocomposite. The XRD pattern of pure PPy shows an obvious broad peak at $2\theta = 20°$, along some very low intensity peaks, suggesting that the PPy conducting polymer is amorphous in nature [16]. XRD pattern of ZrP shows some sharp peaks at $2\theta = 10°, 20°, 25°, 35°$ and some low intensity peaks [24]. In case of PPy–ZrP diffraction peaks lie between 15° and 25°, broaden slightly and low intensity of PPy at 34° and 36° appear in the XRD pattern of PPy–ZrP. These results suggest that PPy is polymerized and deposited on the surface of ZrP millimicron particles and there is the successful incorporation of ZrP nanoparticles in PPy–ZrP nanocomposite. It is also observed that the diffraction pattern of the nanocomposites slightly change as that of ZrP millimicron particles. Thus, we can conclude that PPy has low influence on the crystallization performance of ZrP millimicron particles. The results are also in agreement with the FTIR and TEM studies.

Morphological studies

Figure 4a and c shows the TEM image of ZrP and PPy–ZrP nanocomposite with tubular morphology having an average particle size of ~20–40 and 30–50 nm, respectively. Particle size of ZrP and PPy–TSP lies in nano range, which suggests that the prepared material is nanocomposite. The

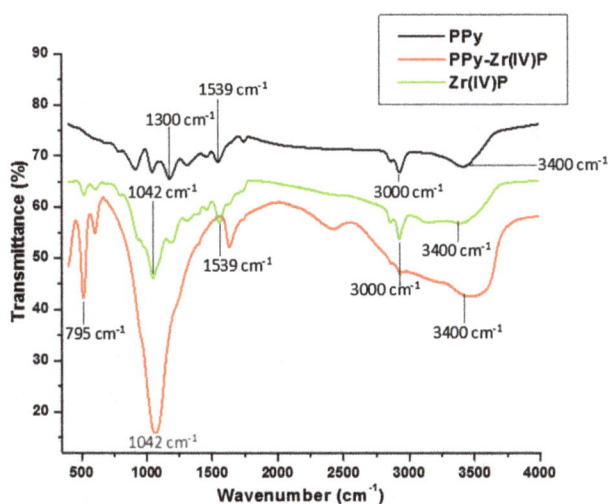

Fig. 2 FTIR spectra of PPy, ZrP and PPy–ZrP cation exchange nanocomposite

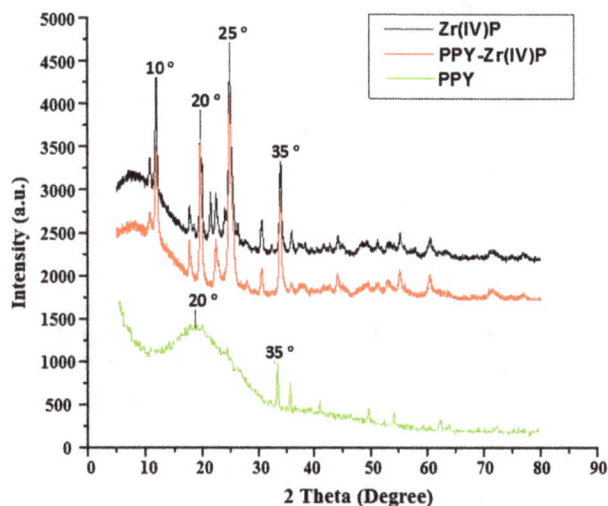

Fig. 3 XRD patterens of PPy, ZrP and PPy–ZrP cation exchange nanocomposite

Fig. 4 TEM image of **a** ZrP, selected area diffraction pattern (SAED) of **b** ZrP, TEM image of **c** PPy–ZrP nanocomposite and selected area diffraction pattern (SAED) of **d** PPy–ZrP nanocomposite

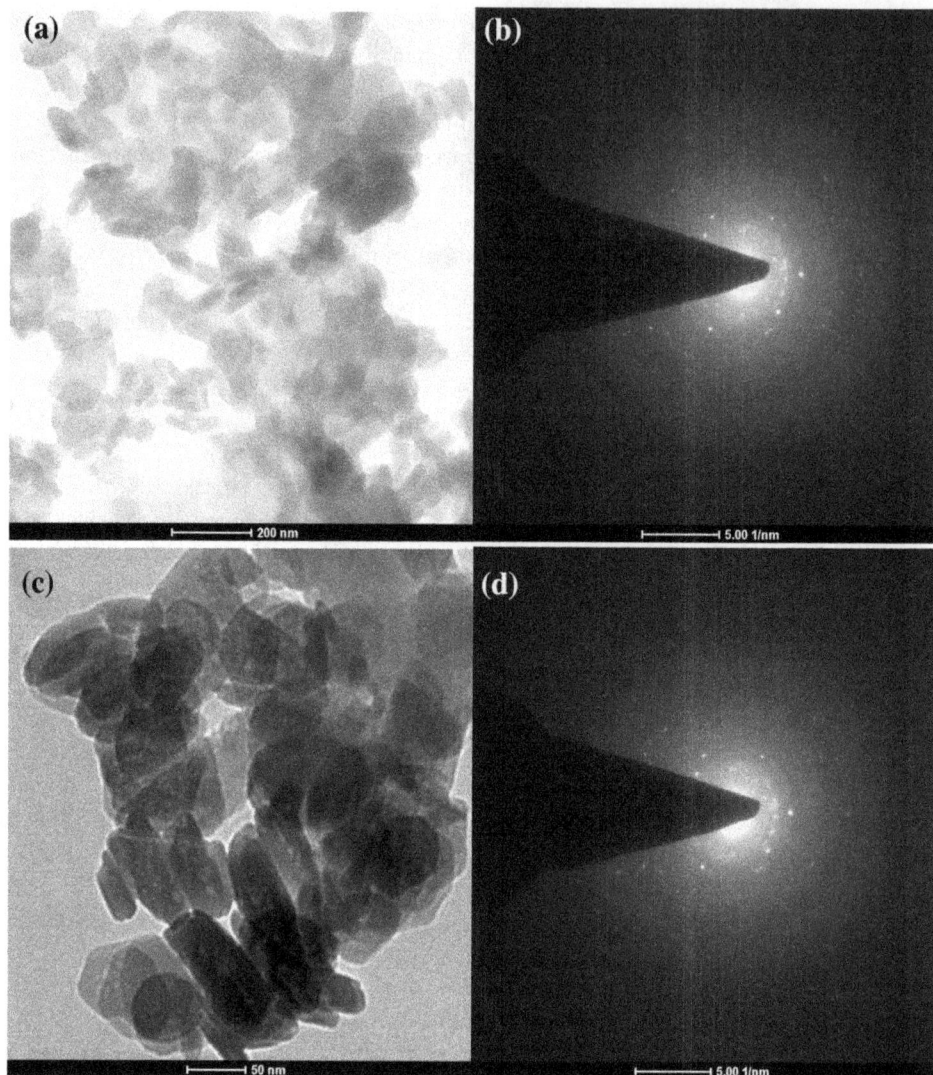

tubular ZrP nanoparticles can be seen as dark spots encapsulated in PPy matrix, which suggests that polymerization of PPy is successfully achieved on the surface ZrP nanoparticles. Figure 4b and d shows selected area diffraction pattern (SAED) of ZrP and PPy–ZrP nanocomposite predicts the crystalline nature of ZrP and semi-crystaline nature of PPyZrP nanocomposite.

The FE-SEM images of PPy, ZrP and PPy–ZrP nanocomposites are shown in Fig. 5a–c at different magnifications. Figure 5a shows globular nanoparticles of PPy, and Fig. 5b shows short tubular nanoparticles of ZrP. The FE-SEM images of PPy–ZrP nanocomposite (Fig. 5c) shows that ZrP nanoparticles are well embedded in the polymer matrix with uniform dispersion. Thus, the results of XRD, FTIR, TEM and SEM studies provided clear evidence that the polymerization of PPy has been successfully achieved on the surface of ZrP nanoparticles. A schematic representation of the formation of PPy and PPy–ZrP nanocomposite is given in Scheme 1.

Thermo gravimetric analysis

The TGA curve of PPy, ZrP and PPy–ZrP nanocomposite is shown in Fig. 6a. In case of PPy, first weight loss was observed at 200 °C (10.27%) due to physisorbed water molecule and volatile impurities. The second weight loss was observed at 400 °C (14.05%) due to degradation of the polymer unsaturated groups. After 250 °C there is gradual weight loss (31.11%) observed up to 1000 °C due to degradation of polymer [25]. In the case of ZrP, the first weight loss was observed at 100 °C (2.39%) due to removal of external water molecules, next on 500 °C (11.32%) due to starting of decomposition of the material and after 600 °C the ZrP was found stable up to 1000 °C. PPy–ZrP nanocomposite shows the first weight loss at 100 °C (6.87% weight loss) due to removal of external water molecules and after that PPyZrP is stable up to 500 °C. The second weight loss of PPy–ZrP appears at 550 °C (12.29% weight loss) because of degradation of

Fig. 5 FE-SEM images of
a PPy **b** ZrP and **c** PPy–ZrP
nanocomposite

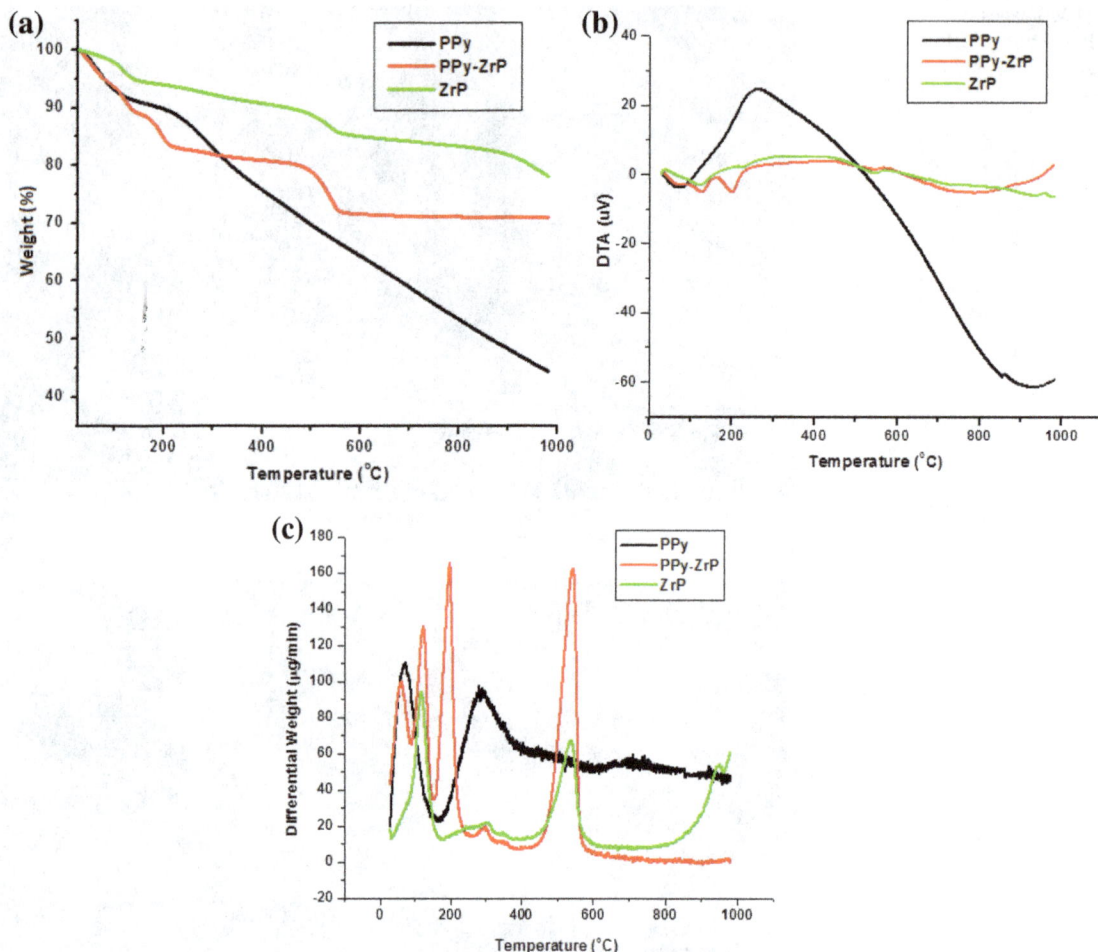

Fig. 6 **a** TGA curves of PPy, ZrP and PPy–ZrP, **b** DTA curves of PPy, ZrP and PPy–ZrP and **c** DTG curves of PPy, ZrP and PPy–ZrP cation exchange nanocomposite

PPy and after 600 °C the PPy–ZrP nanocomposite remains stable up to 1000 °C. The total mass loss up to 1000 °C has been estimated to be about 65.43, 11.62 and 28.68% for PPy, ZrP and PPy–ZrP, respectively. These results confirm that the presence of ZrP in PPy–ZrP nanocomposite is responsible for the higher thermal stability of the composite material in comparison to pristine PPy.

Figure 6b shows the DTA curve of pure PPy, ZrP and PPy–ZrP nanocomposite. DTA of ZrP was found to exhibit two endothermic peaks at 215 °C (2.22 μV) and 532 °C (0.30 μV) and one exothermic peak at 961 °C (−5.62 μV). The endothermic peaks at 215 °C corresponds to decomposition stage between 200 and 300 °C while the endothermic peak at 532 °C corresponds to second decomposition stage (400–600 °C). The exothermic peak corresponds to decomposition stage (900–1000 °C) also indicated in the TGA curve of ZrP (Fig. 6a). DTA of PPy was found to exhibit only one endothermic peak at 250 °C (2.22 μV), corresponds to decomposition stage between 200 and 300 °C as also indicated in TGA of PPy (Fig. 6a).

However, PPy–ZrP exhibited two endothermic peaks at 121 °C (−4.96 μV) and 193 °C (−5.12 μV) corresponds to decomposition stage between (30–150 °C) and (150–300 °C), respectively, and one exothermic peak at 539 °C (1.43 μV) corresponds to decomposition stage between 400 and 650 °C as also indicated in TGA of PPy–ZrP nanocomposite (Fig. 6a).

DTG analysis of pure PPy, ZrP and PPy–ZrP nanocomposite was studied as a function of rate of weight loss (μg min^{-1}) versus temperature (Fig. 6c). In case of pure PPy decomposition at 74 and 290 °C was found with 111 and 96 μg min^{-1} weight loss, respectively, and in case of ZrP decomposition at 109 and 529 °C was found with 94 and 529 μg min^{-1} weight loss, respectively. However, in the case of PPy–ZrP nanocomposite, the decomposition was observed at 64, 121,190, 539 °C with 100, 131, 166, 162 μg min^{-1} weight loss, respectively. Thus, it can be concluded from the DTG analysis that the rate of thermal decomposition is lower in case of PPy–ZrP, whereas in the case of PPy the rate of thermal decomposition was higher.

Fig. 7 EDAX spectra of **a** PPy
b ZrP anc **c** PPy–ZrP cation
exchange nanocomposite

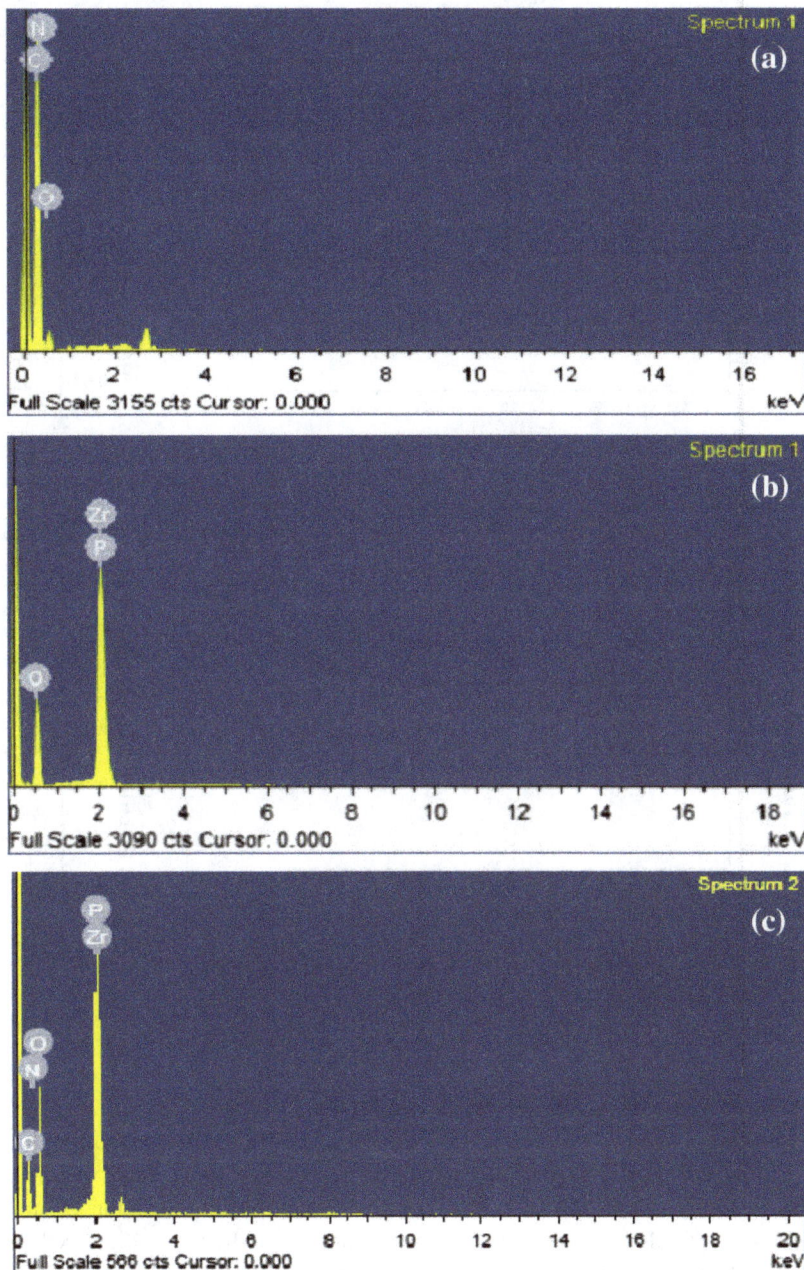

The better thermal resistance of pure PPy–ZrP nanocomposite was due to incorporation of ZrP in the PPy matrix.

Energy-dispersive X-ray analysis

The EDAX patterns of PPy, ZrP and PPy–ZrP cation exchange nanocomposites are shown in Fig. 7. EDAX studies have provided clear evidence that the polymerization of PPy has been successfully achieved on the of the ZrP nanoparticles. The percent composition of elements is given in Table 3.

Table 3 Percent composition of carbon, nitrogen, oxygen, zirconium and phosphorus in PPy–ZrP nanocomposite by EDAX analysis

Element	Weight (%)		
	PPy	ZrP	PPy–ZrP
C	64.73	Nill	23.81
N	19.04	Nill	4.08
O	16.23	44.61	40.61
Zr	Nil	34.09	17.76
P	Nil	44.45	13.74

(a)

(b)

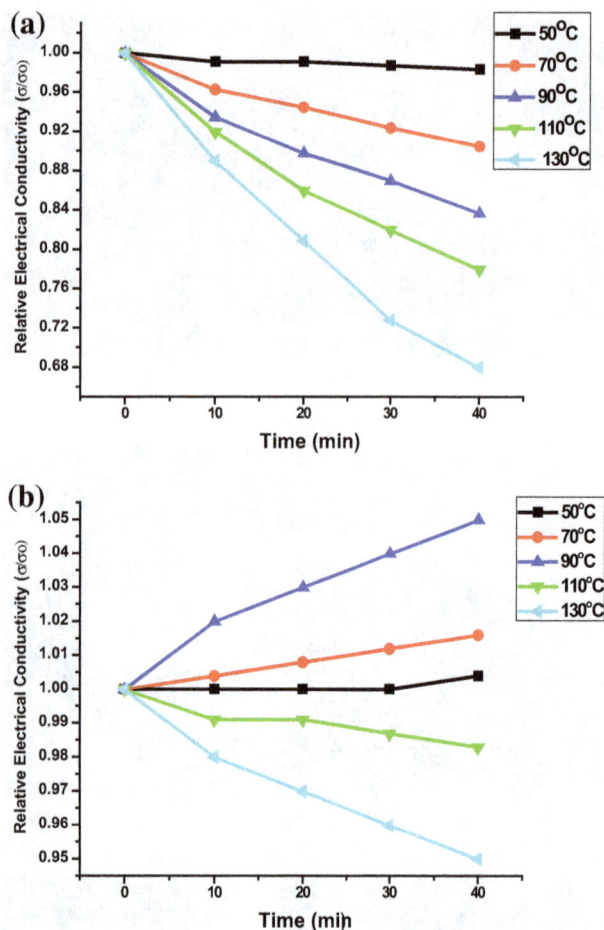

Fig. 8 Isothermal stability of **a** PPy and **b** PPy–ZrP nanocomposite in terms of d.c. electrical conductivity retention at 50, 70, 90, 110 and 130 °C

Stability in terms of DC electrical conductivity retention

The stability of the PPy and PPy–ZrP nanocomposite in terms of DC electrical conductivity retention was studied by isothermal ageing and cyclic ageing conditions in an ambient atmosphere.

DC electrical conductivity retention under isothermal ageing conditions

The isothermal stability of the composite material was examined in terms of DC electrical conductivity retention. In this experiment electrical conductivity was measured five times after an interval of 10 min at a particular temperature, e.g. 50, 70, 90, 110 and 130 °C in an air oven. Figure 8 shows electrical conductivity measurement with respect to time. It was observed that all the composite materials follow Arrhenius equation for the temperature dependence of the electrical conductivity from 50 to 90 °C

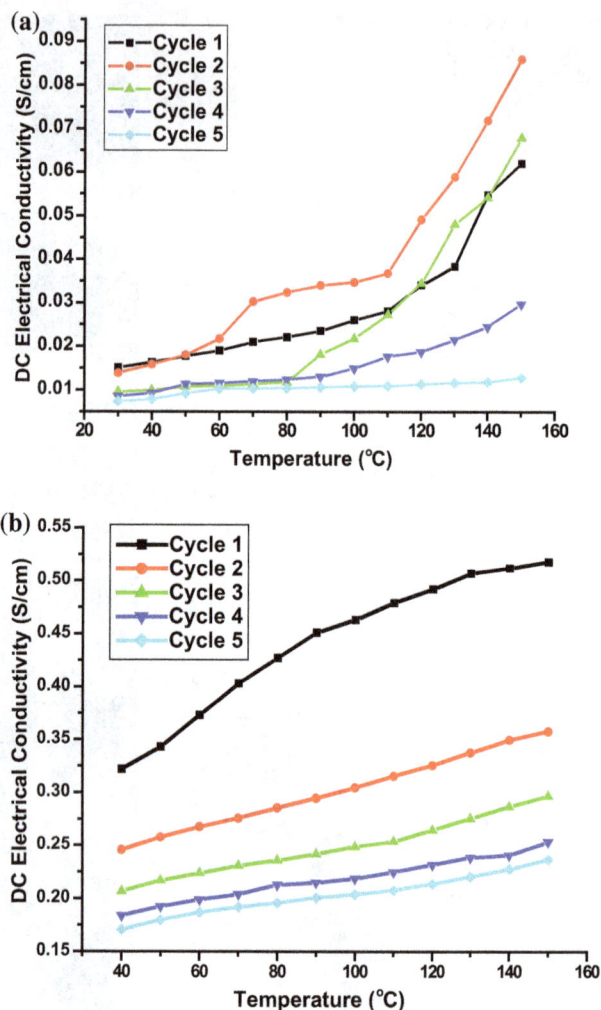

(a)

(b)

Fig. 9 DC electrical conductivity of **a** PPy and **b** PPy–ZrP nanocomposite in terms of cyclic ageing conditions

and after that a deviation in electrical conductivity was observed, it may be due to the loss of dopant and degradation of materials. The isothermal stability of PPy–ZrP cation exchange nanocomposite in terms of DC electrical conductivity retention was found to be better than pristine PPy which suggests that the PPy–ZrP nanocomposite cation exchange may be used in electrical and electronic devices below 100 °C under ambient conditions.

DC electrical conductivity retention under cyclic ageing conditions

The stability of PPy and PPy–ZrP cation in terms of DC electrical conductivity retention exchange nanocomposite was also examined by cyclic ageing technique. It was observed from Fig. 9 that the DC electrical conductivity at the beginning of each cycle was found to be low as compared to previous cycle and which further decreases with the increase in number of cycles for PPy and PPy–

ZrP cation exchange nanocomposite. This may be due to the loss of moisture and polymer degradation during cyclic ageing. From cyclic electrical conductivity study on PPy and PPy–ZrP cation exchange nanocomposite it may be suggested that the electrical conductivity of the PPy–ZrP nanocomposite is more stable than pristine PPy.

Ammonia vapour sensing characteristics of PPy–ZrP nanocomposite

The ammonia vapour sensing performance of PPy–ZrP (PZrP-6) cation exchange nanocomposite was monitored by measuring resistivity changes on exposure to ammonia vapours using laboratory-made assembly designed by using Four-in-line probe electrical conductivity device. The electrical resistance of nanocomposite showed remarkable changes on exposure to 0.2, 0.4, 0.6, 0.8 and 1 M concentrations of aqueous ammonia with vapour concentrations of 0.672, 0.870, 1.020, 1.080, and 1.097%, respectively, at room temperature as a function of time as depicted in Fig. 10a. It was observed that the nanocomposite showed a relatively fast response towards ammonia vapours in the concentration range of 0.2–1 M (vapour concentration 0.672–1.097%) and better resistivity response as compared to PPy (Fig. 10b); however, the change in resistivity on exposure to humidity was also observed as shown in the Fig. 10b. The relative humidity (%RH) inside the glass chamber was calculated as 82.93% by using the following relation [26]:

$$\%RH = E_w(T1)/E_w(T2) \times 100, \tag{5}$$

where E_w (T1) (17.5 mm of Hg) and E_w (T2) (21.1 mm of Hg) are the saturated water vapor pressure at the temperature of water (20 °C) and that of the composite film (23 °C), respectively. The values of the saturated vapor pressure were obtained from Lange's handbook of chemistry by John A. Dean, Fifteenth Edition, Mc Graw-Hill, Inc. [27].Taking into account of some other ammonia-sensing composite materials like Polypyrrole graphitic nanocomposite [28], Polypyrrole/Metal Sulphide nanocomposite [29], poly(3-methythiophene)–titanium(IV) molybdophosphate cation exchange nanocomposite [6] and polyaniline–titanium(IV) phosphate cation exchange nanocomposite [8] reported earlier and compared with PPy–ZrP for ammonia sensing, the following observation can be noted (Fig. 10).

In Polypyrrole graphitic nanocomposite conductivity change in 25 min is 7%; and if it is calculated for PPy–ZrP it is 10.91% in 2 min. In the case of Polypyrrole/Metal Sulphide nanocomposite the response time for ammonia sensing is 20 s; on the other hand response time for PPy–

ZrP is 10 Seconds. In the case of poly(3-methythiophene)–titanium(IV) molybdophosphate cation exchange nanocomposite, the change in resistivity is 0.06 Ω cm in 1.83 min, but in case of PPy–ZrP it is 0.35 Ω cm in 2 min. For polyaniline–titanium(IV) phosphate cation exchange nanocomposite change in resistivity is 0.8 Ω cm in 5 min and in the case of PPy–ZrP it is 0.35 Ω cm in 2 min. All the above results show that PPy–ZrP is a better sensor of Ammonia.

The resistivity was recovered on flushing with the ambient air. The response and recovery time of the sensor was around 10 and 30 s, respectively, for 0.2–1 M aqueous ammonia (vapour concentration 0.672–1.097%). The reversibility of the nanocomposite was also studied and the response of the nanocomposite was found to be highly reversible towards 0.2–0.6 M aqueous ammonia during the test of cyclic measurements as shown in Fig. 11. Further at higher concentration of aqueous ammonia (1 M), the reversible response studies were carried out which showed poor performance. The time taken to regain the resistivity value near to the original one was quite large. This poor performance of nanocomposite at higher concentration may be due to the complete occupying of reacting sites of polymer or because of the insufficient numbers of sites available for ammonia moiety to form the complex structure necessary for obtaining the response behaviour.

The extent of reversibility of the sensor was examined by cyclic measurements using different concentrations of ammonia (0.2, 0.6 and 1 M). The relative standard deviation (RSD %) for 0.2, 0.6 and 1 M was calculated to be 16.60, 1.56 and 1.01%, respectively. From the RSD (%) it can be concluded that the sensor works best in the concentration range from 0.2 to 0.6 M, and at higher concentrations slight irreversibility takes place which may be due to the electrical compensation of the polymer backbone by ammonia.

Second-order kinetics evaluation for ammonia sensing on PPy–ZrP nanocomposite and kinetic parameters

Order of reaction of ammonia vapour sensing on PPy–ZrP nanocomposite was evaluated for the physical interaction of ammonia on PPy–ZrP nanocomposite. The mechanism of interaction may be explained on the basis of the electrostatic interaction of the lone pair of nitrogen of ammonia with carbon in PPy of PPy–ZrP nanocomposite. To ascertain the order of the reaction, the standard equation for first and second were applied as given below:

$$\text{Log } \sigma = \log \sigma_0 - (k_1/2.303)t \tag{6}$$

$$1/\sigma - 1/\sigma_0 = k_1 t, \tag{7}$$

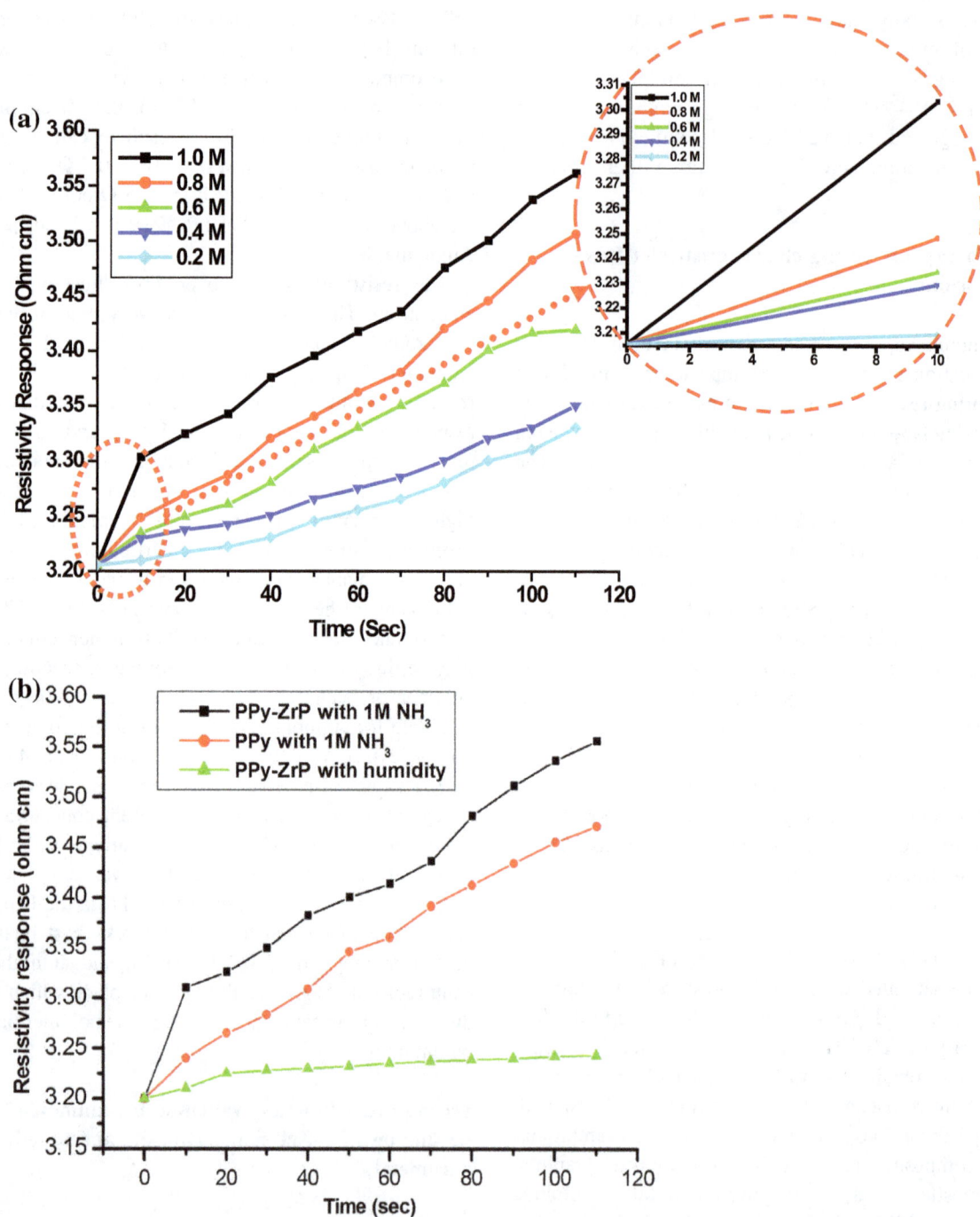

Fig. 10 **a** Effect on the resistivity of PPy–ZrP nanocomposite on exposure to different concentrations of ammonia with respect to time. **b** Effect on the resistivity of PPy and PPy–ZrP nanocomposite on exposure to 1 M concentration of ammonia and in humidity with respect to time

where σ is the conductivity (reverse of resistivity) response recorded during sensing, σ_0 is the conductivity at the start of the sensing, k_1 is the rate constant and t is the time.

A tentative explanation of processes occurring on the surface of PPy–ZrP can be explained as the lone pair of nitrogen of ammonia interacts with the carbon of PPy, which decreases the intensity of positive charge and

hence the mobility of charge carriers decreases resulting in the decrease in conductivity. The mechanistic representation of the electrical compensation of PPy in the PPy–ZrP nanocomposite in the present work is given in Scheme 2.

Conductivity vs. time and reverse of conductivity vs. time graph for sensing of ammonia vapours on PPy–ZrP in

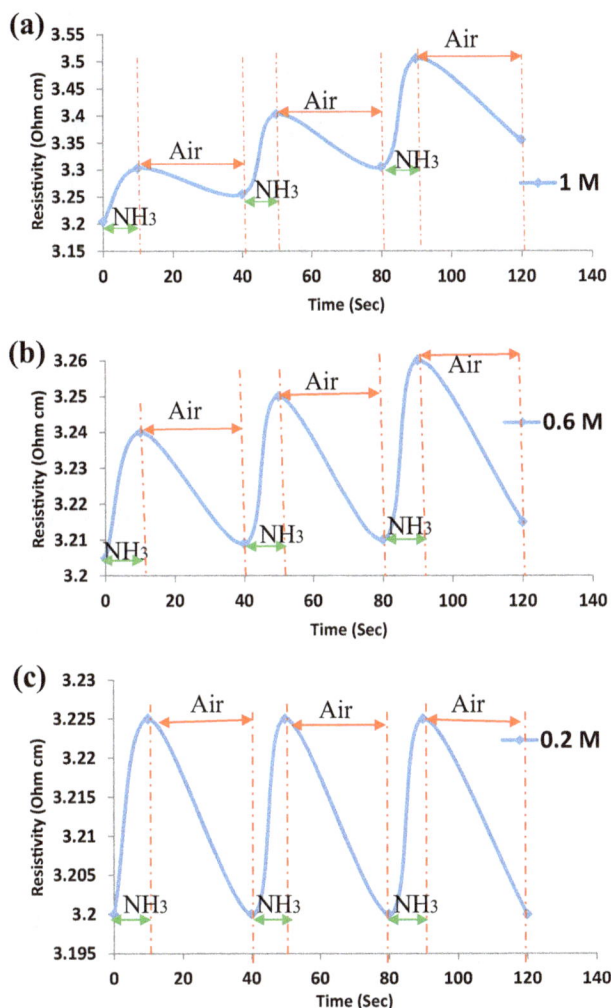

Fig. 11 Reversible resistivity response curves of PPy–ZrP nanocomposite towards **a** 1 M, **b** 0.6 M **c** and 0.2 M concentrations of aqueous ammonia

Fig. 12a and b shows straight line pattern in which Fig. 12b most resembles the second-order reaction.

Rate constants K_1 (0.0027 L Mol^{-1} S^{-1}) and K_2 (0.0029 L Mol^{-1} S^{-1}) at 20 and 25 °C were determined from the slopes of the graph of the inverse of conductivity verse time in Fig. 13.

According to the transition state theory presented by Laidler [30] the rate constant for a process can be written as

$$K = K_B T/h e^{(\Delta S^*/R)} e^{(-\Delta H^*/RT)}, \tag{8}$$

where K_B is Boltzmann's constant, h is Plank's constant, T is absolute temperature, ΔS^* is the entropy of activation, ΔH^* is the enthalpy of activation and R is the gas constant.

The Arrhenius activation energy, E_a is determined from the Arrhenius equation at two different temperatures.

$$\ln K_1/K_2 = E_a/R \left(1/T_1 - 1/T_2\right) \tag{9}$$

The enthalpy of activation, ΔH^*, can be calculated from the relationship:

$$\Delta H^* = E_a - RT \tag{10}$$

The entropy of activation ΔS^* was calculated from the Eq. (8) and the free energy of activation ΔG^* was determined from

$$\Delta G^* = \Delta H^* - T\Delta S^* \tag{11}$$

Results are summarized in Table 4 the negative value of ΔS^* and positive value of ΔH^* indicate the feasibility and endothermic behaviour during sensing process.

Sensing mechanism

The vapour-sensing properties of PPy–ZrP with ammonia vapours can be explained by the interaction of ammonia with PPy in the PPy–ZrP nanocomposite and is almost a reversible process, although a little bit of irreversibility is observed. Under ambient conditions, the value of resistivity increased due to the interaction of lone pair electron of ammonia with the positive charge of PPy in PPy–TSP. The mobility of charge carriers decreases which leads to an increase in resistivity. The interaction between lone pair of ammonia and positive charge of polarons or bipolarons of PPy in PPy–ZRP is electrostatic because it occurs between two opposite charges. Due to the small magnitude of charge, this interaction is weak in nature so it becomes reversible when it comes in contact with air, thus resistivity is restored. The resistivity did not come back to its original value and was always found higher than the previous value for higher concentration of ammonia. Hence, it can be concluded that there are two processes in operation: first, reversible chemisorption of ammonia with PPy occurs and second, compensation or electrical neutralization of the polymer backbone takes place.

In the light of observation by Khan [31] in polypyrrole–titanium(IV) sulphosalicylophosphate nanocomposite cation exchange material, it can be inferred that the lone pair of acetaldehyde interacts with positive charge of PPy, which decreases the intensity of positive charge and hence the mobility of charge carriers decreases resulting in the increase in resistivity. Since the exposure to ammonia was carried out in a closed system and chemical linking is much more complicated process, desorption of ammonia also occurs readily under ambient conditions and thus the resistivity is restored.

On exposure to ammonia for long duration, complete electrical neutralization of the polymer backbone occurred. The mechanistic representation of the electrical compensation of PPy in the PPy–ZrP nanocomposite in the present case is shown in Scheme 2a and b.

Scheme 2 The schematic
diagram showing the
chemisorption (reversible)
interaction of ammonia with
a polaron and **b** bipolaron of
PPy in PPy–ZrP nanocomposite

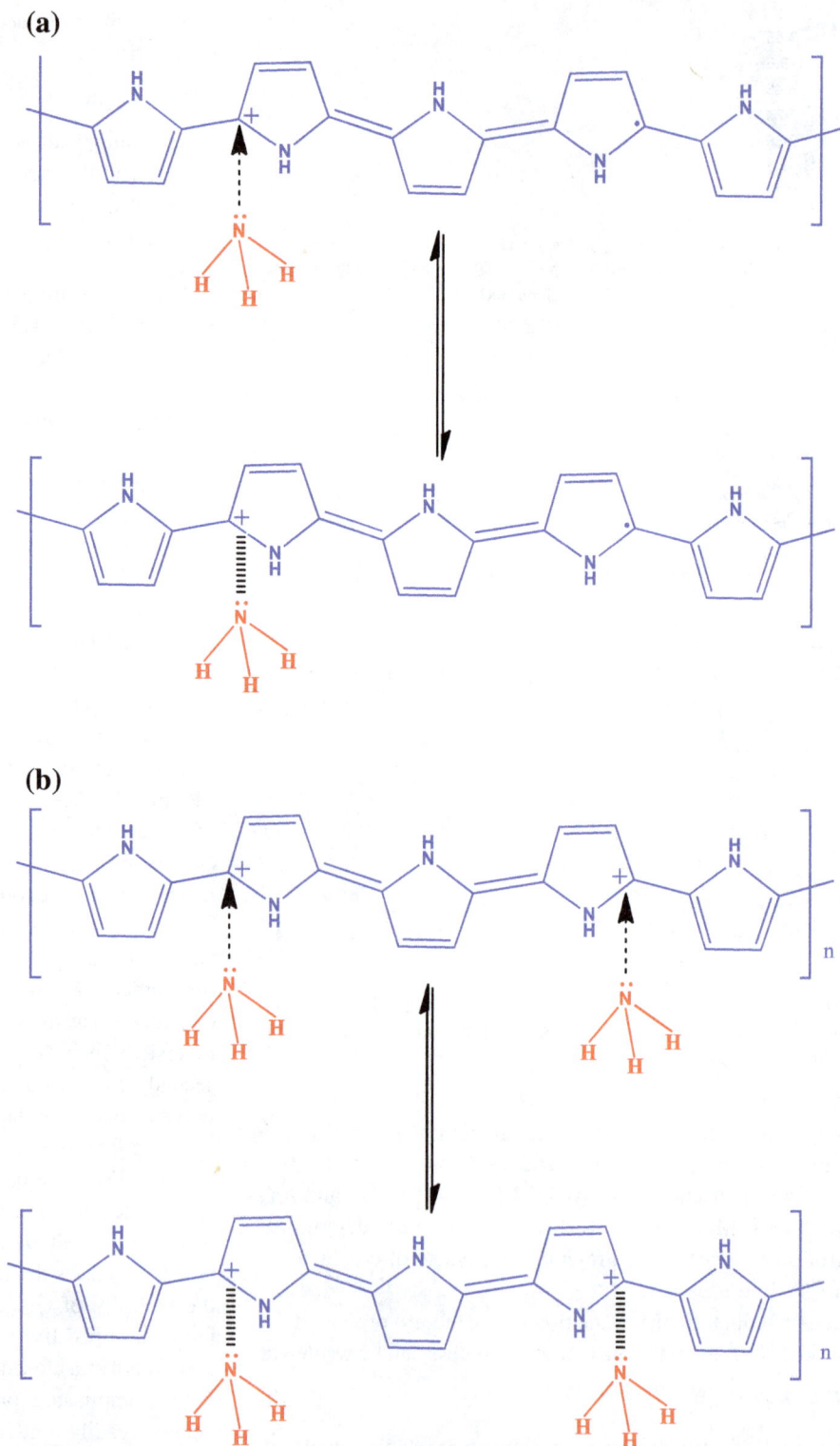

Conclusion

In the present study, the PPy–ZrP cation exchange
nanocomposites have been synthesized by in situ chemical
oxidative polymerization technique. The results of TEM,

SEM, XRD and FTIR studies reveal that the polymerization of pyrrole has been successfully achieved on the surface of the ZrP particles and indicates that there is a strong
interaction between PPy and ZrP nano particles. The PPy–
ZrP cation exchange nanocomposites show improved

Fig. 12 **a** Conductivity versus time graph for evaluating first order of reaction and **b** conductivity versus time graph for evaluating second order of reaction

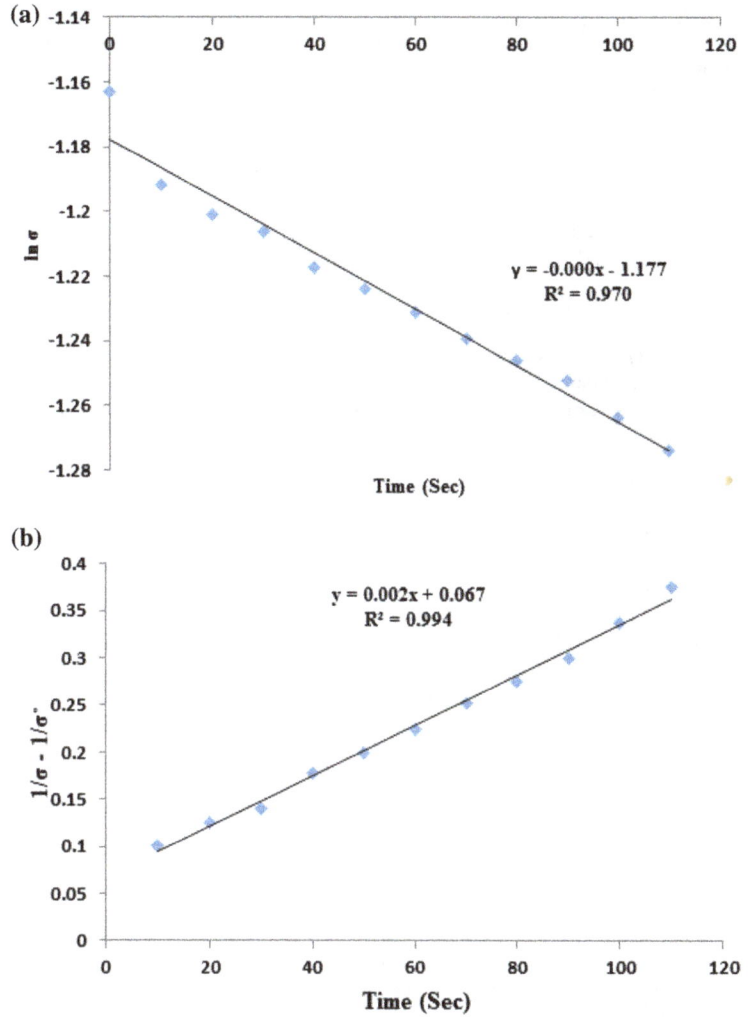

Fig. 13 For calculating rate constant K_1 and K_2 at temperature 20 and 25 °C from the slope of the graph. $K_1 = 0.0027$ 1 mole^{-1} S^{-1}. $K_2 = 0.0029$ 1 mole^{-1} S^{-1}

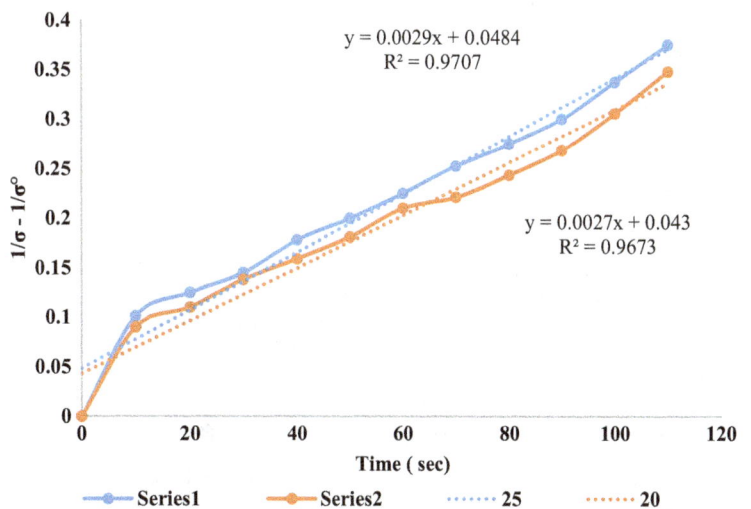

thermal stability, isothermal stability in terms of DC electrical conductivity as well as ion exchange capacity in comparison with the pure PPy. The reproducible ammonia sensing results in the range of 0.2–0.6 M aqueous ammonia indicates that PPy–ZrP nanocomposite can be used in making a sensing device.

Table 4 Kinetic parameters for the ammonia vapour sensing on PPy–ZrP cation exchange nanocomposite

Temperature (°C)	Rate constant K(l mol^{-1} s^{-1})	E_a* (kcal mol^{-1})	ΔH* (kcal mol^{-1})	ΔG* (kcal mol^{-1})	ΔS* (cal K^{-1} mol^{-1})
20	2.7×10^{-3}	2.476	1.893	20.577	−63.766
25	2.9×10^{-3}		1.883	20.902	−63.820

Sensing kinetics of ammonia vapour on PPy–ZrP was also studied and it was found that ammonia vapour sensing on PPy–ZrP follow the second-order of kinetics.

Acknowledgements Authors are thankful to the University Grant Commission (India) for MANF and Department of Applied Chemistry for providing financial support and research facilities.

References

1. Kim BH, Park DH, Joo J, Yu SG, Lee SH (2005) Synthesis, characteristics, and field emission of doped and de-doped polypyrrole, polyaniline, poly (3, 4-ethylenedioxythiophene) nanotubes and nanowires. Synth Met 150:279–284
2. Mahmoudian MR, Alias Y, Basirun WJ (2010) Electrodeposition of (pyrrole-*co*-phenol) on steel surfaces in mixed electrolytes of oxalic acid and DBSA. Mater Chem Phys 124:1022–1028
3. Asan A, Kabasakaloglu M, Aksu ML (2005) The role of oxalate ions in the coverage of mild steel with polypyrrole. Russ J Electrochem 41:175–180
4. Wei L, Hu N, Zhang Y (2010) Synthesis of polymer—mesoporous silica nanocomposites. Materials 3:4066–4079
5. Khan AA, Baig U (2014) Electrical and thermal studies on poly(3-methyl thiophene) and in situ polymerized poly(3-methyl thiophene) cerium(IV) phosphate cation exchange nanocomposite. Compos Part B 56:862–868
6. Khan AA, Baig U (2013) Electrical conductivity and ammonia sensing studies on in situ polymerized poly(3-methythiophene)–titanium(IV) molybdophosphate cation exchange nanocomposite. Sens Actuators, B 177:1089–1097
7. Khan AA, Baig U (2011) Ammonia vapor sensing properties of polyaniline–titanium(IV) phosphate cation exchange nanocomposite. J Hazard Mater 186:2037–2042
8. Shakir M, Iram NE, Khan MS, Al-Resayes SI, Khan AA, Baig U (2014) Electrical conductivity, isothermal stability, and ammonia-sensing performance of newly synthesized and characterized organic-inorganic polycarbazole–titanium dioxide nanocomposite. Ind Eng Chem Res 53:8035–8044
9. Yuvaraj H, Woo MH, Park EJ, Jeong YT, Lim KT (2008) Polypyrrole/gamma Fe$_2$O$_3$ magnetic nanocomposite synthesized in superficial fluide. Eur Polym J 44:637–644
10. Nandi D, Gupta K, Ghosh AK, De A, Ray NR, Ghosh UC (2013) Thermally stable polypyrrole–Mn doped Fe(III) oxide nanocomposite sandwiched in graphene layer: synthesis, characterization with tunable electrical conductivity. Chem Eng J 220:107–116
11. Patil DS, Pawar SA, Devan RS, Gang MG, Ma YR, Kim JH, Patil PS (2013) Electrochemical supercapacitor electrode material based on polyacrylic acid/polypyrrole/silver composite Electrochim Acta 105:569–577
12. Wang S, Jiang SP, White T, Guo J, Wang X (2009) Electrocatalytic activity and interconnectivity of Pt nanoparticles on multiwalled carbon nanotubes for fuel cells. J Phys Chem C 113:18935–18945
13. Wang B, Li C, Pang J, Zhai J, Li Q (2012) Novel polypyrrole-sensitized hollow TiO$_2$/fly ash cenospheres: synthesis, characterization, and photocatalytic ability under visible light. Appl Surf Sci 24:9989–9996
14. Peng S, Wu Y, Zhu P, Thavasi V, Mhaisalkar SG, Ramakrishna S (2011) Facile fabrication of polypyrrole/functionalized multi-walled carbon nanotubes composite as counter electrodes in low-cost dye-sensitized solar cells. J Photochem Photobiol A Chem 223:97–102
15. Cervantes SA, Roca MI, Martinez JG, Olmo LM, Cenis JL, Moraleda JM, Otero TF (2012) Bioelectrochemistry 85:36–43
16. Hong L, Li Y, Yang M (2010) Fabrication and ammonia gas sensing of palladium/polypyrrole nanocomposite. Sens Actuators B Chem 145:25–31
17. Nowicka AM, Fau M, Rapecki T, Donten M (2014) Polypyrrole-Au nanoparticles composite as suitable platform for DNA biosensor with electrochemical impedance spectroscopy detection. Electrochim Acta 140:65–71
18. Deivanayaki S, Ponnuswamy V, Mariappan R, Jayamurugan P (2013) Synthesis and characterization of polypyrrole/TiO$_2$ composites by chemical oxidative method. Optik 124:1089–1091
19. Grover S, Shekhar S, Sharma RK, Singh G (2014) Multiwalled carbon nanotube supported polypyrrole manganese oxide composite supercapacitor electrode: role of manganese oxide dispersion in performance evolution. Electrochim Acta 116:137–145
20. Khan AA, Baig U (2012) Polyacrylonitrile-based organic-inorganic composite anion-exchange membranes: preparation, characterization and its application in making ion-selective membrane electrode for determination of As(V). Desalination 289:21–26
21. Khan AA, Baig U (2012) Electrically conductive membrane of polyaniline–titanium(IV) phosphate cation exchange nanocomposite: applicable for detection of Pb(II) using its ion-selective electrode. J Ind Eng Chem 18:1937–1944
22. Parvatikar N, Jain S, Bhoraskar SV, Prasad MVNA (2006) Spectroscopic and electrical properties of polyaniline/CeO$_2$ composites and their application as humidity sensor. J Appl Polym Sci 102:5533–5537
23. Parvatikar N, Jain S, Khasim S, Revansiddappa M, Bhoraskar SV, Prasad MVNA (2006) Electrical and humidity sensing properties of polyaniline/WO$_3$composites. Sens Actuators B Chem 114:599–603
24. Sun L, Boo WJ, Sue HJ, Clearfield A (2007) Preparation of α-zirconium phosphate nanoplatelets with wide variations in aspect ratios. New J Chem 31:39–43
25. Bose S, Kuila T, Uddin ME, Kim NH, Lau AKT, Lee JH (2010) In-situ synthesis and characterization of electrically conductive polypyrrole/graphene nanocomposites. Polymer 51:5921–5928
26. Khan AA, Khalid M, Niwas R (2010) Humidity and ammonia vapor sensing applications of polyaniline–polyacrylonitrile composite films. Sci Adv Mater 2:474–480
27. Dean JA (1998) Lange's handbook of chemistry, 15th edn. Mc Graw-Hill Inc, pp 528–529
28. Jang WK, Yun J, Kim HI, Lee YS (2013) Improvement of ammonia sensing properties of polypyrrole by nanocomposite with graphitic materials. Colloid Polym Sci 291:1095–1103
29. Yeole B, Sen T, Hansora D, Mishra S (2016) Polypyrrole/metal sulphide hybrid nanocomposites: synthesis, characterization and room temperature gas sensing properties. Mater Res 19:999–1007
30. Laidler KJ (1965) Chemical kinetics. McGraw-Hill, New York, p 556

19

Study of corrosion inhibition of C38 steel in 1 M HCl solution by polyethyleneiminemethylene phosphonic acid

Merah Salah[1,2] · Larabi Lahcène[1] · Abderrahim Omar[3] · Harek Yahia[1]

Abstract A new class of corrosion inhibitors, namely, polyethyleneiminemethylene phosphonic acid (PEIMPA), was synthesized and its inhibiting action on the corrosion of C38 steel in 1 M HCl at 30 °C was investigated by various corrosion monitoring techniques such as weight loss measurements, potentiodynamic polarization, linear polarization resistance (Rp), and surface analysis (SEM and EDX) which are used to characterize the steel surface. Weight loss measurements revealed that the presence of PEIMPA increases the inhibition efficiency by decreasing the corrosion rate. Tafel polarization study showed that the inhibitor acts as a mixed-type inhibitor. Adsorption of PEIMPA on the carbon steel surface was found to obey the Langmuir isotherm. Some thermodynamic functions of dissolution and adsorption processes were also determined and discussed. The SEM results showed the formation of protective film on the mild steel surface in the presence of PEIMPA. The results obtained from different tested techniques were in good agreement.

Keywords Corrosion · Inhibition · C38 steel · Phosphonic acid

✉ Merah Salah
merrah2005@yahoo.fr

[1] Laboratory of Analytical Chemistry and Electrochemistry, Department of Chemistry, Faculty of Science, Tlemcen University, Tlemcen, Algeria

[2] Department of Process Engineering, Faculty of Technology, Saïda University, Saïda, Algeria

[3] Laboratory of Separation and Purification Technology, Department of Chemistry, Faculty of Science, Tlemcen University, Tlemcen, Algeria

Introduction

Study of organic corrosion inhibitor is an attractive field of research due to its usefulness in various industries. Acid is widely used in various industries for the pickling of ferrous alloys and steels. Because of the aggressive nature of the acid medium, the inhibitors are commonly used to reduce acid attack on the substrate metal. Most of the reported corrosion inhibitors are organic compounds containing O, N, S, and P [1–14] in their structures. The phosphoric functions are considered to be the most effective chemical group against corrosion process [15]. The use of organic phosphonic acids to protect carbon steel against corrosion has been the subject of various works [16–26]. Aminomethyl-phosphonic acids are excellent sequestering agents for electroplating, chemical plating, degreasing, and cleaning. It was shown that piperidin-1-yl-phosphonic acids (PPA) and (4-phosphono-piperazin-1-yl) phosphonic acid (PPPA) are used to reduce the corrosion of iron in a NaCl medium, even if PPPA is more efficient than PPA [27]. In the present investigation, the influence of polyethyleneiminemethylene phosphonic acid (PEIMPA) as a corrosion inhibitor of carbon steel in 1 M HCl has been systematically studied by weight loss measurements, potentiodynamic polarization studies, and surface analysis (SEM, EDX). Results are reported and discussed.

Experimental

Polyethyleneiminemethylene phosphonic acid polymer was synthesized (see Scheme 1) from commercially available Lupasol P (polyethylenimine) according to the Moedrizer–Irani reaction [28]. The synthesis was

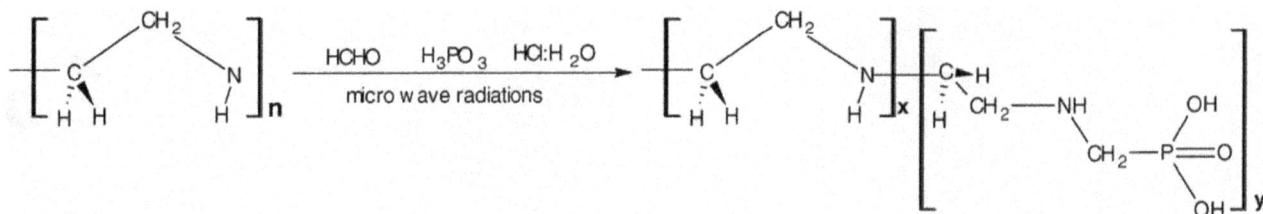

Scheme 1 Synthesis of polyethyleneiminemethylene phosphonic acid from Lupasol P

performed in distilled water under microwave irradiation. In a quartz reactor, a mixture of polyethylenimine (Lupasol P, 80 mmol, 3.44 g), phosphorous acid (80 mmol, 6.68 g), and hydrochloric acid–water (1:1) solution (12 mL) was vigorously stirred and then irradiated (150 W) in a glass cylinder reactor for 1 min. A formaldehyde aqueous solution (160 mmol) was added and irradiated for 8 min.

Then, the precipitation was washed with distilled water to remove unreacted reagents. Finally, phosphonic-modified Lupasol P was washed three times with distilled water and ethanol. After drying, the solid was further pulverized to give a brown powder.

The structure and purity were identified and characterized by elemental microanalysis (Table 1) and ^1H, ^{13}C, and ^{31}P NMR spectroscopy. The spectra showed the expected signals due to the polyethylenimine skeleton and methylene phosphonic units as matched to the proposed structure (Scheme 1).

NMR spectral data: ^1H NMR d (ppm): 4.92 ($N–CH_2$); 2.33 ($CH_2–P$); 1.6 NH. ^{13}C NMR d (ppm): 82.16 ($N–CH_2$), 52.1 ($CH_2–P$). ^{31}P NMR d (ppm): 3.91. The presence of phosphonic acid was confirmed by FTIR measurement: the polymer displays characteristic bonds for P–O–C at 1050 cm^{-1}, P–OH at 2372, and 2338 cm^{-1} 189 and P = O at 1172 cm^{-1}.

Elemental microanalysis suggests the structure made of fragment of the phosphonic acid polymer, corresponding after calculation to $x = 5$ and $y = 9$ (Scheme 1).

A 1 M HCl solution was prepared from an analytical reagent grade of HCl 37% and double-distilled water and was used as corrosion media in the studies. Note that the solubility of polyethyleneiminemethylene phosphonic acid is very high in this medium.

For the weight loss measurements, the experiments were carried out in the solution of 1 M HCl (uninhibited and inhibited) on carbon steel containing 0.30–0.35% C, 0.15–0.35% Si, 0.035% S, 0.5–1.0% Mn, and 0.035% P.

Sheets with dimensions 20 mm × 10 mm × 2 mm were used. They were polished successively with different grades of emery paper up 1200 grade. Each run was carried out in a glass vessel containing 100 ml test solution. A clean weight mild steel sample was completely immersed at an inclined position in the vessel. After 4 h of immersion in 1 M HCl with and without the addition of inhibitor at different concentrations, the specimen was withdrawn, rinsed with double-distilled water, washed with acetone, dried, and weighed. The weight loss was used to calculate the corrosion rate in milligrams per square centimeter per hour.

Electrochemical experiments were carried out in a glass cell (CEC/TH Radiometer) with a capacity of 500 ml. A platinum electrode and a saturated calomel electrode (SCE) were used as a counter electrode and a reference electrode. The working electrode was in the form of a disc cut from mild steel under investigation and was embedded in a Teflon rod with an exposed area of 0.5 cm^2. Potentiodynamic polarizations were conducted in an electrochemical measurement system (VoltaLab 21) which comprises a PGP201 potentiostat, a personal computer, and VoltaMaster4 software. The polarization resistance measurements were performed by applying a controlled potential scan over a small range typically ±15 mV with respect to Ecorr with a scanning rate of 0.5 mV s^{-1}. The resulting current is linearly plotted vs. potential, the slope of this plot at Ecorr being the polarization resistance (Rp). Corrosion current densities were determined by extrapolating the cathodic Tafel regions from the potentiodynamic polarization curves to the corrosion potential. The potentiodynamic current–potential curves were recorded by changing the electrode potential automatically from −700 to −300 mV with the same scanning rate (0.5 mV s^{-1}) under static on the same electrode without any surface treatment. All experiments were carried out in freshly prepared solution at constant temperatures.

Table 1 Elemental microanalysis of polyethyleneiminemethylene phosphonic acid	Microanalysis	%C	%H	%N	%O	%P
	Found	30.8112	7.6250	13.2519	29.3535	18.9574
	Calculated ($x = 5$, $y = 9$)	30.6631	6.6989	13.5359	29.8342	19.2679

Inhibition efficiencies P % were calculated as follows:

Weight loss measurement:

$$P\% = \frac{w - w'}{w} \times 100 \qquad (1)$$

where w and w' are the corrosion rate of steel due to the dissolution in 1 M HCl in the absence and the presence of definite concentrations of inhibitor, respectively.

Linear polarization measurement:

$$P\% = \frac{R'_p - R_p}{R'_p} \times 100 \qquad (2)$$

where R_p and R'_p are the values of linear polarization in the absence and presence of the inhibitor, respectively.

Polarization measurement:

$$P\% = \frac{I_{corr} - I'_{corr}}{I_{corr}} \times 100 \qquad (3)$$

where i_{corr} and i'_{corr} are the corrosion current densities in the absence and the presence of the inhibitor, respectively.

For all methods, the tests were performed in non-deaerated solutions under unstirred conditions.

The surface morphology of the samples before and after adding PEIMPA inhibitor in the medium 1 M HCl after 1 day of immersion was observed by scanning electron microscope (SEM) Quanta 200 FEG coupled with EDX analysis.

Results and discussion

Weight loss measurements

The gravimetric measurements of mild steel in 1 M HCl in the absence and presence of various concentrations of PEIMPA investigated were determined after 4 h of immersion at 30 °C.

Table 2 gives values of the corrosion rates and percentage inhibition efficiency calculated from the weight loss measurements for different concentrations of PEIMPA.

Table 2 Corrosion rates and inhibition efficiencies of PEIMPA at different concentrations in 1 M HCl

Conc. (ppm)	V_{corr} (mg cm^{-2} h^{-1})	P (%)
1 M HCl	0.703	–
100	0.195	72.15
200	0.120	82.93
300	0.104	85.07
400	0.093	86.69
500	0.069	90.11

Inspection of this table shows that the inhibition efficiency increases with increasing inhibitor concentration. The optimum concentration required to achieve this efficiency is found to be 500 ppm. The inhibition of corrosion of carbon steel by the investigated inhibitor can be explained in terms of adsorption on the metal surface. It is generally assumed that the adsorption of the inhibitor at the metal/solution interface is the first step in the mechanism of inhibition in aggressive media.

This compound can be adsorbed on the metal surface by the interaction between lone pair of electrons of hetero atoms and the metal surface. This process is facilitated by the presence of vacant orbitals d of low energy in iron atom, as observed in the transition group metals. Moreover, the formation of positively charged protonated species in acidic solutions facilitates the adsorption of the compound on the metal surface through electrostatic interactions between the organic molecules and the metal surface [29].

Polarization measurements

Figure 1 shows the polarization curves of mild steel in 1 M HCl, blank solution, and in the presence of different concentrations (100–500 ppm) of PEIMPA. With the increase of PEIMPA concentrations, both anodic and cathodic currents were inhibited. This result shows that the addition of PEIMPA inhibitor reduces anodic dissolution and also retards the hydrogen evolution reaction. We note that the corrosion potential varies slightly after the addition of the inhibitor at different concentrations.

Table 3 shows that an increase in inhibitor concentration is resulted in increased inhibition efficiency. It is evident from the results that the I_{corr} values decrease considerably in the presence of inhibitor and that the maximum decrease in I_{corr} coincides with the optimum concentration of

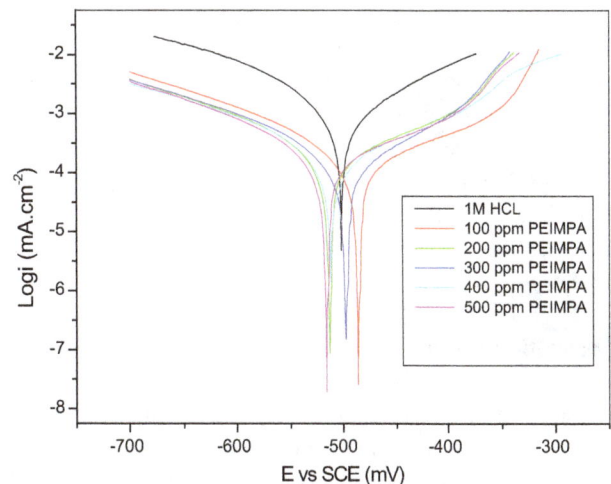

Fig. 1 Polarization curves of carbon steel in 1 M HCl in the presence of different concentrations of PEIMPA at 30 °C

Table 3 Potentiodynamic polarization parameters for corrosion of carbon steel in 1 M HCl with various concentrations of PEIMPA at 30 °C

Conc. (ppm)	E_{corr} vs. SCE (mV)	i_{corr} (mA.cm^{-2})	R_p (Ω.cm^2)	b_c (mV dec^{-1})	P (i_{corr}) (%)	P (R_p) (%)
1 M HCl	−501	1.94	12.98	156	–	–
100	−486	0.567	68.13	143	70.77	80.94
200	−512	0.359	73.54	146	81.49	82.34
300	−497	0.343	81.77	186	82.31	84.12
400	−511	0.336	99.04	134	82.68	86.89
500	−515	0.302	104.50	153	84.43	87.57

inhibitor. Linear polarization technique was performed in 1 M HCl with various concentrations of PEIMPA. The corresponding polarization resistance (R_p) values of carbon steel in the absence and in the presence of different inhibitor concentrations are also given in Table 3. It is apparent that R_p increases with increasing inhibitor concentration. The inhibition percentage (P %) calculated from R_p values is also presented in Table 3. We remark that P % increases with increasing concentration of inhibitor and attains 87% at 500 ppm. The inhibition efficiencies of PEIMPA obtained by potentiodynamic polarization and by polarization resistance methods are in good agreement, particularly, at high concentrations.

For anodic polarization, it can be seen from Fig. 1 that, in the presence of PEIMPA at all concentrations, two linear portions were observed. When the anodic potentials increases, the anodic current increases at a slope of b_{a1} in the low polarization potential region. After passing a certain potential E_u, the anodic current increases rapidly and dissolves at a slope of b_{a2} in the high polarization region. This behavior was already documented for iron in acid solutions [30–33]. The rapid increase of anodic current after E_u may be due to the desorption of PEIMPA molecules adsorbed on the electrode. This means that the inhibition mode of PEIMPA depends on electrode potential. In this case, the observed inhibition phenomenon is generally described as corrosion inhibition of the interface associated with the formation of a bidimensional layer of adsorbed inhibitor species at the electrode surface [34]. Note that the potential E_u is also denoted E_1 in Bartos and Hackerman's paper [30]. Figure 1 shows also that, at potentials higher than E_{corr}, PEIMPA affects the anodic reaction. This result indicates that PEIMPA exhibits both anodic and cathodic inhibition effects.

Adsorption isotherm

The adsorption of the organic compounds can be described by two main types of interaction: physical adsorption and chemisorption that are influenced by the charge nature of the metal, the type of the electrolyte, and the chemical structure of the inhibitor.

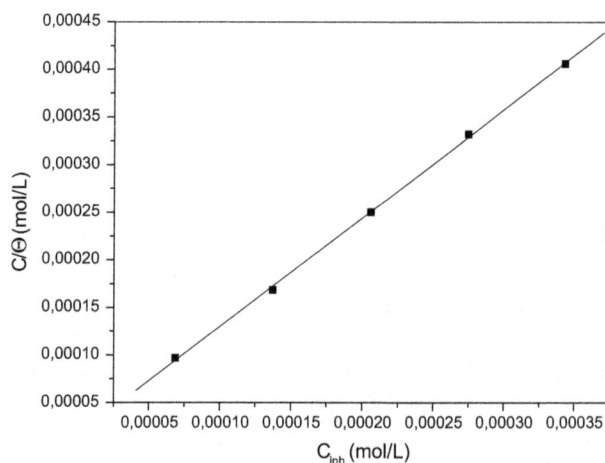

Fig. 2 Langmuir adsorption isotherm of PEIMPA on the carbon steel surface in 1 M HCl from potentiodynamic measurements

The adsorption isotherm can give information on the metal–inhibitor interaction. The adsorption isotherm can be derived from the curve surface coverage against inhibitor concentration. Surface coverage θ was estimated as in Eq. 5. The θ values for different inhibitor concentrations are tested by fitting to various isotherms. So far, the best fit was obtained with the Langmuir isotherm. According to this isotherm, θ is related to concentration inhibitor C via

$$\frac{C}{\theta} = \frac{1}{K} + C \tag{4}$$

where C is the concentration of inhibitor, K is the adsorptive equilibrium constant, and θ is the surface coverage and calculated by the following equation [6, 35, 36]:

$$\theta = 1 - \frac{I_{corr}}{I'_{corr}} \tag{5}$$

where I'_{corr} is the corrosion current density in uninhibited acid and I_{corr} is the corrosion current density in inhibited acid.

Plotting C/θ vs. C yields a straight line, as shown in Fig. 2. The linear correlation coefficient (r) is almost equal to 1 ($r = 0.9999$) and the slight deviation of the slope

(a)

(b)

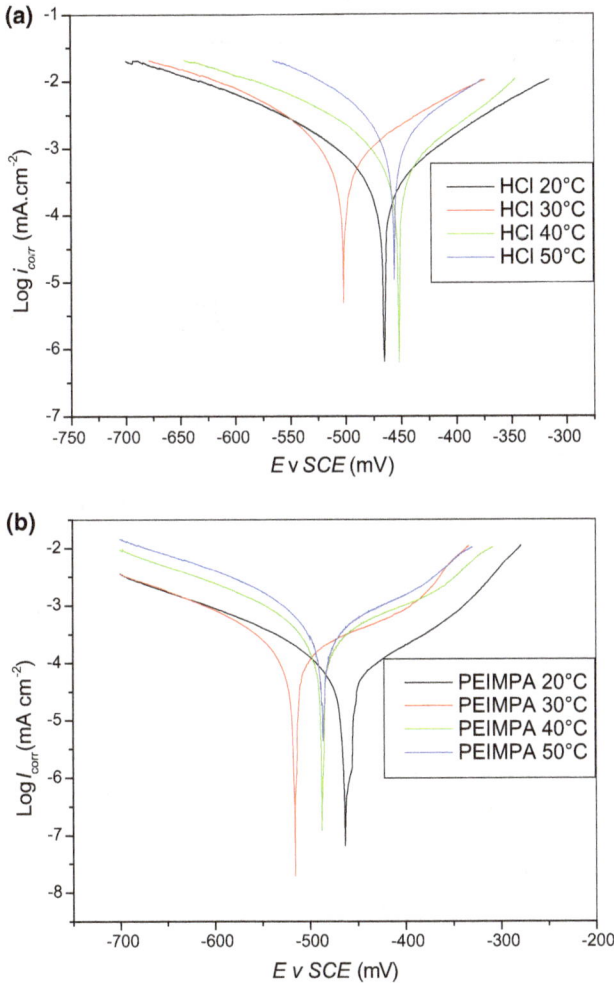

Fig. 3 Polarization curves for C38 steel electrode in 1 M HCl (**a**) and in HCl+ 500 ppm of PEIMPA (**b**) at different temperatures

(1.14) from the unity is attributable to molecular interactions in the adsorbed layer which corresponds to the observed physical adsorption mechanism [37]. The adsorptive equilibrium constant (K) value is 6.22×10^5 L mol^{-1}. The free energy of adsorption ΔG°_{ads} of the inhibitor on mild steel surface can be determined using the following relation:

$$K = \frac{1}{55.5} \exp\left(\frac{\Delta G^\circ_{ads}}{RT}\right) \qquad (6)$$

where R is the gas constant (8.314 J K^{-1} mol^{-1}), T is the absolute temperature (K), and the value 55.5 is the concentration of water in solution expressed in M. The ΔG°_{ads} value calculated is -37.90 kJ mol^{-1}. The negative values of ΔG°_{ads} indicate that the adsorption of inhibitor molecule onto steel surface is a spontaneous process. In general, it is well known that the values of $-\Delta G^\circ_{ads}$ of the order of -20 kJ mol^{-1} or lower indicate a physisorption; those of order of -40 kJ mol^{-1} or higher involve charge sharing or

a transfer from the inhibitor molecules to the metal surface to form a coordinate type of bond (chemisorption). In the present study, the value of ΔG°_{ads} is about between -20 and -40 kJ mol^{-1}, probably mean that the adsorption mechanism of the PEIMPA on steel in 1 M HCl solution is both physisorption and chemisorption.

Noticeably, it is generally accepted that physical adsorption is the preceding stage of chemisorption of inhibitors on metal surface [38]. It was reported that in this domain, the surface charge of steel at E_{corr} in HCl solution is expected to be positive. Thus, the anions are first adsorbed on the steel surface creating an excess negative charge, which, in turn, facilitates physical adsorption of the inhibitor cations [39]. Accordingly, the Cl$^-$ and phosphonate ions adsorb and the surface becomes negatively charged. Due to the electrostatic attraction, the protonated PEIMPA molecules are adsorbed on carbon steel surface (physisorption).

Along with electrostatic force of attraction, inhibitor also adsorbs on the carbon steel surface through chemical adsorption. The adsorption of free molecules could take place via interaction of the unshared pairs of electrons of nitrogen and oxygen atoms of the –PO(OH)$_2$ group and the vacant d-orbitals of iron atoms.

Therefore, inhibition takes place through both physisorption and chemisorption. However, chemisorption has no substantial contribution [12]. Indeed, the PEIMPA molecules are easily protonated to form ionic forms in acid solution. It is logical to assume that in this case, the electrostatic cation adsorption is mainly responsible for the protective properties of this compound.

Effect of temperature

To investigate the mechanism of inhibition and to determine the activation energy of the corrosion process, polarization curves of steel in 1 M HCl were determined at various temperatures (303–333 K) in the absence and presence of 500 ppm of PEIMPA. Representative Tafel polarization curves for C38 steel electrode in 1 M HCl without and with 500 ppm at different temperatures are shown in Fig. 3a, b. Similar polarization curves were obtained in the case of the other concentrations of PEIMPA (not given). The analysis of these figures reveals that raising the temperature increases both anodic and cathodic current densities, and consequently, the corrosion rate of C38 steel increases. The corresponding data are given in Table 4. In the studied temperature range (303–333 K), the corrosion current density increases with increasing temperature both in uninhibited and inhibited solutions and the values of the inhibition efficiency of PEIMPA decrease with the increase of temperature.

Table 4 Electrochemical parameters and the corresponding inhibition efficiencies for the corrosion of C38 in 1 M HCl and 500 ppm PEIMPA at various temperatures

Milieu	$T/(°C)$	$E_{corr}/(mVvSCE)$	$i_{corr}/(mA/cm^2)$	$b_c/(mV/dec)$	$P/(\%)$
1 M HCl	20	−466	1.46	198	–
	30	−501	1.94	156	–
	40	−459	2.83	193	–
	50	−457	4.01	196	–
500 ppm PEIMPA	20	−462	0.160	180	89.04
	30	−515	0.302	134	84.34
	40	−486	0.562	175	80.15
	50	−485	1.052	188	73.81

Figure 4 and 5 present the Arrhenius plots of the natural logarithm of the current density vs. 1/T, for 1 M solution of hydrochloride acid, without and with the addition of PEIMPA and $\ln(i_{corr}/T)$ with reciprocal of the absolute temperature, respectively. Straight lines with coefficients of correlation (c.c.) high to 0.99 are obtained for the supporting electrolyte and PEIMPA.

The values of the slopes of these straight lines permit the calculation of the Arrhenius activation energy, E_a, according to

$$\ln I_{corr} = -\frac{E_a}{RT} + \ln A \qquad (7)$$

where R is the universal gas constant and A is the Arrhenius factor.

In addition, from transition-state plot according to the following equation:

$$\ln\left(\frac{I_{corr}}{T}\right) = \frac{-\Delta H_a^°}{RT} + B \qquad (8)$$

where $\Delta H_a^°$ is the enthalpy of activation and B is a constant.

The E_a and $\Delta H_a^°$ values were determined from the slopes of these plots. The calculated values of E_a and ΔH_a

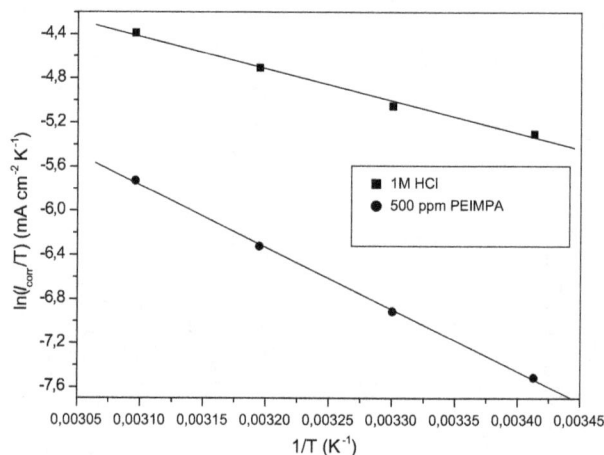

Fig. 5 $\ln(i_{corr}/T)$ vs. 1/T for C38 steel dissolution in 1 M HCl in the presence of PEIMPA at 500 ppm

in the absence and the presence of PEIMPA at 500 ppm are given in Table 5. Inspection of these data reveals that the increase in E_a in the presence of the inhibitor may be interpreted as physical adsorption. Indeed, a higher energy barrier for the corrosion process in the inhibited solution is associated with physical adsorption or weak chemical bonding between the inhibitor species and the steel surface [22, 40]. Szauer and Brand explained that the increase in activation energy can be attributed to an appreciable decrease in the adsorption of the inhibitor on the carbon steel surface with the increase in temperature. A corresponding increase in the corrosion rate occurs because of the greater area of metal that is consequently exposed to the acid environment [41].

The enthalpy of activation values is found to be positive in the absence and presence of inhibitor and reflects the endothermic mild steel dissolution process. It is evident from Table 5 that the value of $\Delta H_a^°$ increased in the presence of PEIMPA than the uninhibited solution indicating protection efficiency. This suggested the slow dissolution and hence lower corrosion rate of mild steel [2, 7]. This result permits verifying the known thermodynamic equation between E_a and $\Delta H_a^°$ [42]:

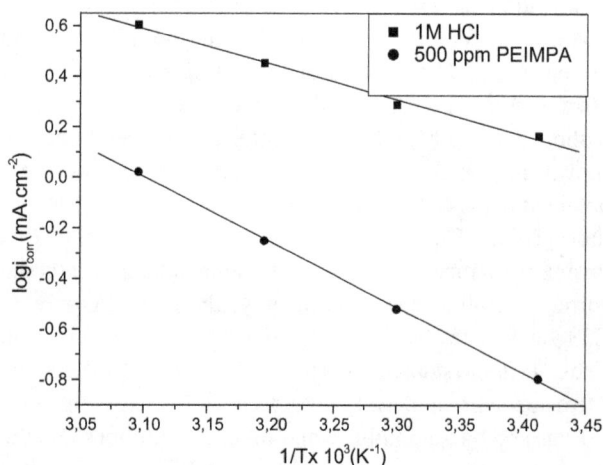

Fig. 4 Arrhenius *plots* of log icorr vs. 1/T without and with 500 ppm of PEIMPA

Table 5 Apparent activation energy E_a and activation enthalpy ΔH°_a of dissolution of C38 steel in 1 M HCl in the absence and presence of 500 ppm PEIMPA

Milieu	E_a (kJ mol^{-1})	$\Delta H^\circ a$ (kJ mol^{-1})	$E_a - \Delta H^\circ a$ (kJ mol^{-1})
1 M HCl	26.82	24.21	2.61
500 ppm PEIMPA	49.32	46.71	2.61

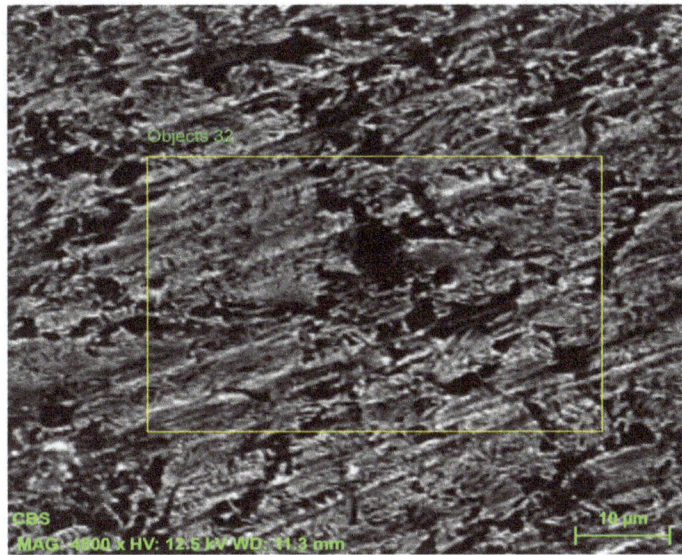

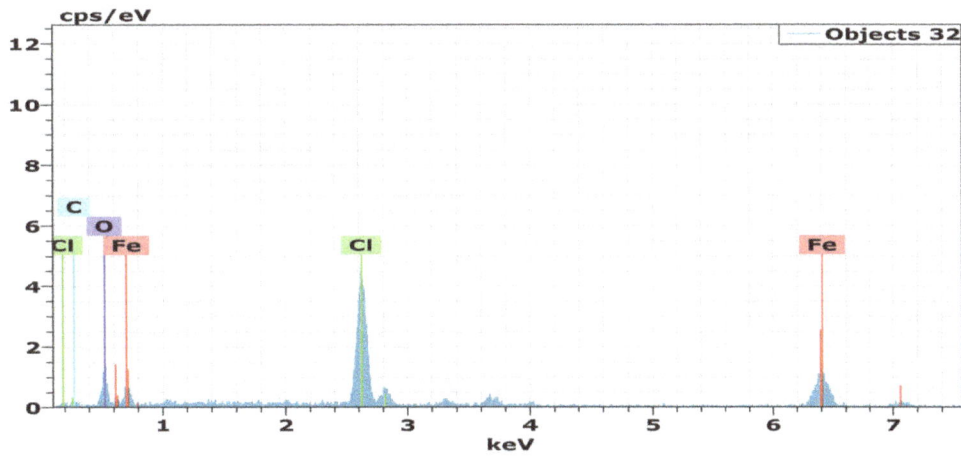

element	% masses
iron	88.74
carbon	4.60
oxygen	4.13
chlorine	0.41

Fig. 6 SEM and EDX data of C38 steel immersed in 1 M HCl solution for 24 h

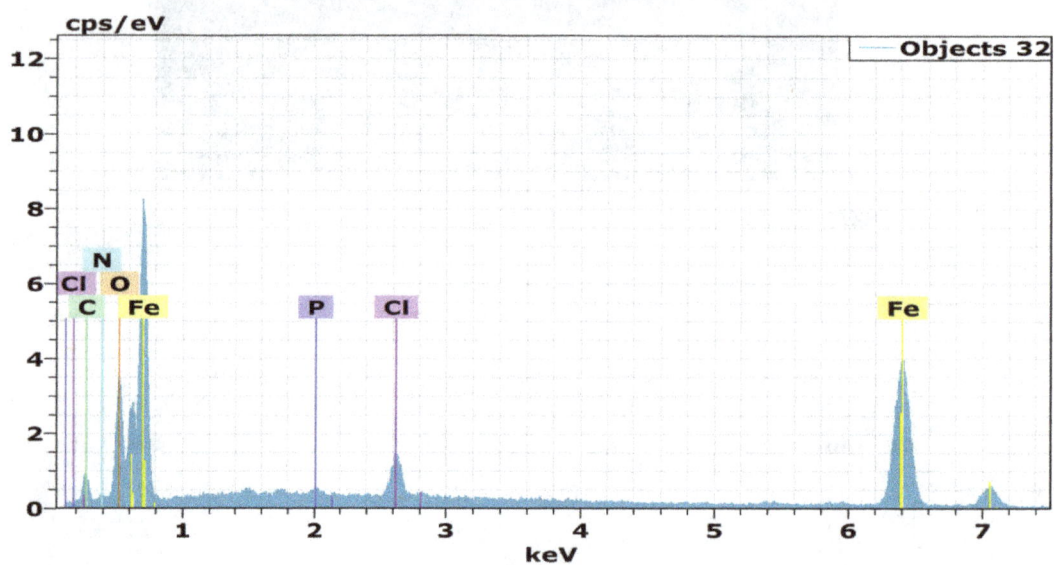

element	% masses
iron	76.36
oxygen	7.20
carbon	4.01
chlorine	2.76
Phosphor	0.30
nitrogen	0.50

Fig. 7 SEM and EDX data of C38 steel immersed in 1 M HCl+ 500 ppm PEIMPA for 24 h

$$E_a - \Delta H \circ a = RT. \tag{9}$$

Surface examination by SEM/EDX

Figure 6 shows the scanning electron micrographs of C38 after immersion for 24 h in 1 M HCl solution. The specimen surface in Fig. 6 appears to be roughened extensively by the corrosive environment and the porous layer of corrosion product is present. The EDX spectra show the characteristics peaks of some of the elements constituting of the steel sample after 24 h immersion in 1 M HCl without inhibitor, which reveals the presence of oxygen and iron, suggesting, therefore, the presence of iron oxide/hydroxide.

When PEIMPA was added into the corrosion test solution (Fig. 7), a smooth surface was noticed traducing a good protection effect of the corrosion inhibitor by a formation of a thick and compact film. This may be intercepted by the adsorption of these inhibitors on the electrode surface.

In the presence of PEIMPA inhibitor, the EDX spectra show additional lines of nitrogen and phosphorus, due to the adsorbed layer of inhibitor that covered the electrode surface. In addition, the Fe peaks are decreasing in relation to the uninhibited steel surface sample. This diminution of the Fe lines occurs because of the overlying inhibitor film.

These results confirm those from weight loss and polarization measurements, which suggest that a protective film was formed over the metal surface and hence retarded both anodic and cathodic reactions.

Conclusion

On the basis of this study, the following conclusions can be drawn:

- PEIMPA shows a very good activity of preventing corrosion of carbon steel in 1 M HCl.
- Inhibition efficiency of PEIMPA varies directly with the concentration and inversely with the temperature.
- Potentiodynamic polarization experiments reveal that PEIMPA acts as mixed-type inhibitor.
- The adsorption of PEIMPA on the metal surface follows Langmuir isotherm.
- The corrosion inhibition is probably due to the adsorption of PEIMPA on the metal surface and blocking its active sites by the phenomenon of physical and chemical adsorptions.
- The weight loss, linear polarization, and polarization curves are in good agreement.

- SEM and EDX techniques reveal that the inhibitor molecules form a good protective film on the steel surface and confirm the result obtained by gravimetric and polarization methods.

References

1. Larabi L, Harek Y, Benali O, Ghalem S (2005) Hydrazide derivatives as corrosion inhibitors for mild steel in 1 M HCl. Prog Org Coat 54:256–262
2. Benali O, Larabi L, Tabti B, Harek Y (2005) Influence of 1-methyl 2-mercapto imidazole on corrosion inhibition of carbon steel in 0.5 M H$_2$SO$_4$. Anti Corros Method Mater 52:280–285
3. Benali O, Larabi L, Mekelleche SMB, Harek Y (2006) Influence of substitution of phenyl group by naphthyl in a diphenylthiourea molecule on corrosion inhibition of cold-rolled steel in 0.5 M H$_2$SO$_4$. J Mat Sci 41:7064–7073
4. Larabi L, Benali O, Harek Y (2007) Corrosion inhibition of cold rolled steel in 1 M HClO$_4$ solutions by N-naphtyl N'-phenylthiourea. Mat Lett 61:3287–3291
5. Benali O, Larabi L, Traisnel M, Gengembre L, Harek Y (2007) Electrochemical, theoretical and XPS studies of 2-mercapto-1-methylimidazole adsorption on carbon steel in 1 M HClO$_4$. Appl Surf Sci 253:6130–6139
6. Benali O, Larabi L, Merah S, Harek Y (2011) Influence of the methylene blue dye (MBD) on the corrosion inhibition of mild steel in 0.5 M sulphuric acid, part I: weight loss and electrochemical studies. J Mater Environ Sci 2:39–48
7. Sabirneeza AAF, Subhashini S (2014) Poly(vinyl alcohol–proline) as corrosion inhibitor for mild steel in 1 M hydrochloric acid. Int J Ind Chem 5:111–120
8. Touhami F, Aouniti A, Kertit S, Abed Y, Hammouti B, Ramdani A, El-Kacemi K (2009) Corrosion inhibition of armco iron in 1 M HCl media by new bipyrazolic derivatives. Corros Sci 42:929–940
9. Bouklah M, Hammouti B, Aouniti A, Benhadda T (2004) Thiophene derivatives as effective inhibitors for the corrosion of steel in 0.5 M H$_2$SO$_4$. Prog Org Coat 47:225–228
10. Bouklah M, Hammouti B, Lagrenée M, Bentiss F (2006) Thermodynamic properties of 2,5-bis(4-methoxyphenyl)-1,3,4-oxadiazole as a corrosion inhibitor for mild steel in normal sulfuric acid medium. Corros Sci 48:2831–2842
11. Kertit S, Essoufi H, Hammouti B, Benkaddour B (1998) 1-phenyl-5-mercapto-1,2,3,4-tétrazole (PMT): un nouvel inhibiteur de corrosion de l'alliage Cu-Zn efficace à très faible concentration. J Chim Phys 95:2072–2082
12. Labjar N, Lebrini M, Bentiss F, Chihib NE, Hajjaji SE, Jama C (2010) Corrosion inhibition of carbon steel and antibacterial properties of aminotris-(methylenephosphonic) acid. Mater Chem Phys 119:330–336
13. Amar H, Benzakour J, Derja A, Villemin D, Moreau B, Braisaz T (2006) Piperidin-1-yl-phosphonic acid and (4-phosphono-piperazin-1-yl) phosphonic acid: a new class of iron corrosion inhibitors in sodium chloride 3% media. Appl Surf Sci 252:6162–6172
14. Amar H, Benzakour J, Derja A, Villemin D, Moreau BJ (2003) A corrosion inhibition study of iron by phosphonic acids in sodium chloride solution. J Electroanal Chem 558:131–139
15. Truc TA, Pebere N, Hang TTX, Hervaud Y, Boutevin B (2002) Study of the synergistic effect observed for the corrosion protection of a carbon steel by an association of phosphates. Corros Sci 44:2055–2071

16. Andijani I, Turgoose S (2003) Studies on corrosion of carbon steel in deaerated saline solutions in presence of scale inhibitor. Desalination 123:223–231

17. Rajendran S, Reenkala SM, Anthony N, Ramaraj R (2002) Synergistic corrosion inhibition by the sodium dodecylsulphate–Zn^{2+} system. Corros Sci 44:2243–2252

18. To XH, Pebere N, Pelaprat N, Boutevin B, Hervaud Y (1997) A corrosion-protective film formed on a carbon steel by an organic phosphonate. Corros Sci 39:1925–1934

19. Rajendran S, Apparao BV, Palaniswamy N (1999) Synergistic effect of 1-hydroxyethane-1, 1-diphosphonic acid and Zn2+ on the inhibition of corrosion of mild steel in neutral aqueous environment. Anti Corros Method Mater 46:23–28

20. Pech MA, Chi-Canul LP (1999) Investigation of the Inhibitive Effect of N-phosphono-methyl-glycine on the corrosion of carbon steel in neutral solutions by electrochemical techniques. Corrosion 55:948–956

21. Nakayama N (2000) Inhibitory effects of nitrilotris(methylenephosphonic acid) on cathodic reactions of steels in saturated $Ca(OH)_2$ solutions. Corros Sci 42:1897–1920

22. Gunasekharan G, Natarajan R, Palaniswamy N (2001) The role of tartrate ions in the phosphonate based inhibitor system. Corros Sci 43:1615–1626

23. Rajendran S, Apparao BV, Palaniswamy N, Periasamy V, Karthikeyan G (2001) Corrosion inhibition by strainless complexes. Corros Sci 43:1345–1354

24. Bouklah M, Krim O, Messali M, Hammouti B, Elidrissi A (2011) A pyrrolidine phosphonate derivative as corrosion inhibitor for steel in H2SO4 solution. Warad I Der Pharma Chim 3:283–293

25. Dkhireche N, Abdelhadi R, Ebn Touhami M, Oudda H, Touir R, Elbakri M, Sfaira M, Hammouti B, Senhaji O, Taouil R (2012) Elucidation of dimethyldodecylphosphonate and CTAB synergism on corrosion and scale inhibition of mild steel in simulated cooling water system. Int J Electrochem Sci 7:5314–5330

26. Kharbach Y, Haoudi A, Skalli MK, Kandri Rodi Y, Aouniti A, Hammouti B, Senhaji O, Zarrouk A (2015) The role of new phosphonate derivatives on the corrosion inhibition of mild steel in 1 M H2SO4 media. J Mater Environ Sci 6:2906–2916

27. Villemin D, Didi MA (2015) Aminomethylenephosphonic acids syntheses and applications (A Review). Orient J Chem 31:01–12

28. Ferrah N, Abderrahim O, Didi MA, Villemin D (2011) Removal of copper ions from aqueous solutions by a new sorbent: polyethyleneiminemethylene phosphonic acid. Desalination 269:17–24

29. Merah S, Larabi L, Benali O, Harek Y (2008) Synergistic effect of methyl red dye and potassium iodide on inhibition of corrosion of carbon steel in 0.5 M H2SO4. Pigm Resin Technol 37:291–298

30. Bartos M, Hackerman N (1992) A Study of inhibition action of propargyl alcohol during anodic dissolution of iron in hydrochloric acid. J Electrochem Soc 139:3428–3433

31. Kuo HC, Nobe KJ (1978) Electrodissolution kinetics of iron in chloride solutionsVI. Concentrated acidic solutions. J Electrochem Soc 125:853–860

32. Mac Farlane DR, Smedley SI (1986) The dissolution mechanism of iron in chloride solutions. J Electrochem Soc 133:2240–2244

33. Feng Y, Siow KS, Teo WK, Hseieh AK (1999) The synergistic effects of propargyl alcohol and potassium iodide on the inhibition of mild steel in 0.5 M sulfuric acid solution. Corros Sci 41:829–852

34. Lorentz WJ, Mansfeld F (1986) Interface and interphase corrosion inhibition. Corros Sci 31:467–476

35. Tsuru T, Haruyama S, Gijutsu B (1978) Corrosion inhibition of iron by amphoteric surfactants in 2 M HCl. J Jpn Soc Corros Eng 27:573–581

36. Xianghong L, Shuduan D, Hui F (2010) Blue tetrazolium as a novel corrosion inhibitor for cold rolled steel in hydrochloric acid solution. Corros Sci 52:2786–2792

37. Malki Alaoui L, Kertit S, Bellaouchou A, Guenbour A, Benbachir A, Hammouti B (2008) Phosphate of aluminum as corrosion inhibitor for steel in H3PO4. Portug Electroch Acta 26:339–347

38. Wang FP, Kang WL, Jin HM (2008) Corrosion electrochemistry mechanism, methods and applications. Chemical Industrial Engineering Press, Beijing, p 242

39. Popova A, Sokolova E, Raicheva S, Christov M (2003) AC and DC study of the temperature effect on mild steel corrosion in acid media in the presence of benzimidazole derivatives. Corros Sci 45:33–58

40. Bentiss F, Lebrini M, Lagrenee M (2005) Thermodynamic characterization of metal dissolution and inhibitor adsorption processes in mild steel/2,5-bis(n-thienyl)-1,3,4-thiadiazoles/hydrochloric acid system. Corros Sci 47:2915–2931

41. Szauer T, Brandt A (1981) On the role of fatty acid in adsorption and corrosion inhibition of iron by amine-fatty acid salts in acidic solution. Electrochim Acta 26:1219–1224

42. Laidler KJ (1963) Reaction kinetics, vol 1, 1st edn. Pergamon Press, New York

20

Properties of complex ammonium nitrate-based fertilizers depending on the degree of phosphoric acid ammoniation

Konstantin Gorbovskiy[1] · Anatoly Kazakov[2] · Andrey Norov[1] · Andrey Malyavin[1] ·Anatoly Mikhaylichenko[3]

Abstract Complex ammonium nitrate-based NP and NPK fertilizers are multicomponent salt systems prone to high hygroscopicity, caking and explosive thermal decomposition. The slurries that used in the production of these fertilizers can also exhibit insufficient thermal stability. One of the most important issues for such slurries is their viscosity, which determines the energy costs for transportation and processing into the final product. Increasing the degree of phosphoric acid ammoniation helps to reduce the ammonium nitrate's content in the product, but the main question remains about the properties of such fertilizers. This article is devoted to studying properties of complex NP and NPK ammonium nitrate-based fertilizers and their intermediates with increasing the degree of phosphoric acid ammoniation.

Keywords Ammonium nitrate-based fertilizer · Hygroscopicity · Caking · Microcalorimetry · Thermal decomposition · Slurry viscosity

✉ Konstantin Gorbovskiy
sulfur32@bk.ru

[1]   The Research Institute for Fertilizers and Insecto-Fungicides Named after Professor Y. Samoilov, 162622 Cherepovets, Vologda Region, Russia

[2]   Institute of Problems of Chemical Physics of the Russian Academy of Sciences, 142432 Chernogolovka, Moscow Region, Russia

[3]   D. Mendeleev University of Chemical Technology of Russia, 125047 Moscow, Russia

Introduction

Ammonium nitrate (AN) is one of the most common commercially available nitrogen fertilizers, the content of nitrogen in which amounts up to 35% by mass. The main agrochemical advantage of AN compared to other simple nitrogen fertilizers is to present nitrogen both in ammonia and nitrate forms. Herewith, the high content of this component enables to mix it with other types of fertilizers and obtain complex fertilizer with the high content of basic nutrients—nitrogen, phosphorus and potassium. The main disadvantages of such types of fertilizers are their high hygroscopicity, caking [1] and the increased requirements for fire and explosion safety [2]. All the above-mentioned factors, and in particular the last, are the main disadvantages limiting the production of complex AN-based fertilizers.

Cases of explosion of AN and complex AN-based fertilizers are well known: in 1921 in the warehouse in Oppau (Germany), in 1947 in the warehouse in the bay in Texas City (USA), in 2001 in the warehouse in Toulouse (France), in 2013 in the warehouse in West (USA). The largest explosion of technological installations was recorded in 1952 in Nagoya (Japan), in 1978—in Chirchik (Uzbekistan) in 1981—in Cherepovets (Russia), in 1994— in Port Neil (USA), in 2009—in Kirovo-Chepetsk (Russia).

Ammonium phosphates $NH_4H_2PO_4$ and $(NH_4)_2HPO_4$, ammonium sulfate and potassium chloride are also used in the production of complex AN-based NPK fertilizers. Herewith, the following reactions take place:

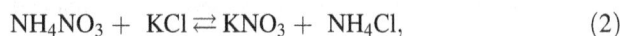

$$NH_4H_2PO_4 + KCl \rightleftharpoons KH_2PO_4 + NH_4Cl, \tag{1}$$

$$NH_4NO_3 + KCl \rightleftharpoons KNO_3 + NH_4Cl, \tag{2}$$

$$(NH_4)_2SO_4 + 2KCl \rightleftharpoons K_2SO_4 + 2NH_4Cl. \tag{3}$$

KH_2PO_4, KNO_3 and K_2SO_4 in combination with unreacted $NH_4H_2PO_4$, NH_4NO_3 and $(NH_4)_2SO_4$ (accordingly) form solid solutions—compounds of isomorphic-substituted type.

The composition of the solid solutions is determined by the extent of the conversion of the reactions (1–3). $(NH_4)_2HPO_4$ does not react with KCl. Moreover, AN can form various double salts: $NH_4NO_3 \cdot 2KNO_3$, $(NH_4)_2SO_4 \cdot 2NH_4NO_3$, $(NH_4)_2SO_4 \cdot 3NH_4NO_3$. Formation of $NH_4NO_3 \cdot 2KNO_3$ depends on the extent of the conversion of the reaction (2) [3]. The double salts $(NH_4)_2SO_4 \cdot 2NH_4NO_3$ and $(NH_4)_2SO_4 \cdot 3NH_4NO_3$ in the presence of KCl can decompose with the formation of solid solutions [4].

Thus, complex AN-based fertilizers are complex salt systems, whose composition is defined by the ratio of initial components.

The presence of all the above-mentioned compounds can variously affect the decomposition of complex AN-based fertilizers and their propensity for detonation. The presence of $NH_4H_2PO_4$, $(NH_4)_2HPO_4$ and $(NH_4)_2SO_4$ reduces the rate of AN decomposition [5, 6], and chloride-anions Cl^-, on the contrary, act as catalysts for AN decomposition [7–9].

Despite this, increasing demands of the agrochemical sector leads to the necessity to develop new grades of the fertilizers, the production of which is possible only when using concentrated nitrogen fertilizers, especially ammonium nitrate and urea. However, considerable difficulties emerge in case of urea used, which consist in high hygroscopicity and caking, reduction of the amide nitrogen proportion in the product due to decomposition of urea at relatively low temperatures during granulation and drying, and complexity of the technological process because of heavy clogging of equipment [10, 11].

One of the ways to improve the quality of complex AN-based fertilizers and reduce the risk of explosion is to increase the ammoniation degree of wet-process phosphoric acid, which reduces the AN portion in the product. Such way can improve the properties of the final product (decrease hygroscopicity and caking), increase its thermal stability, decrease the amount of different compounds in exhaust gases (nitrous gases, chlorine and fluorine compounds) during thermal decomposition, increase fire and explosion safety, and also decrease viscosity of ammonium phosphate–nitrate slurries produced during the production of fertilizer that can decrease energy cost for their transportation. However, information on influence of the degree of phosphoric acid ammoniation on the above-mentioned properties of complex AN-based fertilizers and their intermediates is absent in the literature.

Thus, the purpose of this work is to study the properties of complex AN-based fertilizers and intermediates in their

production depending on the degree of phosphoric acid ammoniation.

Experimental section

Preparation of the samples

To produce complex fertilizers, concentrated hemihydrate phosphoric acid, nitric acid, ammonium sulfate and potassium chloride (mineral concentrate "Silvin") were used. Wet-process phosphoric acid was obtained from the Khibiny apatite concentrate (the Cola Peninsula, Russia) of composition: P_2O_5—51.72, CaO—0.67, MgO—0.23, F—1.33, SO_3—4.53, Fe_2O_3—0.55, Al_2O_3—0.90, SiO_2—0.43% by mass by sulfuric acid attack. Phosphoric and nitric acids were mixed in a certain ratio and ammoniated in a reactor equipped with the agitator device, the reflux condenser and the water jacket, which allowed ammoniation to be carried out under near-isothermal conditions at 70 ± 2 °C.

The degree of ammoniation of phosphoric acid $NH_3:H_3PO_4$ (M) was determined by pH value of the 1% by mass aqueous solution of the slurry obtained and using the reference source [12]. Ammonium sulfate and potassium chloride were introduced into the slurry in an amount necessary to obtain the desired grade, mixed thoroughly and dried at 65 °C. Then, the charge mixture was crushed and put in a pan granulator with diameter of 300 mm and length of 150 mm. Granules of 2–4 mm were finally dried at 65 °C to reach the required humidity. The product obtained was analyzed for content of basic elements.

X-ray diffraction analysis

X-ray diffraction analysis of the investigated samples was performed when used powder diffractometer «STADI-MP» (STOE, Germany) with curved Ge (111) monochromator and radiation of CuK_α ($\lambda = 1.54056$ Å). The data acquisition was carried out in stepwise overlapping of scanning area mode by means of position-sensitive linear detector, the capture angle of which amounted 5° over 2θ with channel width of 0.02°. The reliability and accuracy of compounds in X-ray patterns obtained were established by means of database of 2013 International Centre for Diffraction Data.

Derivatographic analysis

Derivatographic analysis was carried out when used Paulik–Erdei derivatograph (MOM, Hungary) of Q-1500 series while heating in the air at atmospheric pressure in open quartz crucibles with heating rate of 2.5°/min. Al_2O_3 precalcinated at 1000 °C was used as a reference. The sample weight amounted 0.2 g. The thermocouple was Pt/Pt–Pd. The interpretations of the dependencies obtained were carried out in compliance with the literature data [13–16].

Hygroscopicity

Hygroscopicity (K) of the samples obtained was determined by means of climatic chamber BINDER KBF 115 (BINDER, Germany) with internal circulation. The value of K was determined by means of conditioning of granule samples with the diameter of 3–4 mm with the mass of 3.500 ± 0.006 g in the chamber at 25 °C and the relative air humidity (φ) of 80% for 1 h. Granules were uniformly distributed in a cup with the diameter of 50 mm and height of 10 mm in a single layer. The value of K was determined as the amount of water absorbed with a sample of unit mass for 1 h.

Caking

Determination of caking (σ) of samples obtained was conducted by means of climatic chamber with internal circulation BINDER KBF 115 (BINDER, Germany) at temperature of 45 °C, $\varphi = 40\%$, and special presses equipped with calibrated spring. The spring load for each sample was 340 kPa. The samples detention time in the chamber was 6 h. Caking was determined as averaged maximum force required for breaking of formed cylindrical pellet divided by its cross-section area (pellet size: diameter 33 mm, height 40 mm).

Static strength

Determining the static strength, P was conducted by means of IPG-1M (Urals Scientific Research Institute of Chemistry with Experiment Plant, Russia) according to the formula:

$$P = \frac{\sum_{i=1}^{N} F_i}{\frac{\pi d_m^2}{4} N}, \qquad (4)$$

where F is the mean force required for breaking of one granule, d_m is the mean diameter of one granule equal to 3.5 mm, and N is the number of measured granules.

Microcalorimetry

The microcalorimetric studies of the thermal decomposition kinetics were conducted by measuring the heat release rate in the samples under study with differential automatic calorimeter DAC-1-2 [17]. Tests were carried out in the vacuum-sealed glass ampoules with inner volume of about

2 cm^3, a mass of each tested mixture sample was 1 g. The free inner volume after putting each sample and sealing an ampoule was in the range 0.7–1.2 cm^3 per 1 g of the mixture tested. These ampoules were entirely put into the calorimeter and had no cold surfaces, and reaction products could not leave the boundaries of the reaction space.

Gravimetric study of the thermal decomposition

Studies of mass loss in the thermal decomposition were conducted by maintaining granulated samples with mass of 20.00 ± 0.05 g in the electric oven without forced convection at the given temperature for a given period of time. The content of ammonium and nitrate nitrogen, chlorine, fluorine and sulfur was determined in products of the thermal decomposition. The fraction of these elements that have been released into the gas phase was calculated according to the formula:

$$X_A = \frac{\omega_0(A)m_0 - \omega_t(A)m_t}{m_0}, \qquad (5)$$

where X_A is the fraction of A ($A = N_{amm}$, N_{nitr}, Cl, F) released into the gas phase per the unit mass of the initial sample; $\omega_0(A)$ is the mass fraction of A in the initial sample; m_0 is the mass of the initial sample; $\omega_t(A)$ is the mass fraction of A in the sample after the decomposition for time t; m_t is the mass of the sample after the decomposition for a time t.

Dynamic viscosity

The dynamic viscosity of slurries was determined by means of rotation viscometer HAAKE VT 74 Plus (Thermo Scientific, USA). In order to do that, the slurry obtained was placed in the cylindrical vessel provided with a thermostatic jacket and connected to circulation bath in which a polysilicon oil was circulated. After viscosity measurements, the slurry humidity was measured.

Processing experimental data obtained and the determination of confidence intervals for 95% confidence probability were conducted with the mathematical statistics methods by means of software application of origin.

Results and discussion

The composition of the fertilizer samples and X-ray diffraction analysis

Table 1 shows the results of analyses of fertilizer samples.

Figure 1 shows X-ray patterns for samples 1 and 2 (grade 26:13:0), 3 and 4 (grade 22:11:11), 5 and 6 (grade 16:16:16).

Table 1 The composition of
the fertilizer samples (%mass.)

Sample no.	Grade	N_{amm}	N_{nitr}	P_2O_5	S	K_2O	M	H_2O
1	26:13:0	18.5	7.8	13.8	8.4	–	1.68	0.55
2		15.8	10.1	13.3	4.2	–	1.06	0.42
3	22:11:11	14.9	7.6	11.4	5.6	11.4	1.71	0.59
4		13.6	10.8	11.7	4.0	11.4	1.04	0.55
5	16:16:16	13.8	2.2	15.9	8.2	16.5	1.65	0.52
6		12.3	4.0	16.5	4.0	16.4	1.07	0.48
7	20:10:10	16.0	3.9	9.9	11.0	10.3	1.70	0.53
8		14.9	5.2	10.5	10.1	10.5	1.03	0.52
9	19:9:19	12.6	6.6	9.3	8.0	20.0	1.67	0.44
10		11.6	8.2	9.3	2.8	20.3	1.03	0.51
11	27:6:6	16.4	10.9	6.4	2.6	6.5	1.66	0.49
12		15.8	11.9	6.3	2.6	6.4	1.06	0.50

Fig. 1 X-ray patterns of the
fertilizer samples: **a**—*1*, **b**—*2*,
c—*3*, **d**—*4*, **e**—*5*, **f**—*6*; *1*
$(NH_4)_2HPO_4$, *2* $NH_4H_2PO_4$, *3*
NH_4NO_3, *4* $(NH_4)_2SO_4$, *5*
$2NH_4NO_3\cdot(NH_4)_2SO_4$, *6*
$3NH_4NO_3\cdot(NH_4)_2SO_4$, *7*
$(NH_4,K)H_2PO_4$, *8* $(NH_4,K)NO_3$,
9 $(NH_4,K)_2SO_4$, *10* KCl, *11*
NH_4Cl, *2θ* Bragg angle (degree)

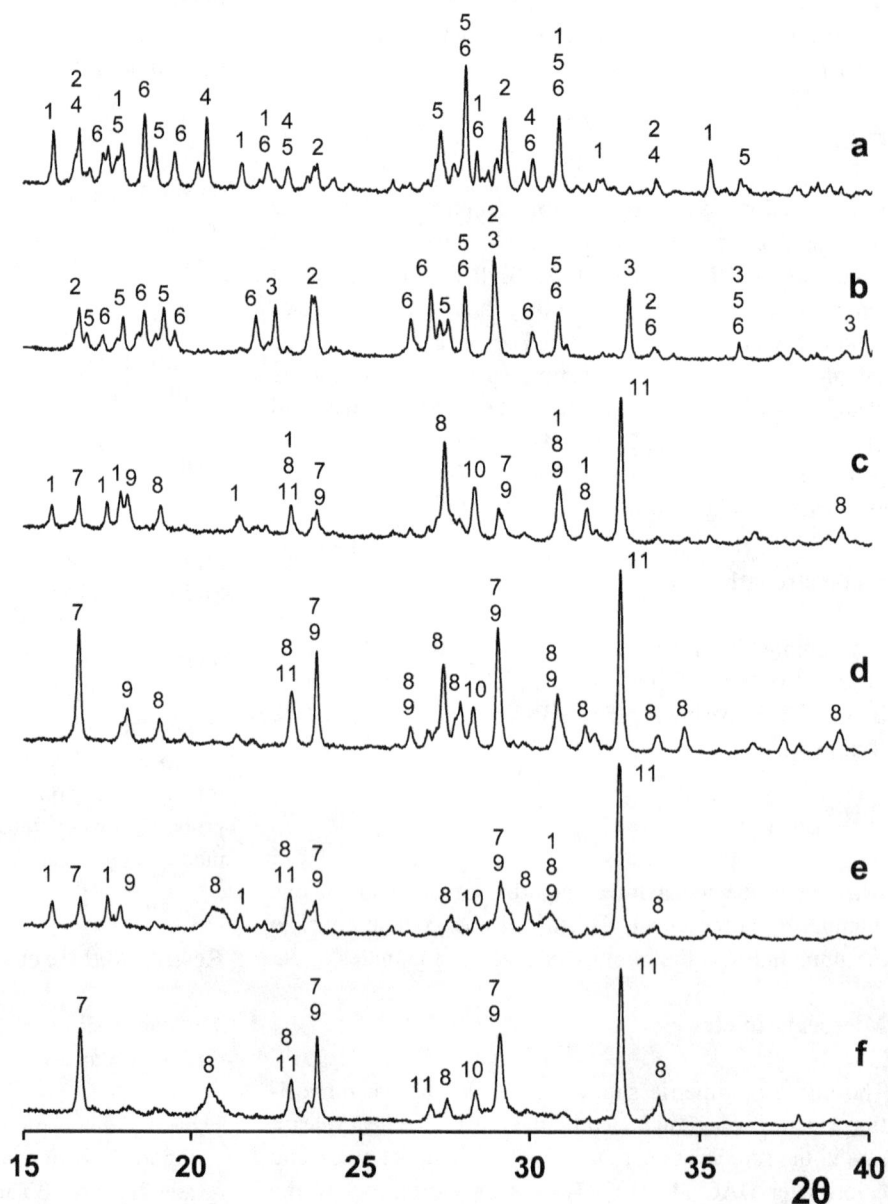

X-ray patterns for the samples of grades 16:16:16 and 22:11:11 demonstrate the presence of solid solutions $(NH_4,K)NO_3$, $(NH_4,K)H_2PO_4$ and $(NH_4,K)_2SO_4$, as well as of NH_4Cl and KCl. For samples 3 and 5, the presence of $(NH_4)_2HPO_4$ was established.

Comparing X-ray patterns for the samples 3 and 4 of grade 16:16:16 and 5 and 6 of grade 22:11:11 shows that the intensity of the main diffraction peak of NH_4Cl decreases with a higher degree of ammoniation. This is associated with a reduction of the original content of AN in the composition of samples that results in reducing the amount of NH_4Cl produced in reaction (2).

X-ray patterns for sample 1 of 26:13:0 grade demonstrate the presence of $(NH_4)_2HPO_4$, $NH_4H_2PO_4$, $(NH_4)_2SO_4$, $2NH_4NO_3 \cdot (NH_4)_2SO_4$ and $3NH_4NO_3 \cdot (NH_4)_2SO_4$ and for the sample 2 the presence of NH_4NO_3, $NH_4H_2PO_4$, $2NH_4NO_3 \cdot (NH_4)_2SO_4$ and $3NH_4NO_3 \cdot (NH_4)_2SO_4$.

Comparing X-ray patterns for samples 1 and 2 of grade 26:13:0 demonstrates that the composition of sample 2 has the unbound AN, which could not fully converted to $2NH_4NO_3 \cdot (NH_4)_2SO_4$ and $3NH_4NO_3 \cdot (NH_4)_2SO_4$ due to a high content of AN and a low content of $(NH_4)_2SO_4$ in the composition of the fertilizer. This may lead to significant deterioration of the properties of sample 2 compared with sample 1.

All these compounds are typical for complex AN-based fertilizers that is noted in [1, 3, 4, 12].

Derivatographic analysis

Figures 2, 3, 4 and 5 show the results of the derivatographic analysis for samples 1, 2, 3 and 4. Analysis of curves of the differential thermal analysis (DTA) and of the differential thermogravimetric analysis (DTG) confirms the data of X-ray diffraction analysis.

Curves of the differential thermal analysis (DTA) for 22:11:11 samples are characterized by the following peaks: the reverse phase transition of $(NH_4,K)NO_3$ in

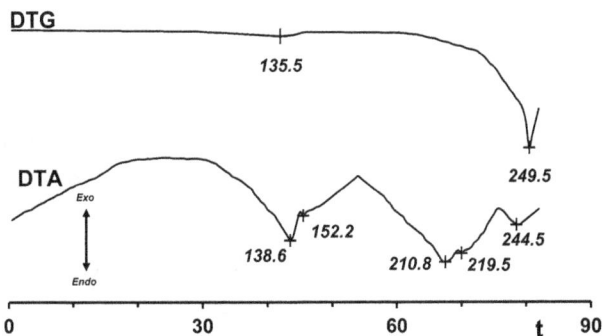

Fig. 3 *Curves* of the differential thermal analysis (DTA) and of the differential thermogravimetric analysis (DTG) of sample 2: *t* time (min)

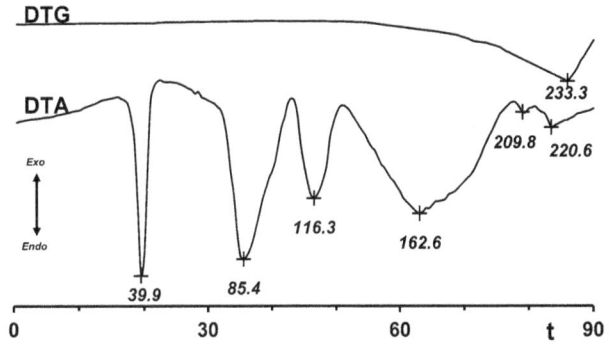

Fig. 4 *Curves* of the differential thermal analysis (DTA) and of the differential thermogravimetric analysis (DTG) of sample 3: *t* time (min)

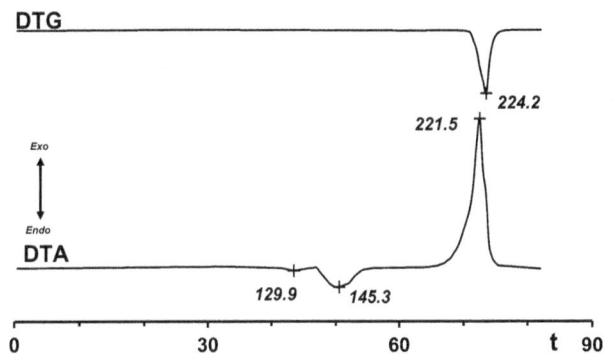

Fig. 5 *Curves* of the differential thermal analysis (DTA) and of the differential thermogravimetric analysis (DTG) of sample 4: *t* time (min)

$NH_4NO_3 \cdot 2KNO_3$ (113.1 and 129.9 °C, respectively) [15]; melting (132.8 and 145.3 °C) [15]; the exothermal decomposition of the product including the decomposition of NH_4NO_3 [15], the polycondensation of $(NH_4,K)H_2PO_4$ and the decomposition of $(NH_4)_2HPO_4$ for samples 3 and 4 (197.2 and 221.5 °C [13].

It can be concluded by comparing the DTG and DTA curves that sample 3 has the higher thermal stability as

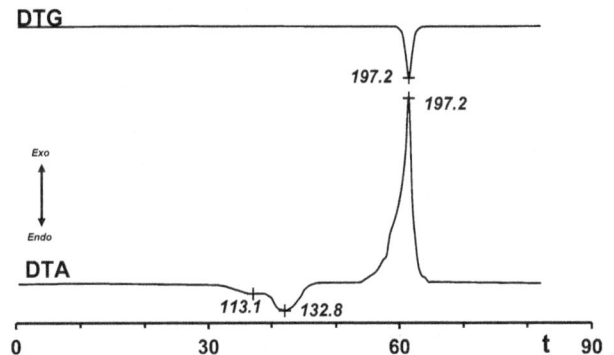

Fig. 2 *Curves* of the differential thermal analysis (DTA) and of the differential thermogravimetric analysis (DTG) of sample 1: *t* time (min)

compared to sample 4, which may be related to a lower content of AN and a higher content of $(NH_4)_2HPO_4$. It is also worth noting that there is no peak characteristic for the $(NH_4)_2HPO_4$ decomposition in the DTA and DTG curves of sample 3, which would be in the range of 120–200 °C. It may be assumed that its absence is due to the interaction between $(NH_4)_2HPO_4$ and HNO_3, which is formed as the result of the partial dissociation of NH_4NO_3, according to the reaction:

$$(NH_4)_2HPO_4 + HNO_3 \rightarrow NH_4H_2PO_4 + NH_4NO_3. \quad (6)$$

The decomposition of $(NH_4)_2HPO_4$ is apparently to occur at higher temperatures due to the course of reaction (6). In the case of sample 3, this process takes place in the intensive exothermal decomposition of the product.

At heating sample 2, the peaks are observed on the DTA curve, which is related to the following phenomena: the reverse phase transition of AN IV → III (39.9 °C) [13]; the reverse phase transition of AN III → II (85.4 °C) [13]; the reverse phase transition of AN II → I (116.3 °C) [13]; melting and partial decomposition of adducts $2NH_4NO_3$-$(NH_4)_2SO_4$ and $3NH_4NO_3 \cdot (NH_4)_2SO_4$ (162.9 °C) [16]; the polycondensation of $NH_4H_2PO_4$ (209.8 °C) [13]; the AN decomposition (220.6 °C) [13].

There are no peaks, which are characteristic to AN in the DTA curve of sample 1. The thermal decomposition of this sample is characterized by the following processes: the decomposition of $(NH_4)_2HPO_4$ (138.6 °C) [13]; the melting and partial decomposition of adducts $2NH_4NO_3$-$(NH_4)_2SO_4$ and $3NH_4NO_3 \cdot (NH_4)_2SO_4$ (152.2 °C) [16]; the polycondensation of $NH_4H_2PO_4$ (210.8 °C) [13]; the AN decomposition (219.5 °C) [13]; the $(NH_4)_2SO_4$ decomposition (244.5 °C) [14].

It can be concluded when compared the DTG and DTA curves that the presence of $(NH_4)_2HPO_4$ as a part of sample 1 leads to the fact that at temperatures over 100 °C $(NH_4)_2HPO_4$ decomposes to $NH_4H_2PO_4$ to release NH_3

into a gas phase. However, sample 2 exhibits the higher thermo-stability than sample 1 when further heated.

It should also be noted that the decomposition of samples 1 and 2 takes place endo-thermally as opposed to samples 3 and 4, whose decomposition proceeds with the release of the large amount of heat. This is due to the absence of chlorine compounds in the composition of samples 1 and 2, which are capable to accelerate the exothermal AN and complex AN-based fertilizers decomposition [7–9, 15].

Hygroscopicity, caking and static strength

Table 2 presents the results of studying hygroscopicity, caking and static strength of the fertilizer samples obtained.

The presented data show that for the same grade of the fertilizer the increase of M reduces the hygroscopicity and caking; however, the static strength of granules decreases also. The reduction of hygroscopicity can be associated with a reduced content of AN, which is highly hygroscopic. The reduction of caking can also be associated with a reduced ammonium chloride content with increasing M, which is apparent from intensity of peaks for NH_4Cl in the presented X-ray patterns [1, 18]. The reduction of static strength of granules can be the result of lower strength of phase contacts between granules with increase of M in the granulation process [19].

The maximum difference in hygroscopicity and caking is observed for 26:13:0 grade, which can be due to the presence of AN in sample 2, whereas in sample 1 AN is connected in double salts $(NH_4)_2SO_4 \cdot 2NH_4NO_3$ and $(NH_4)_2SO_4 \cdot 3NH_4NO_3$. The minimum difference in hygroscopicity and caking is observed for 27:6:6 grade, which can be explained by the high content of nitrate nitrogen in both samples and the small difference in its content between them.

Table 2 Hygroscopicity, caking and static strength of granulated fertilizer samples

Sample no.	Grade	K, mmole g^{-1} h^{-1}	$\sigma \times 10^{-2}$, kPa	P, MPa
1	26:13:0	3.21 ± 0.13	3.00 ± 0.13	2.44 ± 0.14
2		5.30 ± 0.20	4.47 ± 0.18	3.70 ± 0.20
3	22:11:11	4.04 ± 0.19	3.54 ± 0.19	3.16 ± 0.19
4		5.00 ± 0.20	$4,10 \pm 0.30$	4.40 ± 0.30
5	16:16:16	3.04 ± 0.12	1.76 ± 0.16	5.00 ± 0.30
6		3.51 ± 0.17	3.10 ± 0.30	5.10 ± 0.30
7	20:10:10	3.74 ± 0.17	2.97 ± 0.15	2.39 ± 0.15
8		4.06 ± 0.15	3.90 ± 0.20	3.80 ± 0.20
9	19:9:19	3.22 ± 0.15	2.59 ± 0.10	3.28 ± 0.19
10		3.96 ± 0.11	3.36 ± 0.16	4.40 ± 0.20
11	27:6:6	5.00 ± 0.10	3.90 ± 0.30	$3.90 \pm 0,20$
12		5.16 ± 0.12	4.40 ± 0.30	4.90 ± 0.30

dQ/dt

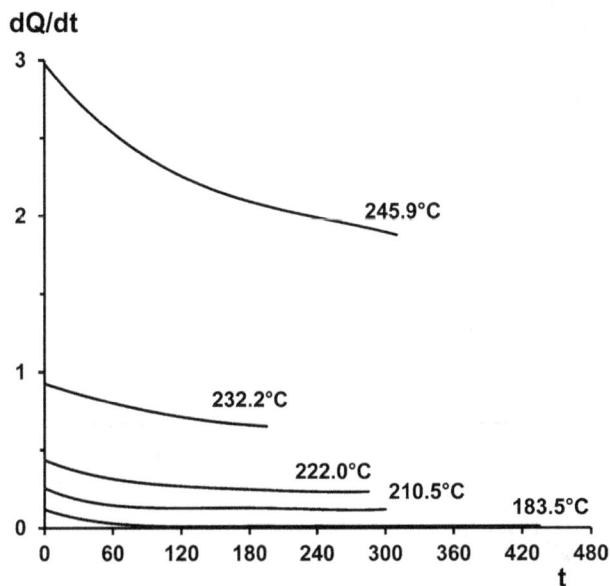

Fig. 6 Dependence of the heat release rate dQ/dt (mW g^{-1}) on time t (min) in the thermal decomposition of sample 3

dQ/dt

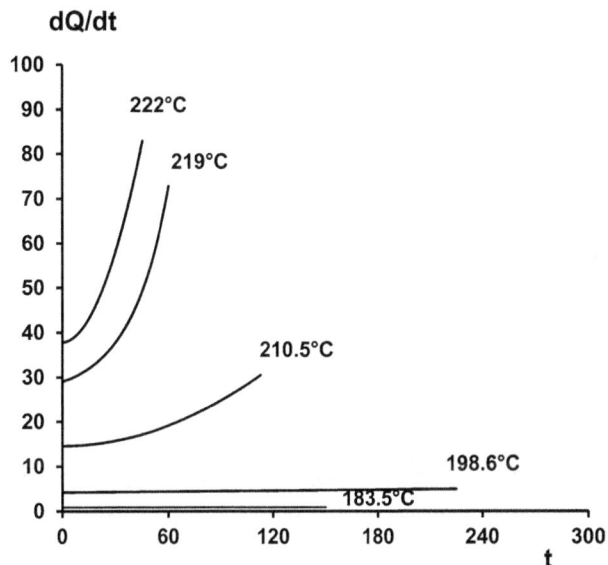

Fig. 7 Dependence of the heat release rate dQ/dt (mW g^{-1}) on time t (min) in the thermal decomposition of sample 4

It should also be noted that the highest increase of the caking was observed for 16:16:16 grade ($\sigma_6/\sigma_5 = 1.76$), whereas for the other grades this ratio is much lower. This is possible due to the high ratio of the content of NH$_4$Cl in two samples of 16:16:16 grade and almost twofold increase in the content of AN in sample 6 when M simultaneously reduced. The closest value to this one is $\sigma_2/\sigma_1 = 1.49$ for 26:13:0 grade. The high ratio σ_2/σ_1 for 26:13:0 grade is apparently due to the fact that in sample 2 the part of NA presents in the free form, while in sample 1 NA is fully bound in double salts.

Microcalorimetry

Figures 6 and 7 show the curves of the heat release rate dependence on time in the thermal decomposition of samples 3 and 4 in the temperature range of 183.5–245.9 °C.

As indicated above, chloride-anions Cl$^-$ contained in samples under study are catalysts of the AN decomposition, and their catalytic effect increases with the increase of the content of nitric acid in the system and virtually does not occur when its content is low. The accelerating action of Cl$^-$ in the AN decomposition is related to accumulation of nitryl chloride NO$_2$Cl, nitrosyl chloride NOCl and chlorine Cl$_2$ in the system, being more effective oxidizers of ammonium cation NH$_4^+$ and ammonia as compared to nitric acid. The presence of NH$_4$H$_2$PO$_4$, (NH$_4$)$_2$HPO$_4$ and (NH$_4$)$_2$SO$_4$ together with Cl$^-$ reduces Cl$^-$ catalytic effect in AN decomposition.

The study of the heat release rate for sample 4 revealed its low thermal stability. In the decomposition of sample 4

Cl$^-$ accelerating action prevails over decreasing the AN decomposition rate in response to H$_2$PO$_4^-$, HPO$_4^{2-}$ and SO$_4^{2-}$ anions and, therefore, the decomposition of this sample occurs with the self-acceleration.

Sample 3 has a lower content of AN as compared to sample 4, herewith in its composition a large portion of H$_2$PO$_4^-$ is substituted with HPO$_4^{2-}$. Anion of HPO$_4^{2-}$ is capable to a higher degree to reduce the concentration of undissociated nitric acid, and so to increase the thermal stability of sample 3. Besides, the content of (NH$_4$)$_2$SO$_4$ in sample 3 is also higher than in sample 4. All this contributes to the fact that the accelerating action of Cl$^-$ is not detected, and the decomposition occurs without self-acceleration. Thus, sample 3 has significantly higher thermal stability as compared to sample 4.

Besides the study of the fertilizer samples, the heat release rate was also measured as a function of time in the thermal decomposition of nitrate–phosphate–ammonium slurries at obtaining sample 3 with $M = 1.0$ (sample 3a) and $M = 1.4$ (sample 3b) with humidity of about 8% mass in the temperature range of 243.5–277.0 °C (Figs. 8, 9).

The study of the heat release rate for these samples revealed their high thermal stability, while sample 3b was more thermally stable than sample 3a, which can be explained by the higher content of (NH$_4$)$_2$HPO$_4$ in it.

Figure 10 shows the temperature dependencies of the initial heat release rates $(dQ/dt)_{t=0}$ in the thermal decomposition of samples 3, 4, 3a and 3b in Arrhenius coordinates. For comparison, Fig. 5 also illustrates the temperature dependence of the initial heat release rates in the AN thermal decomposition studied previously [20].

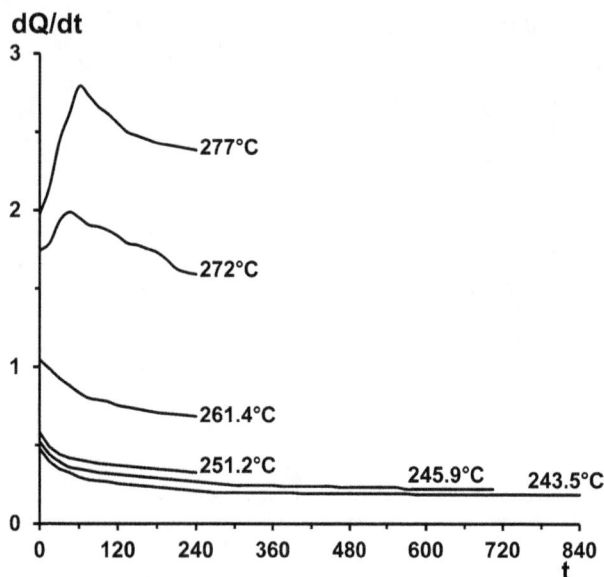

Fig. 8 Dependence of the heat release rate dQ/dt (mW g^{-1}) versus time t (min) in the thermal decomposition of sample 3a

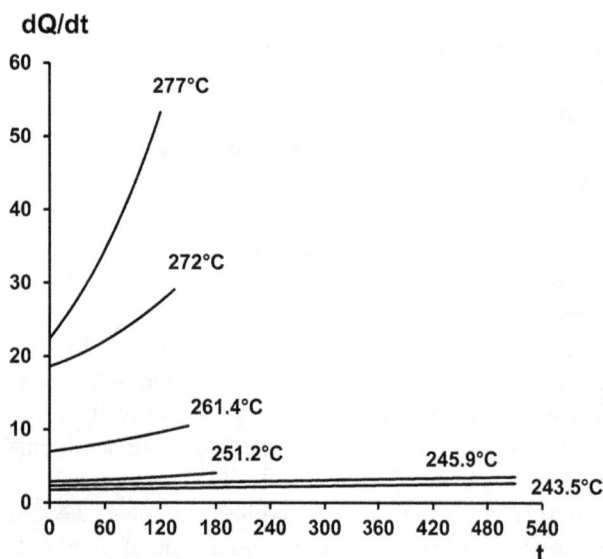

Fig. 9 Dependence of the heat release rate dQ/dt (mW g^{-1}) versus time t (min) in the thermal decomposition of sample 3b

The equations of the obtained dependence of $(dQ/dt)_{t=0}$ (mW g^{-1}) on temperature (K) are as follows:

for sample 3

$$\left(\frac{dQ}{dt}\right)_{t=0} = 10^{11.7\pm0.7}\exp\left(-\frac{(17.2\pm0.8)\times10^3}{T}\right), \quad (7)$$

for sample 4

$$\left(\frac{dQ}{dt}\right)_{t=0} = 10^{18.1\pm0.7}\exp\left(-\frac{(22.3\pm0.9)\times10^3}{T}\right), \quad (8)$$

for sample 3a

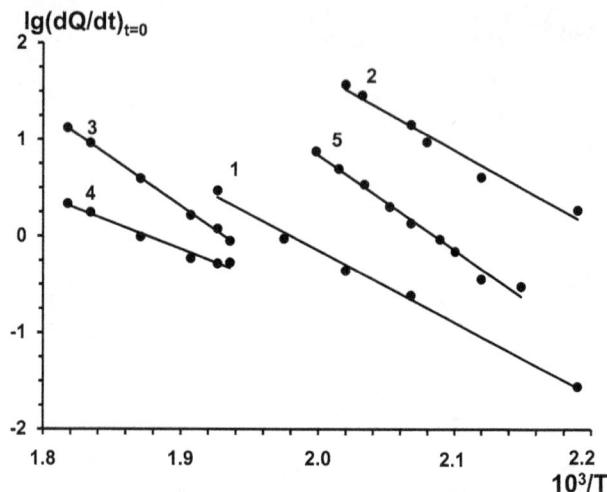

Fig. 10 Dependence of $\lg[dQ/dt \text{ (mW g}^{-1})]_{t=0}$ on $10^3/T$ (K^{-1}) for samples 3 (1), 4 (2), 3a (3), 3b (4) and ammonium nitrate (5)

$$\left(\frac{dQ}{dt}\right)_{t=0} = 10^{16.1\pm0.3}\exp\left(-\frac{(22.8\pm0.4)\times10^3}{T}\right), \quad (9)$$

for sample 3b

$$\left(\frac{dQ}{dt}\right)_{t=0} = 10^{7.3\pm0.8}\exp\left(-\frac{(12.7\pm0.9)\times10^3}{T}\right). \quad (10)$$

The dependencies presented in Figs. 6, 7, 8, 9 and 10 show that the initial heat release rate of sample 4 is on average by 1–2 orders higher than that for sample 3. Herewith the initial heat release rate of sample 4 significantly exceeds that of AN, while for sample 3 the situation is inverse. Samples 3a and 3b have even higher thermal stability as compared to sample 3, which may be explained by lack of Cl$^-$ in their composition and the high water content.

In any real conditions of conducting the discussed reaction, the thermal explosion is only possible when the values of external parameters of the process exceed the critical ones for the thermal explosion, but calculation of the critical conditions for a real complex production process is a very time-consuming task, and the adiabatic induction period of thermal explosion τ_{ad} is calculated simply. If the value τ_{ad} is much greater than the real time of the production process at an appropriate temperature, then the thermal explosion will not occur, and in any real process conditions the induction period may only be greater than under adiabatic conditions. However, if the value τ_{ad} and process real time are close enough or if τ_{ad} is even less, it is necessary to calculate the critical conditions of the thermal explosion (the critical temperature for the actual size of the unit and the conditions of heat transfer from it). Only these calculations can give final decision on possibility of the thermal explosion in the process considered.

Calculation of the adiabatic induction period is the most simple and available method to assess the possibility of the thermal explosion for any particular composition. In the complete absence of heat removal (adiabatic conditions) and at a sufficiently high value of the process heat, the thermal explosion will always occur; besides, the degree of conversion in the reaction discussed during induction period will be very small, because all the heat is used for heating a substance. As far as there is no heat removal, the adiabatic induction period is independent of the sample mass and heat removal conditions and it is considered as a characteristic for a substance or mixture discussed. In the theory of thermal explosion because of the weak influence of the process acceleration, the exact quantitative equation for calculating the adiabatic induction period was obtained only for zero-order reaction, and the reaction rate change in the subsequent stages is assumed to have a very small action on the adiabatic induction period [21]:

$$\tau_{ad} = \frac{c_p}{Q_0 k_0} \cdot \frac{RT_0^2}{E} \cdot \exp\left(\frac{E_c}{RT_0}\right), \tag{11}$$

where c_p is the heat capacity of the sample; Q_0 is the total process heat; k_0 and E_c are the pre-exponential factors and the activation energy of the decomposition rate constant; T_0 is the absolute temperature of the decomposition; $R = 8.314$ J mole^{-1} K^{-1} is the universal gas constant.

When $\left(\frac{dQ}{dt}\right)_{t=0} = Q_0 k_0 \exp\left(-\frac{E}{RT_0}\right)$ Eq. (11) takes the following form:

$$\tau_{ad} = \frac{RT_0^2}{E} \cdot \frac{c_p}{\left(\frac{dQ}{dt}\right)_{t=0}}. \tag{12}$$

The results from paper [12] were used to determine the heat capacity of samples under study, provided that in a first approximation the heat capacities of samples 3, 4 and 3a, 3b are equal in pairs. The value of the AN heat capacity was taken according to the data in [2]. The τ_{ad} values obtained are given in Table 3. The τ_{ad} values for the same initial temperature may be considered as the characteristics

of a relative explosion risk of a substance. The adiabatic induction periods of the thermal explosion for samples 3, 3a and 3b are greater than that for AN, and for sample 4 they are almost by an order less, which reveals the potential danger of thermal spontaneous ignition of the sample during production operations at high temperatures.

Gravimetric study of the thermal decomposition

The study of the mass loss in the thermal decomposition was carried out for samples 3 and 4 at temperatures of 170, 180, 190 and 200 °C. In addition to the study of the mass loss, the release of ammonium nitrogen, nitrate nitrogen, chlorine and fluorine to the gas phase was also evaluated. The research results are presented in Figs. 11, 12 and 13.

The decomposition intensity for sample 4 is much higher than that for sample 3. The release of chlorine, fluorine, ammonium nitrogen, and nitrate nitrogen from sample 4 to the gas phase is also much more intensive than that from sample 3. Ammonium nitrogen in the initial decomposition stage is released from sample 3 in a greater quantity than that from sample 4. It is related to the higher content of $(NH_4)_2HPO_4$, which starts to decompose in NH_3 and $NH_4H_2PO_4$ at low temperatures.

It is also worth mentioning that the maximum amount of fluorine released into the gas phase for both samples is almost the same. It is related to the fact that fluorine in both samples according to [22] is present in the form of compounds $(NH_4)_2SiF_6$, NH_4F, $NH_4NO_3 \cdot (NH_4)_2SiF_6$, KNO_3-K_2SiF_6, $(NH_4)_2SiF_6 \cdot NH_4F$, etc., the decomposition of which depends only on the process temperature. The higher fluorine release rate for sample 4 is related to the more intense exothermal decomposition of this sample.

For chlorine, the release into the gas phase depends on the content of AN, so for sample 4 a significantly greater amount of chlorine is released into the gas phase than for sample 3.

The release of chlorine, fluorine, nitrous gases and ammonium compounds into the gas phase leads to the essential complication and more expensive purification of

	T, K	τ_{ad}, h				
		Sample 3	Sample 4	Sample 3a	Sample 3b	Ammonium nitrate
Table 3 Adiabatic induction period of the thermal explosion τ_{ad} of samples 3, 4, 3a and 3b and AN depending on temperature T	473	94.64	1.50	226.83	232.00	11.38
	478	68.04	0.97	146.60	180.92	7.00
	483	49.29	0.63	95.69	142.00	4.35
	488	35.97	0.42	63.05	112.02	2.73
	493	26.45	0.28	41.94	88.90	1.73
	498	15.58	0.19	28.14	70.94	1.11
	503	14.59	0.13	19.05	56.90	0.72
	508	10.94	0.09	13.01	45.88	0.47

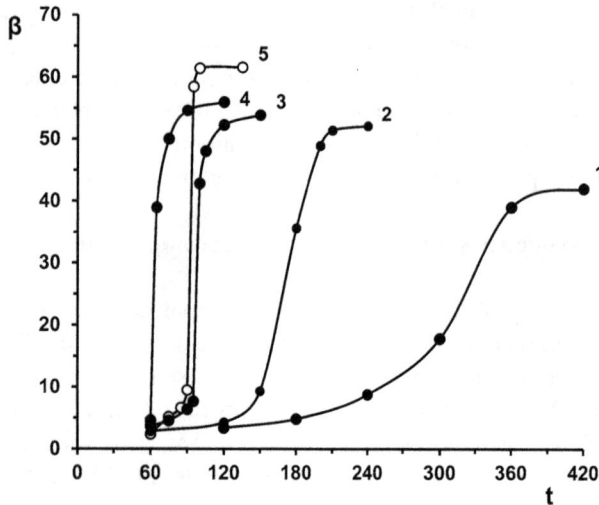

Fig. 11 Dependence of thermal decomposition degree $\beta = (m - m_0)/m_0 \cdot 100$ (%) versus time t (min) for samples 3 and 4 at constant temperature; sample 3: *1* 170 °C, *2* 180 °C, *3* 190 °C, *4* 200 °C; sample 4: *5* 180 °C

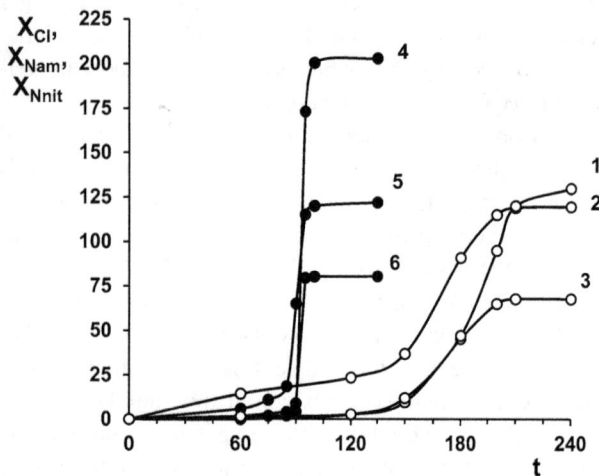

Fig. 12 Release of chlorine X_{Cl}, ammonium X_{Namm} and nitrate nitrogen X_{Nnitr} (g kg^{-1}) into the gas phase in the thermal decomposition of samples 3 and 4 at temperature of 180 °C versus time t (min); sample 3: curve *1* Cl, *2* N_{am}, *3* N_{nit}; sample 4: curve *4* Cl, *5* N_{am}, *6* N_{nit}

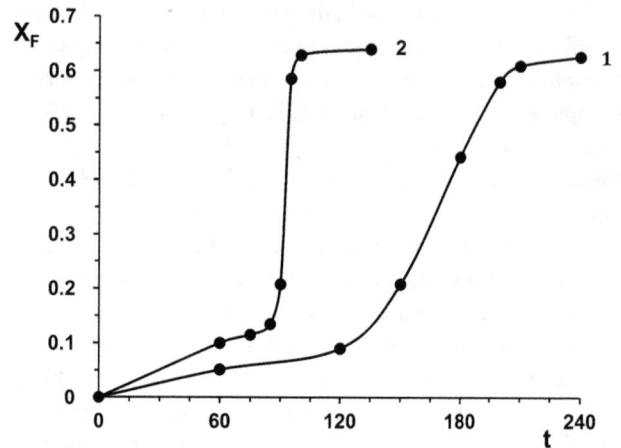

Fig. 13 Release of fluorine into the gas phase X_F (g kg^{-1}) in the thermal decomposition of samples 3 (*curve 1*) and 4 (*curve 2*) at temperature 180 °C versus time t (min)

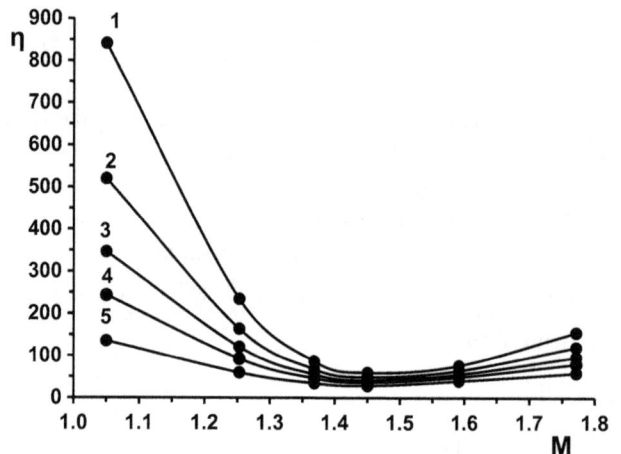

Fig. 14 The dependence of dynamic viscosity η (mPa s) of ammonium–phosphate–nitrate slurry on M at temperature 110 °C and for different values of humidity: *1* 5% mass., *2* 6% mass., *3* 7% mass., *4* 8% mass., *5* 10% mass

Figure 14 shows the dependence of dynamic viscosity of such slurry on M for different values of humidity at 110 °C. The slurry viscosity is apparent to reach the minimum value at $M = 1.45$ for all the values of humidity.

It should be mentioned that the same behavior of viscosity was observed for phosphate ammonia slurries obtained from various types of a phosphate raw [3, 23]. The presence of minimum in the viscosity curve is probably due to the high solubility of ammonium phosphates at $M = 1.4$–1.5. The presence in the slurry of impurities of iron, aluminum, magnesium, fluorine, silicon, etc. leads to increasing viscosity due to the formation of poorly soluble compounds [3, 24, 25].

Figures 15 and 16 show the dependences of the slurry dynamic viscosity (for $M = 1.05$ and $M = 1.45$) on humidity at different temperatures. The figures show that

exhaust gases from them, as well as to the more intense corrosion of equipment.

Dynamic viscosity

The study of dynamic viscosity was performed for ammonium–phosphate–nitrate slurries obtained at production of 22:11:11 grade when $M = 1.7$. To obtain such slurries, phosphoric and nitric acids were mixed in the ratio P_2O_5:$HNO_3 = 0.36$:1 (by mass.) and ammoniated up to the specified value of M.

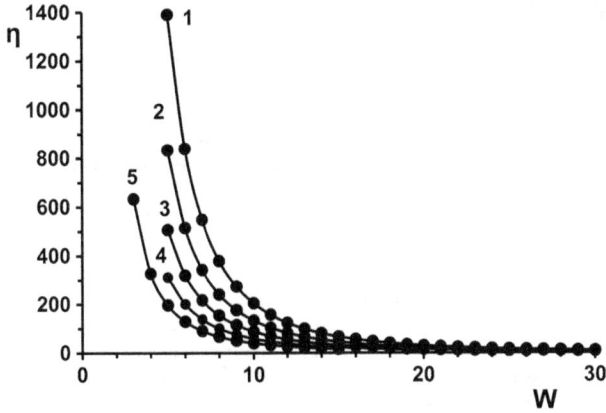

Fig. 15 Dependence of dynamic viscosity η (mPa s) of ammonium–phosphate–nitrate slurry for $M = 1.05$ on humidity (%mass.) for different values of temperature: *1* 100 °C, *2* 105 °C, *3* 110 °C, *4* 115 °C, *5* 120 °C

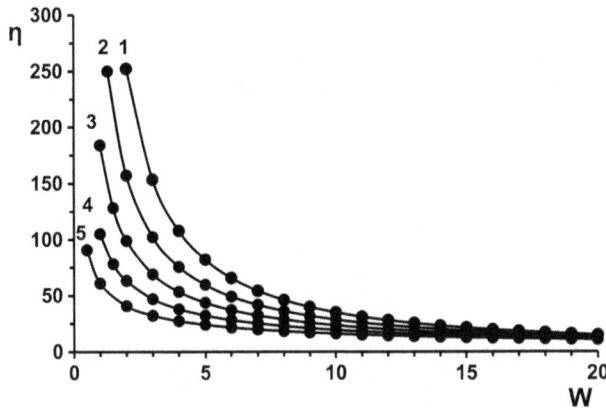

Fig. 16 Dependence of dynamic viscosity η (mPa s) of ammonium–phosphate–nitrate slurry for $M = 1.45$ on humidity (%mass.) for different values of temperature: *1* 100 °C, *2* 105 °C, *3* 110 °C, *4* 115 °C, *5* 120 °C

the slurry viscosity increases with decreasing humidity. Herewith for the slurries at $M = 1.05$, a more rapid increase of viscosity with decreasing humidity is observed.

Temperature influence on dynamic viscosity of the slurries obeys the law of Arrhenius–Andrade [26]:

$$\eta = A_{\exp}\left(\frac{E_v}{RT}\right), \qquad (13)$$

where A is the pre-exponential factor and E_v is the activation energy for viscous flow.

Table 4 presents the equations of dynamic viscosity dependence on temperature at the different values of humidity for slurries studied. As can be seen from the equations presented, the activation energy of a viscous flow, the values of pre-exponential factor and the value of dynamic viscosity of the slurry for $M = 1.45$ are substantially less than for $M = 1.05$.

Using the slurries having higher mobility and flowability during their processing in the granular product can significantly reduce the energy costs for the removal of moisture from the granules and reduce the amount of the recirculated product obtainable by a recycle method.

Conclusions

On the basis of the studies performed, it has been found that increasing the degree of phosphoric acid ammoniation with $M = 1.0$–1.1 to $M = 1.6$–1.7 influences on the properties of the complex AN-based NP and NPK fertilizers.

It has been shown by X-ray diffraction and derivatographic analysis that the composition of NPK fertilizer (16:16:16 and 22:11:11) contains $(NH_4,K)H_2PO_4$, $(NH_4,K)_2SO_4$, $(NH_4,K)NO_3$, KCl and NH_4Cl. When $M = 1.6$–1.7, $(NH_4)_2HPO_4$ presents also additionally. The composition of NP fertilizer (26:13:0) contains $NH_4H_2PO_4$, $2NH_4NO_3 \cdot (NH_4)_2SO_4$ and $3NH_4NO_3 \cdot (NH_4)_2SO_4$. When $M = 1.0$–1.1, NH_4NO_3 presents additionally in the system, when $M = 1.6$–1.7, $(NH_4)_2HPO_4$ and $(NH_4)_2SO_4$ present additionally.

It is found that the decomposition of NPK fertilizers occurs with the strong exothermal effect and NP fertilizers decomposition occurs with the endothermal effect. The strong exothermal effect of the thermal NPK fertilizer

	Table 4 Equations of dynamic viscosity (mPa·s) dependence for ammonium phosphate nitrate slurries for $M = 1.05$ and $M = 1.45$ for the various humidity values W (% mass.)	W	$M = 1.05$	$M = 1.45$
		5	$\eta = 10^{-13.6 \pm 0.6} \exp\left(\frac{(14.7 \pm 0.6) \times 10^3}{T}\right)$	$\eta = 10^{-8.5 \pm 0.4} \exp\left(\frac{(9.0 \pm 0.4) \times 10^3}{T}\right)$
		10	$\eta = 10^{-11.7 \pm 0.6} \exp\left(\frac{(11.9 \pm 0.6) \times 10^3}{T}\right)$	$\eta = 10^{-5.1 \pm 0.2} \exp\left(\frac{(5.7 \pm 0.3) \times 10^3}{T}\right)$
		15	$\eta = 10^{-10.6 \pm 0.5} \exp\left(\frac{(10.5 \pm 0.4) \times 10^3}{T}\right)$	$\eta = 10^{-3.03 \pm 0.13} \exp\left(\frac{(3.7 \pm 0.1) \times 10^3}{T}\right)$
		20	$\eta = 10^{-9.8 \pm 0.3} \exp\left(\frac{(9.6 \pm 0.3) \times 10^3}{T}\right)$	$\eta = 10^{-1.59 \pm 0.02} \exp\left(\frac{(2.4 \pm 0.1) \times 10^3}{T}\right)$
		25	$\eta = 10^{-9.2 \pm 0.3} \exp\left(\frac{(9.0 \pm 0.3) \times 10^3}{T}\right)$	$\eta = 10^{-0.47 \pm 0.01} \exp\left(\frac{(1.30 \pm 0.05) \times 10^3}{T}\right)$

decomposition is associated with the presence of chlorine-contained compounds.

It has been shown that hygroscopicity and caking for 26:13:0, 22:11:11, 16:16:16, 20:10:10, 19:9:19 and 27:6:6 grades decrease by increasing M from 1.0–1.1 to 1.6–1.7.

The study of the thermal decomposition by the example of 22:11:11 grade has demonstrated that increasing the degree of ammonization up to the specified values increases the thermal stability and reduces the intensity of the release of compounds of chlorine, fluorine and nitrous gases into the gas phase.

The study of thermal and rheological properties of ammonium–phosphate–nitrate slurries has allowed to set their high thermal stability, which increases with the increase of the phosphoric acid ammoniation degree. The viscosity of the slurries changes extremely having the minimum value at $M = 1.4–1.5$ and the maximum value at $M = 1.0$. The viscosity of the slurries increases with decreasing moisture content and decreases with increasing temperature according to the law of Arrhenius–Andrade.

Compliance with ethical standards

Conflict of interest The authors declare no competing financial interest.

References

1. Kuvshinnikov IM (1987) Mineral fertilizers and salts: properties and methods of their improvement. Khimiya, Moscow
2. Olevskiy VM (1978) Ammonium nitrate technology. Khimiya, Moscow
3. Kononov AV, Sterlin VN, Evdokimova LI (1988) Principles of technology of complex fertilizers. Khimiya, Moscow
4. Shmulyan EK, Portnova NL, Doroshina TV, Abashkina TF, Vinnik MM (1975) Determination of ammonium nitrate phosphate fertilizers and intermediate products composition in the process of nitric-sulfuric decomposition of Karatau rock phosphates. Byulleten Tekhniko-Ekonomicheskoi Informacii NIITE-KHIMa 8:18–23
5. Rubtsov YI, Strizhevsky II, Kazakov AI, Andrienko LP, Moshkovich EB (1989) Possibility of reduction of thermal decomposition rate for ammonium nitrate. J Appl Chem-USSR 62:2169–2174
6. Kazakov AI, Ivanova OG, Kurochkina LS, Plishkin NA (2011) Kinetics and mechanism of thermal decomposition of ammonium nitrate and sulfate mixtures. Russ J Appl Chem 84:1516–1523. doi:10.1134/S1070427211090102
7. Keenan AG, Dimitriades B (1962) Mechanism for the chloride-catalyzed thermal decomposition of ammonium nitrate. J Chem Phys 37:1583–1586. doi:10.1063/1.1733343
8. Rubtsov YI, Strizhevsky II, Kazakov AI, Moshkovich EB, Andrienko LP (1989) Kinetic mechanism of influence of Cl⁻ on thermal decomposition of ammonium nitrate. J Appl Chem-USSR 62:2417–2422
9. Rubtsov YI, Kazakov AI, Nedelko VV, Shastin AV, Larikova TS, Sorokina TV, Korsounskii BL (2008) Thermolysis of ammonium nitrate/potential donor of active chlorine compositions. J Therm Anal Calorim 93:301–309. doi:10.1007/s10973-007-8868-z
10. Chatterjee SK (1990) Experience with production of urea-based high-grade NPK fertilizers, urea-based NPK plant design and operating alternative: workshop proceedings. International Development Centre, Muscle Shoals, pp 14–20
11. Ranadurai S (1990) Operation experiences with NP-NPK granulation of coromandel fertilizers. Urea-based NPK plant design and operating alternative: workshop proceedings. International Development Centre, Muscle Shoals, pp 21–26
12. Borisov VM, Azhikina YV, Galtsov AV (1983) Physics and chemistry of phosphoric fertilizers production. Reference book, Khimiya
13. Zaitsev PM, Tavrovskaya AY, Podlesskaya AV, Portnova NL (1982) Thermal stability of mineral fertilizer components. Report 1. Nitrates, chlorides, fluorides, fluorine silicates, ammonium, potassium, calcium, aluminum and iron phosphates. Trudy NIUIFa 240:154–167
14. Tavrovskaya AY, Podlesskaya AV, Portnova NL (1982) Thermal stability of mineral fertilizer components. Report 2. Ammonium, potassium, calcium, aluminum and iron phosphates. Magnesium compounds. Trudy NIUIFa 240:168–185
15. Tavrovskaya AY, Portnova NL, Abashkina TF (1976) Thermographic study of ammonium nitrate phosphate fertilizer. Byulleten Tekhniko-Ekonomicheskoi Informacii NIITEKHIMa 7:10–14
16. Babkina TS, Golovina NB, Bogachev AG, Olenev AV, Shevelkov AV, Uspenskaya IA (2012) Crystal structures and physicochemical properties of mixed salts of ammonium nitrate and sulfate. Russ Chem B 61:3339. doi:10.1007/s11172-012-0005-x
17. Galperin LN, Kolesov YR, Mashkinov LB, Terner YE (1973) Differential automatic calorimeters (DAC) of different purpose. Book of reports from VI USSR conference on calorimetry. Inorganic Chemistry and Electrochemistry Institute of GSSR Academy of Sciences, Tbilisi, pp 539–543
18. Walker GM, Magee TRA, Holland CR, Ahmad MN, Fox JN, Moffat NA, Kells AG (1998) Caking process in granular NPK fertilizer. Ind Eng Chem Res 37:435–438. doi:10.1021/ie970387n
19. Shchukin ED, Amelina EA (2003) Surface modification and contact interaction of particles. J Dispers Sci Technol 24:377–395. doi:10.1081/DIS-120021796
20. Rubtsov YI, Kazakov AI, Morozkin SY, Andrienko LP (1984) Kinetics of heat release at thermal decomposition of commercial ammonium nitrate. J Appl Chem-USSR 57:1926–1929
21. Frank-Kamenetskiy DA (1987) Diffusion and heat exchange in chemical kinetics. Nauka, Moscow
22. Tavrovskaya AY, Portnova NL, Abashkina TF, Zaitsev PM (1977) Thermographic study of fluorine silicate compounds contained in ammonium nitrate phosphate fertilizer. Trudy NIUIFa 231:184–194
23. Akiyama T, Ando J (1972) Constituents and properties of ammoniated slurry from wet-process phosphoric acid. B Chem Soc Jpn 45:2915–2920. doi:10.1246/bcsj.45.2915
24. Zhong B, Li J, Xiang Zhang Y, Liang B (1999) Principle and technology of ammonium phosphate production from middle-quality phosphate ore by a slurry concentration process. Ind Eng Chem Res 38:4504–4506. doi:10.1021/ie980419m
25. Campbell GR, Leong YK, Berndt CC, Liow JL (2006) Ammonium phosphate slurry rheology and particle properties—the influence of Fe(III) and Al(III) impurities, solid concentration and degree of neutralization. Chem Eng Sci 61:5856–5866. doi:10.1016/j.ces.2006.05.010
26. Barnes HA (2000) Handbook of elementary rheology. University of Wales Institute of Non-Newtonian Fluid Mechanicsm, Aberystwyth

Permissions

All chapters in this book were first published in IJIC, by Springer International Publishing AG.; hereby published with permission under the Creative Commons Attribution License or equivalent. Every chapter published in this book has been scrutinized by our experts. Their significance has been extensively debated. The topics covered herein carry significant findings which will fuel the growth of the discipline. They may even be implemented as practical applications or may be referred to as a beginning point for another development.

The contributors of this book come from diverse backgrounds, making this book a truly international effort. This book will bring forth new frontiers with its revolutionizing research information and detailed analysis of the nascent developments around the world.

We would like to thank all the contributing authors for lending their expertise to make the book truly unique. They have played a crucial role in the development of this book. Without their invaluable contributions this book wouldn't have been possible. They have made vital efforts to compile up to date information on the varied aspects of this subject to make this book a valuable addition to the collection of many professionals and students.

This book was conceptualized with the vision of imparting up-to-date information and advanced data in this field. To ensure the same, a matchless editorial board was set up. Every individual on the board went through rigorous rounds of assessment to prove their worth. After which they invested a large part of their time researching and compiling the most relevant data for our readers.

The editorial board has been involved in producing this book since its inception. They have spent rigorous hours researching and exploring the diverse topics which have resulted in the successful publishing of this book. They have passed on their knowledge of decades through this book. To expedite this challenging task, the publisher supported the team at every step. A small team of assistant editors was also appointed to further simplify the editing procedure and attain best results for the readers.

Apart from the editorial board, the designing team has also invested a significant amount of their time in understanding the subject and creating the most relevant covers. They scrutinized every image to scout for the most suitable representation of the subject and create an appropriate cover for the book.

The publishing team has been an ardent support to the editorial, designing and production team. Their endless efforts to recruit the best for this project, has resulted in the accomplishment of this book. They are a veteran in the field of academics and their pool of knowledge is as vast as their experience in printing. Their expertise and guidance has proved useful at every step. Their uncompromising quality standards have made this book an exceptional effort. Their encouragement from time to time has been an inspiration for everyone.

The publisher and the editorial board hope that this book will prove to be a valuable piece of knowledge for researchers, students, practitioners and scholars across the globe.

List of Contributors

Abdelrahman A. Badawy
Physical Chemistry Department, National Research Centre, Dokki, Cairo, Egypt

Shaimaa M. Ibrahim
Chemistry Department, Faculty of Education, Ain Shams University, Cairo, Egypt
Department of Chemistry, Faculty of Science, Qassim University, Buraidah, Saudi Arabia

Neba F. Abunde and Ahmad Addo
Department of Agricultural Engineering, College of Engineering, Kwame Nkrumah University of Science and Technology, Kumasi, Ghana

N. Asiedu
Department of Chemical Engineering, College of Engineering, Kwame Nkrumah University of Science and Technology, Kumasi, Ghana

Abdelmajid Regti, My Rachid Laamari and Mohammadine El Haddad
Equipe de Chimie Analytique and Environnement, Faculté Poly-disciplinaire, Université Cadi Ayyad, BP 4162, 46000 Safi, Morocco

Salah-Eddine Stiriba
Equipe de Chimie Moléculaire, Matériaux et Modélisation, Faculté Poly-disciplinaire, Université Cadi Ayyad, BP 4162, 46000 Safi, Morocco
Instituto de Ciencia Molecular/ICMol, Universidad de Valencia, C/. Catedrático José Beltrán 2, Paterna, 46980 Valencia, Spain

M. Jannathul Firdhouse and P. Lalitha
Department of Chemistry, Avinashilingam Institute for Home Science and Higher Education for Women University, Coimbatore 641043, Tamil Nadu, India

Omer El-Amin Ahmed Adam
Chemistry Department, University of Kassala, Kassala 31111, Sudan
Chemistry Department, Faculty of Science and Arts in Baljurashi, Al baha University, Baljurashi 65635, Saudi Arabia

Akl M. Awwad
Royal Scientific Society, Al-Jubaiha, Amman 11941, Jordan

Vajjiravel Murugesan
Department of Chemistry, B. S. Abdur Rahman University, Vandalur, Chennai 600 048, India

M. J. Umapathy
Department of Chemistry, College of Engineering Guindy, Anna University, Chennai 600 025, India

Namrata Chaubey and Vinod Kumar Singh
Department of Chemistry, Udai Pratap Autonomous College, Varanasi 221002, India

M. A. Quraishi
Department of Chemistry, Indian Institute of Technology, Banaras Hindu University, Varanasi 221005, India

Riza A. Magbitang and Rheo B. Lamorena
Natural Sciences Research Institute, College of Science, University of the Philippines Diliman, 1101 Quezon City, Philippines
Institute of Chemistry, College of Science, University of the Philippines Diliman, 1101 Quezon City, Philippines

Matthew C. Menkiti
Civil, Environmental and Construction Engineering Department, Texas Tech University, Lubbock, TX, USA
Chemical Engineering Department, Nnamdi Azikiwe University, Awka, Nigeria

Ocholi Ocheje and Chinedu M. Agu
Chemical Engineering Department, Nnamdi Azikiwe University, Awka, Nigeria

A. Kalla, M. Benahmed and N. Djeddi
Laboratoire des Molécules Bioactives et Applications, Université Larbi Tébessi, Route de Constantine, 12000 Tébessa, Algeria

S. Akkal
Laboratoire de Phytochimie et Analyses physicochimiques et Biologiques, Département de Chimie, Faculté de Sciences exactes, Université Mentouri Constantine, Route d'Ain el Bey, 25000 Constantine, Algeria

H. Laouer
Laboratoire de Valorisation des Ressources Naturelles Biologiques, Département de Biologie et d'écologie végétale, Université Ferhat Abbas de Sétif 1, Sétif, Algérie

Robert Schmidt, H. Martin Scholze and Achim Stolle
Institute for Technical Chemistry and Environmental Chemistry (ITUC), Friedrich-Schiller University Jena, Lessingstr. 12, 07743 Jena, Germany

Sameh A. S. Alariqi, Niyazi A. S. Al-Areqi and Elyas Sadeq Alaghbari
Department of Chemistry, Faculty of Applied Science, University of Taiz, Taiz, Yemen

R. P. Singh
Division of Polymer Science and Engineering, National Chemical Laboratory, Dr. Homi Bhabha Road, Pune 411008, India

J. Siame
Department of Chemical Engineering, School of Mines and Mineral Sciences, Copperbelt University, Kitwe, Zambia

H. Kasaini
US Metals Refining Group, Inc.,, Colorado, USA

Sunday E. Elaigwu
Department of Chemistry, University of Hull, Cottingham Road, Hull HU6 7RX, UK
Department of Chemistry, University of Ilorin, PMB 1515, Ilorin, Kwara, Nigeria

Gillian M. Greenway
Department of Chemistry, University of Hull, Cottingham Road, Hull HU6 7RX, UK

Vinod Raphael Palayoor and Shaju Shanmughan Kanimangalath
Department of Chemistry, Government Engineering College, Thrissur 680009, Kerala, India

Joby Thomas Kakkassery and Sini Varghese
Research Division, Department of Chemistry, St. Thomas' College (Autonomous), Thrissur 680001, Kerala, India

I. O. Otunniyi, M. Oabile, A. A. Adeleke and P. Mendonidis
Metallurgical Engineering, Vaal University of Technology, Vanderbijlpark, South Africa

Pranab Ghosh, Mainul Hoque and Gobinda Karmakar
Natural Product and Polymer Chemistry Laboratory, Department of Chemistry, University of North Bengal, Darjeeling 734013, India

Malay Kr. Das
Department of Physics, University of North Bengal, Darjeeling 734013, India

Asif Ali Khan, Rizwan Hussain and Umair Baig
Analytical and Polymer Research Laboratory, Department of Applied Chemistry, Faculty of Engineering and Technology, Aligarh Muslim University, Aligarh, UP 202002, India

Merah Salah
Laboratory of Analytical Chemistry and Electrochemistry, Department of Chemistry, Faculty of Science, Tlemcen University, Tlemcen, Algeria
Department of Process Engineering, Faculty of Technology, Sai‹da University, Sai‹da, Algeria

Larabi Lahcène and Harek Yahia
Laboratory of Analytical Chemistry and Electrochemistry, Department of Chemistry, Faculty of Science, Tlemcen University, Tlemcen, Algeria

Abderrahim Omar
Laboratory of Separation and Purification Technology, Department of Chemistry, Faculty of Science, Tlemcen University, Tlemcen, Algeria

Konstantin Gorbovskiy, Andrey Norov and Andrey Malyavin
The Research Institute for Fertilizers and Insecto-Fungicides Named after Professor Y. Samoilov, 162622 Cherepovets, Vologda Region, Russia

Anatoly Kazakov
Institute of Problems of Chemical Physics of the Russian Academy of Sciences, 142432 Chernogolovka, Moscow Region, Russia

Anatoly Mikhaylichenko
D. Mendeleev University of Chemical Technology of Russia, 125047 Moscow, Russia

Index

www.ingramcontent.com/pod-product-compliance
Lightning Source LLC
Chambersburg PA
CBHW08202519032 6
41458CB00010B/3280